Indigenous Environmental Knowledge

John Edington

Indigenous Environmental Knowledge

Reappraisal

Springer

John Edington
Environmental Studies Programme
University of Cardiff
Wales, UK

ISBN 978-3-319-62490-7 ISBN 978-3-319-62491-4 (eBook)
DOI 10.1007/978-3-319-62491-4

Library of Congress Control Number: 2017948393

Printed on acid-free paper

This Springer imprint is published by Springer Nature
The registered company is Springer International Publishing AG
The registered company address is: Gewerbestrasse 11, 6330 Cham, Switzerland

This book is dedicated to the memory of my wife Ann, my constant companion on our village forays.

Synopsis

Opinions about the credibility of indigenous environmental knowledge have been highly polarised. Whilst an attitude of general denigration was the norm during colonial times, the pendulum of opinion has recently tended to swing in the opposite direction, with indigenous rural peoples now being widely regarded as environmentally well-aware. In a survey spanning over 20 countries, the author demonstrates that in many respects this reputation is well deserved. In spheres such as agriculture and architecture, traditional concepts anticipate modern ones and may indeed surpass them in having relevance to current environmental issues. In certain other matters however, local environmental perceptions are less well-founded. The appreciation of disease hazards is often noticeably incomplete, as so too is the local capacity to manage wild populations sustainably. These two blind spots stem mainly from a lack of access to suitable investigative techniques and a tendency to make unsubstantiated generalisations. Coupled with hostile external influences – some of them geographical whilst others human in origin – indigenous peoples have thus been widely prevented from achieving the quality of life standards they would wish for themselves and which we would wish for on their behalf. The author first offers a series of detailed reviews of various aspects of village life from these perspectives and then goes on to suggest how some of the remaining difficulties might be resolved as part of the development process. A final chapter explores the contrasting notion that the better-founded facets of village knowledge systems could hold the key to certain difficulties currently facing the Western world. This is the first time all these issues have been reviewed together in a comprehensive way.

Acknowledgements

My thanks are due to the Inter-University Council for Higher Education Overseas and to the British Council for sponsoring my initial visits to Nigeria, Malaysia, Nepal, Thailand, the Philippines and Vietnam. Subsequently various overseas universities and institutions have acted as my hosts (usually in exchange for inputs to teaching programmes), and this provided me with bases from which to make the other village investigations. I gratefully acknowledge their assistance. The organisations and persons who helped are listed below in alphabetical order according to country:

Belize – Toledo Ecotourism Association, Mr. C. Schmidt

Brazil – Mamirauá Sustainable Development Institute, Ms. C. D. V. Koike

China and *Hong Kong* – Mrs. S.-W. Sim

Ethiopia – Plant Genetic Resource Centre, Addis Ababa, Mr. H. Mekbib

India – Central Arid Zone Research Institute, Jodhpur, Dr. I. Prakash; Indian Wildlife Institute, Dehradun, Dr. G. S. Rawat; Central Himalayan Environment Association, Nainital, Professor A. K. Pant; Indian Institute of Technology, Roorkee, Appropriate Ecotechnology Development Group, Mr. Mahendra Pant; Karnataka State Forest Department, Mr. P. D. Gaonkar; Museum of Mankind in Bhopal

Japan – Osaka Prefecture University, Professor K. Tanida

Jordan – Jordanian Royal Society for the Conservation of Nature

Malaysia – Universiti Pertanian Malaysia (now Universiti Putra Malaysia) in Serdang, Universiti Sains Malaysia in Penang and the Institute for Medical Research in Kuala Lumpur

Nepal – Tribhuvan University, Dr. T. C. Majupuria and Dr. K. Joshi; National Planning Commission, Dr. A. R. Joshi; Department of Medicinal Plants, Kathmandu, Mr. D. P. Joshi; UNICEF Nepal; Ms. Sarah Hawes, Jambu Sherpa and Nima Sardar

Nigeria – University of Benin, Professor J. C. Ene; Bayero University, Kano, Professor R. Shotter

Oman – Shankland Cox Ltd., Mr. R. H. Daly

Philippines – University of the Philippines at Los Baños, Dr. P. Payawal, Professor L. de Padua, Mr. H. Salmon (British Council)

South Africa – Witwatersrand University, Professor H. E. H. Paterson, Conservation Corporation of Africa

Thailand – Forest Industry Organization, Kasetsart University, Bangkok, Dr. S. Chettamart

Uganda – Makerere University, Dr. J. Obua; Kibale Association for Rural Environmental Development, Mr. R. Tooro

USA – University of California, Berkeley, Professor J. R. McBride

Vietnam – University of Hanoi, Professor M. D. Yen

Zimbabwe – National Herbarium, Mr. R. B. Drummond; Zimbabwe Women's Bureau, Mrs. H. Ponde and Mrs. A. Chonyera; National Training Centre for Rural Women, Mrs. L. Kamanga

Amongst these many collaborators, two people need to be singled out for particular mention. One is Dr. Ananda Raj Joshi who, whilst working as my doctoral student studying village communities in Nepal, helped to crystallise many of the themes developed in this book. I am also grateful to him for the permission to refer to unpublished material contained in his doctoral thesis. The other is Professor Joe McBride of the Department of Forestry and Resource Management, University of California, Berkeley, who not only participated in the firewood studies in Nepal but has also been a source of constant encouragement and useful comment throughout the course of the work. Graham Morris and Mary I. Anderson deserve my thanks for acting as "lay" readers of the manuscript. Unless indicated otherwise, the line drawings are my own work. Likewise, apart from exceptions noted in the captions, the photographs are all ones taken in situ by my late wife or myself.

I also need to thank some organisations for the permission to use copyright material. These include the Royal Society of Chemistry for the food composition values in Table 3.2 which are derived from *McCance and Widdowson's the Composition of Foods* and the International Mountain Society for the inventory of tree uses contained in Table 5.2, which is based on material first published in the journal *Mountain Research and Development*.

Contents

About the Author

John Edington who may be known to fellow zoologists for his pioneer work on caddis flies, has also spent many years studying the application of *ecology* to human affairs. This latter interest has already generated two well-regarded texts: *Ecology and Environmental Planning* (Chapman and Hall) and *Ecology, Recreation and Tourism* (Cambridge University Press). In this, the third of the trilogy, he explores the controversial issue of the credence to be placed on indigenous environmental knowledge He finds much to admire in these knowledge systems but also has occasion to voice some criticisms. This evaluation is based on first-hand investigations in over 20 countries in the developing world. The author has been based in Cardiff University throughout his working life first as a lecturer in zoology and subsequently as the director of the Environmental Studies Programme.

Chapter 1
Introduction

1.1 Changing Perceptions

Arguments about the merits of indigenous peoples have a long and respectable history. As far back as the seventeenth century the English philosopher Thomas Hobbes famously proclaimed the lives of such peoples to be "nasty, brutish and short" and suggested moreover that they had only a very limited understanding of the world around them. Sometime later his fellow philosopher, the Frenchman Jean-Jacques Rousseau, took a diametrically opposite view and was loud in praise of their virtues. Both these commentators concentrated mainly on demeanour and general behaviour, but later, the discussion turned increasingly to the issue of environmental awareness. During colonial times, traditional communities were not highly rated on this score (Jiggins 1989). They gained a reputation for being backward and unenterprising, perhaps because they had shown themselves reluctant to be swept along by European schemes for agricultural modernisation and forest exploitation. More recently public opinion has tended to swing in the opposite direction, with the notion gaining ground that traditional communities may not only know what is best for themselves, but also be capable of offering the Western world some useful pointers (Goldsmith et al. 1972).

1.2 Hunter-Gatherers

The popular perception of indigenous peoples has been greatly coloured by accounts of the groups known as "hunter-gatherers". These peoples have no fixed bases (barring the occasional cave), lack any kind of domesticated animals or crop plants, and need to roam widely to obtain the wild plant and animal materials they require for everyday existence. One group, the Bushmen of the Kalahari Desert of Southern Africa, has aroused particular interest largely because of the engaging accounts of

J. Edington, *Indigenous Environmental Knowledge*,
DOI 10.1007/978-3-319-62491-4_1

their lives penned by the writer and explorer Laurens van der Post (1958, 1961). Subsequently our knowledge of them has been greatly expanded by a succession of detailed studies by professional anthropologists (Marshall 1964; Lee 1972, 1976, 1979; Lee and DeVore 1976). These duly confirm the Bushmen as remarkable people, adept at harnessing the sparse resources of the dry environment they occupy. A typical band would be familiar with 200 species of wild plants and know how to use at least 105 of these as food (Lee 1972, 1979). Likewise they can recognize 54 species of wild animals, and habitually hunt some 17 of these (Lee 1972). They are especially skilled at using materials from local wild species to fashion the various items of equipment they need for daily life (Fig. 1.1). Thus they make the shafts of tools and weapons from the straight branches of local trees and use root-bark cylinders from the umbrella thorn *(Acacia tortilis)* to form quivers to hold their arrows. Bow strings are made from antelope sinews and the arrows themselves from reed-stems tipped with bone points. A special feature of the arrows is that they have detachable tips coated with a toxin obtained from the chrysomelid beetle *Diaphidia nigro-ornata.* Typical carrying devices for women take the form of a bag-cum-sleeping blanket made from antelope hide and for men, an intricately knotted net fashioned from antelope sinews. Water is critically important to the Bushmen and supplies are variously obtained from pools, accumulations in tree boles, water-holding sediments (the water being drawn up through a filtration straw) and certain plants such as *Fockea monroi* that contain water in their tissues. Famously, ostrich eggshells are used to store and carry water, each shell containing about a litre (Lee 1979). Bushmen exhibit the keynote feature of all hunter-gatherers namely that, in order to tap resources that are dispersed and fluctuating, they are obliged to keep constantly on the move (Fig. 1.2).

Other hunting and gathering groups show similar ingenuity in adapting local biological materials to their needs (Tindale 1972). When the aboriginal *Pitjandjara* of Central Australia have to carry water they use, not egg shells, but large bowls fashioned from *Eucalyptus* trunk bases and lined with *mulga* grass to stop the water slopping out. Spears for hunting are fashioned from the woody stems of the shrub *Pandorea doratoxylon,* with the heads made conveniently detachable by being temporarily fixed to the shafts with plant resins. This group is also notable for its very detailed knowledge of desert food chain relationships For example they know that the seasonal rains will stimulate not only the growth of directly edible species such as pigweed (*Portulaca oleracea*), but also plants on which various useful insects thrive. Thus they know that when the creeping vine *Boerhavia diffusa* spreads over the sand it will become colonized by the much-prized edible caterpillars of the hawk moth *Celerio lineata.* They are also aware that other types of caterpillars that appear seasonally on shrubs, whilst not themselves edible, attract the attention of large *Varanus* lizards which certainly are (Tindale 1972).

A remarkable fact about hunting and gathering is that it was once the only mode of life pursued by mankind, a state of affairs that persisted for the first 140,000 years of human existence (Stringer 2002; Cela-Conde and Ayala 2007). It was only as relatively recently as 10,000 years ago that most hunter-gatherers started to settle down as farmers or to disappear as viable groups. From once representing 100% of

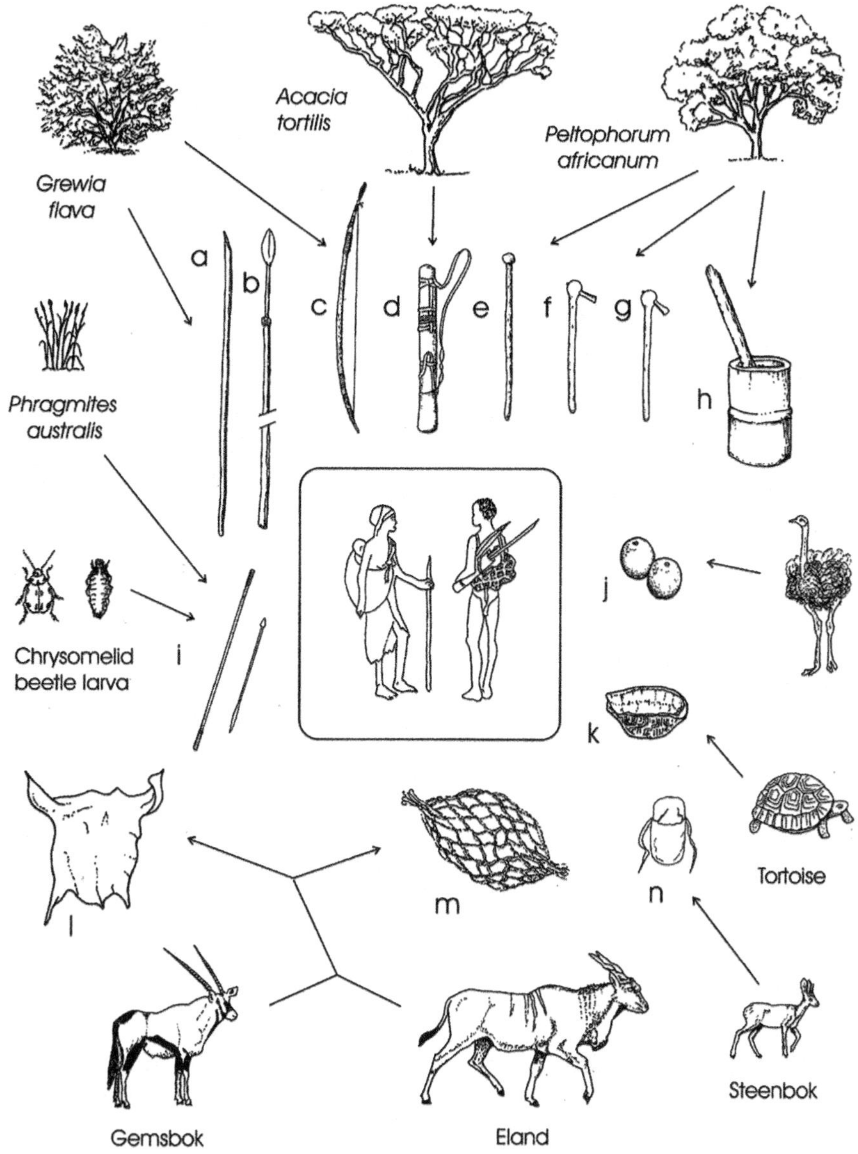

Fig. 1.1 Everyday equipment fashioned by !Kung Bushmen from local plant and animal materials. (**a**) Digging stick, (**b**) spear, (**c**) bow, (**d**) quiver, (**e**) club, (**f**) axe, (**g**) adze, (**h**) pestle and mortar, (**i**) arrow (tipped with poison from the chrysomelid beetle *Diaphidia nigro-ornata*), (**j**) ostrich eggs for carrying water, (**k**) tortoise carapace for use as a bowl, (**l**) woman's carrying bag, (**m**) man's carrying net, (**n**) handbag, (**l–n**), all made from antelope hide and sinews (Developed from Lee 1979)

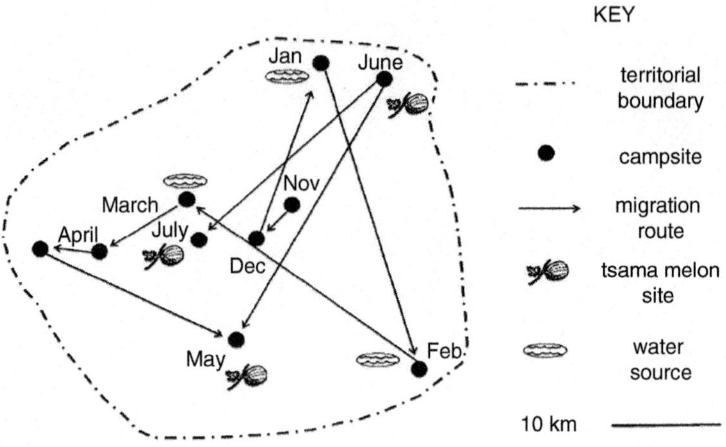

Fig. 1.2 Annual migrations of a Bushmen band. Movements are dictated largely by the seasonal availability of water, game animals and plant foods such as the *tsama* melon *(Citrullus lanatus)* (Based on Silberbauer 1972)

the world's population, they have by today dramatically declined to only 0.001% of it (Lee and DeVore 1968). We will be discussing later (p.12) what factors might have triggered this dramatic change in circumstances. The survivors include, in addition to the Bushmen and the Australian aborigines, such groups such as the African "pygmies" and the northern *Inuit* (Turnbull 1961; Lee and DeVore 1968).

When the key hunter-gatherer specialists met together in Chicago in 1966 to compare notes, they were loud in their praise of the various surviving groups, considering their virtues to include acute environmental awareness, frugal resource-use and a gentle disposition (Lee and DeVore 1968). However, this positive appraisal has not entirely stood the test of time. When prehistoric hunter gatherers are also taken into the equation, evidence emerges for some very wasteful practices. For example, prehistoric bison hunters in North America were evidently in the habit of stampeding entire herds over cliff tops, subsequently allowing most of the carcasses to go to waste (Wheat 1967; Reeves 1983). Indeed it appears that during prehistoric times, a number of large mammals were hunted so relentlessly as to drive them to extinction (Martin and Wright 1967). Postulated casualties include various species of mammoths, camels, horses and bison (Fig. 1.3) (Martin 1984; McDonald 1984).

Furthermore, a fresh look at even some of the surviving groups reveals them in a somewhat less flattering light. For instance when the *Pitjandjara* used fire as a hunting tool in the Australian desert, they seem to have been very wasteful, with many more animals, in this case lizards, being killed than could ever have been used (Tindale 1933). Likewise hunting decisions by the Bushmen seem generally to be made on an entirely mystical basis (Bleek and Lloyd 1911), without any attempt to match harvesting levels with available resources. In this sense, the apparent sustainability of contemporary hunting and gathering operations may be largely an illusion

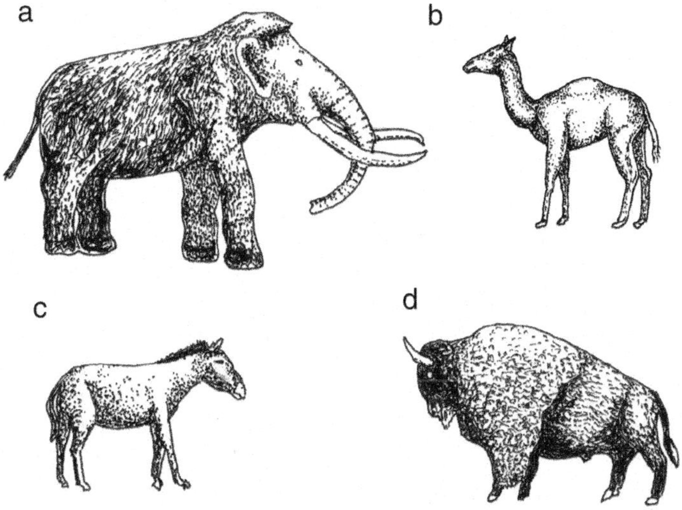

Fig. 1.3 North American species believed to have been driven to extinction by early hunters. (**a**) Mammoth *(Mammuthus)*, (**b**) camel *(Camelops)*, (**c**) horse *(Equus)* (**d**) bison *(Bison)* (developed from Martin and Wright 1967; Martin and Klein 1984)

created by a combination such factors as small band sizes, simple weaponry, low human reproductive rates and mobile life styles. Even the hunter-gatherers' reputation for gentleness (Thomas 1959; Marshall 1961) has recently come under fire, with evidence emerging for a history of seriously violent interactions both within and between groups (LeBlanc and Register 2004). One can speculate on the reasons for the early emphasis on the virtues of these groups rather than their shortcomings, but one possibility is that, in the aftermath of two destructive world wars, there was a particular desire to identify groups capable of being presented as positive role-models for mankind.

1.3 The Challenges of Permanent Settlement

Something of the same ambiguities are apparent when we move from hunter-gatherers to settled farmers. Virtually all rural inhabitants in the developing world now come under this latter heading and exploring the levels of environmental awareness they display in their different spheres of activity is the principal aim of this book. Extending over virtually every climatic zone and natural habitat, they offer a particularly varied and fruitful field for investigation. Compared with an itinerant life, settlement brings a mixture of benefits and challenges. For example the construction of permanent dwellings (as opposed to the flimsy shelters of hunter-gatherers) allows designs to be made more sophisticated, both in the matter of selecting suitable raw materials and of creating comfortable indoor living conditions. Unfortunately, some of the insects that attack plants in the wild state continue their depredations after these plant materials have been incorporated into buildings. In the tropics, timber components are seriously at risk from termite attacks and roof-thatching materials from the ravages of various caterpillars.

Another category of unwelcome organisms that gravitate towards permanent dwellings are those carrying disease. Malarial mosquitoes made a significant evolutionary step when some species developed the habit of entering dwelling houses at night in search of blood meals. Indeed the estimated emergence of malaria as a serious disease neatly coincides with the dates suggested by archaeologists for the beginnings of settled agriculture (Volkman et al. 2001; Tishkoff et al. 2001). Especially favourable conditions would have been created by the juxtaposition of dwellings with irrigated farm plots capable of harbouring mosquito larvae. Another category of health hazard is posed by the various insects and ticks that find the mud-brick walls of houses an acceptable substitute for the earth-lined animal burrows they previously occupied (Smith 1973). Notable health problems generated in such circumstances include Chagas' disease and tick-borne relapsing fever.

With farming, being able to concentrate on a circumscribed area close to a village centre, not only carries logistical benefits, it also allows expertise to accumulate over successive generations about how best to manage this particular plot of land. Difficulties can arise however from the fact that continually harvesting crops from the same ground will cause soil fertility to decline unless decisive steps are taken to stop this from happening. Also, crops bred to maximize their nutritional qualities, are highly attractive to pests and such organisms evolve to become ever more efficient at exploiting this resource. This problem is usually aggravated by the fact that human selection for other characteristics often inadvertently undermines the crop's natural defensive mechanisms. These may be chemical or structural in character. A good example of a loss in the latter category is where the disappearance of spiky awns from the heads of cereals makes them more vulnerable to bird attack. The mere act of cultivating soil prior to planting brings another kind of problem in its wake, because it favours that class of wild plant we call "weeds", all them rapid colonizers of exposed soil and active competitors with crop plants.

A process essential to the functioning of most villages is the importation of wild plant materials from the surrounding countryside. These materials serve an extremely wide variety of functions (Table 1.1). In a typical village such as that portrayed in Fig. 1.4, these materials come mainly from surrounding forests and grazing areas, but with a few originating from path sides and waste ground. With items that are highly valued or ones that are required in bulk, there is an ever-present risk of overexploitation. In this particular village for example, some important medicinal plants have started to become scarce as a result of over-collecting, usually by outsiders and procuring sufficient supplies of firewood and leaf fodder has become a constant preoccupation. Both these latter materials are required in large quantities, with daily *per capita* use running at 2 kg and 10 kg respectively. The temptation is to concentrate first on supplies closest to the village, but inevitably this has the effect of extending the length of foraging journeys, with the extra physical burdens usually falling on the shoulders of the village women and girls who do most

Table 1.1 Use of wild-plants by three traditional village communities (individual species may be used for as many as four different purposes)

	Ethnic group and locality		
	Sherpa	*Tewa*	*Mödö*
	Nepal (Joshi 1987)	New Mexico (Robbins et al. 1916)	Sudan (Persson 1986)
	Number of species used for particular purposes		
Herbal medicines and cosmetics	48	27	11
Foods and beverages	47	34	20
Firewood	27	3	10
Animal fodder	27	1	0
House construction materials	10	6	23
Agricultural implements	3	4	10
Domestic equipment	11	13	18
Clothing	0	4	7
Paints and dyestuffs	1	5	0
Mineral salts	0	0	3
Hunting devices	0	11	19
Fish poisons	3	0	6
Natural insecticides	6	0	0
Ritual purposes	0	9	3
Total number of species used	95	70	70

of the collecting work. A common aggravating factor is the plundering of firewood by members of larger adjacent villages and by urban firewood collectors. All of which sets a premium on developing strategic plans for plant resource use and on finding ways to match harvesting levels with regeneration rates.

Villages are also commonly beset with water supply and waste disposal issues. Whilst hunter-gatherers frequently programme water procurement into their itineraries, settled villages have to develop *in situ* strategies to meet their needs. Strangely, this does not invariably mean locating the settlement immediately adjacent to an accessible water source. Quite commonly, proximity to flat land suitable for cultivation and house building is given a higher priority and this can considerably lengthen water supply lines, adding another major chore if the transport is done manually. The disposal of human wastes from permanent settlements is another challenge and a failure to perform this function effectively can have serious health implications. Hunter-gatherers have the advantage of being able to walk away from their wastes; the process of desiccation having an additional sanitizing effect in dry environments. In consequence the incidence of gastro-intestinal parasites in such communities is generally much less than in settled agricultural ones (Dunn 1968). Indeed archaeological evidence shows that the process of settlement coincided with a marked upsurge in the incidence of parasitic infections of all kinds (Fischer 1980; Cockburn and Cockburn 1980).

a

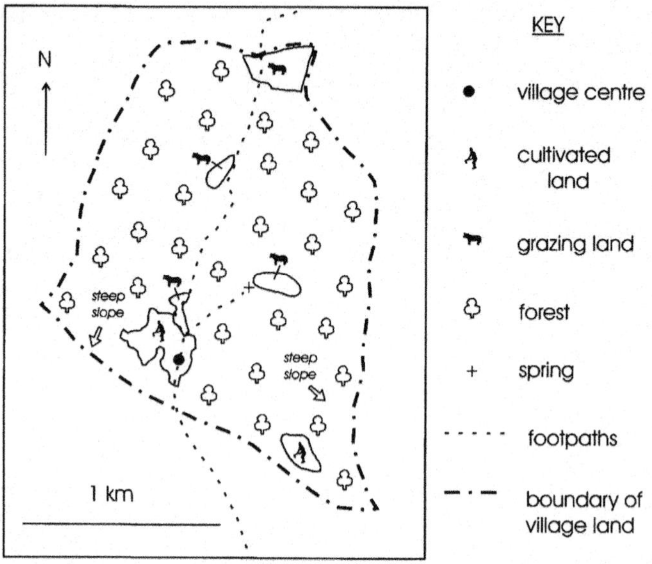

b

Fig. 1.4 A typical village structure (from Nepal) showing a village core and cultivated area situated on flatter land, a quite distant water source and extensive peripheral forest and grazing areas – the source of a wide variety of useful wild plants. (**a**) Village map, (**b**) view of the village from the South East (Developed from Joshi 1987)

1.4 Local Investigative Methods

Moving from this list of challenges to the investigatory means locally available to address them, many of the matters in question lend themselves to direct examination by straightforward trial and error techniques. If the timbers used on a building are unsuitable, this becomes clearly evident when the structure fails. Likewise with farming operations, over a series of successive seasons it is relatively easy to distinguish the strategies that produce good crop yields from those that don't. Indeed this is a sphere where village investigators are known additionally to carry out procedures that have all the hallmarks of formal scientific experiments. For example rice farmers in West Africa have been observed to plant rice varieties in experimental plots in order to gauge their performance; the varieties having been obtained either through exchange with other villages or by the collection of mutant strains from the fields. The experimental plots are chosen to represent particular soil types or sites at specific points on the flood profile and the aim of the experiments is to assess relative performance. Yields are quantified using home-made measuring devices such as calabashes and tobacco tins (Richards 1985).

Some matters however, are rendered at least partly opaque by an absence of suitable investigative techniques. For example, a lack of access to microscopy has hindered the local development of anything approaching the modern "germ theory" of disease. This has seriously hampered efforts to bring about health improvements (p.158). Likewise a lack of familiarity with population census methods has hindered the development of effective strategies for managing wild plant and animal resources (p.183). In certain spheres, and herbal medicine is one of them, much valid information has been collected by village observers, but such is the human desire for completeness (Glasser 1988) that all too often the gaps have been filled using unsubstantiated generalizations (p.135). The "hot" and "cold" classification system and the so-called "law of signatures" rule for matching remedies with diseases are two such examples (Arber 1953; Helman 1990).

1.5 Invoking the Supernatural

Another complicating factor is the widespread local adherence to a "world view" in which supernatural agencies are routinely considered to involve themselves in everyday life. This can have some positive aspects. It affords people the psychological reassurance that potential misfortunes can be averted or at least minimized by making representations to the relevant deities (unsubstantiated though these expectations may be) (Jahoda 1969). Praying for the sick is in a somewhat different category because there is at the very least the possibility of benefitting from *placebo* effects (p.130). It is also the case that worthwhile conservation measures can often be strengthened by placing them in a religious framework. For instance off the coast of Indonesia, useful restrictions on fishing in the vicinity of fish-spawning grounds are strengthened by the belief that such sea-areas come under the protection of the sea goddess *Loro Kudal* (Polunin 1985). Likewise in the Ivory Coast, prohibitions on tree cutting on erosion-prone slopes are strengthened by the claim that these areas fall within the jurisdiction of the goddess *Assie* and are therefore sacrosanct (Shepherd 1992). Indeed throughout Africa and Asia there are forest remnants that owe their continued existence to a proximity to religious shrines (Gadgil and Vartak

1976; Ingles 1995). Supernatural postulations can however also have seriously negative consequences. For instance, invoking witchcraft as a cause of illness may not only divert attention from authenticated and workable remedies, it can also have seriously disruptive social consequences (p.139).

In exploring these many interactions our approach initially is to take each village activity in turn, with the aim of describing the completeness or otherwise of the relevant village knowledge systems. It is fair to say even at this point that the exercise reveals some striking environmental insights. Following this, a penultimate chapter considers the scope for using Western inputs to address any conspicuous deficiencies revealed; this review being placed in the context of overall community development. In a final chapter, attention is turned to the opposite and less commonly discussed proposition, namely that some traditional concepts and technologies are sufficiently relevant to modern life to warrant re-adoption.

References

Arber A (1953) Herbals, their origin and evolution, *A chapter in the history of botany 1470–1670.* Cambridge University Press, Cambridge

Bleek WHI, Lloyd LC (1911) Specimens of bushman folklore. George Allen, London

Cela-Conde C, Ayala FJ (2007) Human evolution. Trails from the past. Oxford University Press, Oxford

Dunn FL (1968) Epidemiological factors: health and disease in hunter-gatherers. In: Lee RB, DeVore I (eds) Man the hunter. Aldine Publishing Company, Chicago, pp 221–228

Fischer C (1980) Bog bodies of Denmark. In: Cockburn A, Cockburn E (eds) Mummies, disease and ancient cultures. Cambridge University Press, Cambridge, pp 177–193

Gadgil M, Vartak VD (1976) The sacred groves of Western Ghats in India. Econ Bot 30:152–160

Glasser M (1988) Accountability of anthropologists, indigenous healers, and their governments: a plea for reasonable medicine. Soc Sci Med 27:1461–1465

Goldsmith E, Allen R, Allaby M, Davoll J, Lawrence S (1972) A blueprint for survival. Penguin Books, Harmondswith

Helman CG (1990) Culture, health and illness. An introduction for health professionals. Wright, London

Ingles AA (1995) Religious beliefs and rituals in Nepal. Their influence on forest conservation. In: Halladay P, Gilmour DA (eds) Conserving biodiversity outside protected areas. The role of traditional agro-ecosystems. IUCN, Gland/Cambridge, pp 205–224

Jahoda G (1969) The psychology of superstition. Allen Lane, Penguin, London

Jiggins J (1989) An examination of the impact of colonialism in establishing negative values and attitudes towards indigenous agricultural knowledge. In: Warren DM, Slikkerveer LJ, Titiola SO (eds) Indigenous knowledge systems: implications for agriculture and international development. Iowa State University, Amco, Iowa, pp 68–78

Joshi AR (1987) A study of the environmental relationships of certain village communities in the central development region of Nepal. Doctoral Thesis, University of Wales, Cardiff

LeBlanc SA, Register KE (2004) Constant battles. Why we fight. St. Martin's Griffin, New York

Lee RB (1972) The !Kung bushmen of Botswana. In: Bicchieri MG (ed) Hunters and gatherers today. Holt, Rinehart and Winston, New York, pp 326–368

Lee RB (1976) !Kung spatial organization, an ecological and historical perspective. In: Lee RB, DeVore I (eds) Kalahari hunter-gatherers. Studies of the !Kung San and their neighbours. Harvard University Press, Cambridge, MA, pp 73–97

Lee RB (1979) The !Kung San: men, women and work in a foraging society. Cambridge University Press, Cambridge
Lee RB, DeVore I (eds) (1968) Man the hunter. Aldine Publishing Company, Chicago
Lee RB, DeVore I (eds) (1976) Kalahari hunter-gatherers. Studies of the !Kung San and their neighbours. Harvard University Press, Cambridge, MA
Marshall L (1961) Sharing, taking and giving: relief of social tensions among !Kung bushmen. Africa 31:231–249
Marshall L (1964) The !Kung Bushmen of the Kalahari Desert. In: Gibbs JL (ed) Peoples of Africa. Holt, Rinehart and Winston, New York, pp 241–278
Martin PS (1984) Prehistoric overkill: the global model. In: Martin PS, Klein RG (eds) Quaternary extinctions, a prehistoric revolution. University of Arizona Press, Tucson, pp 354–403
Martin PS, Klein RG (eds) (1984) Quaternary extinctions. A prehistoric revolution. University of Arizona Press, Tucson
Martin PS, Wright HE (eds) (1967) Pleistocene extinctions: the search for a cause. Yale University Press, New Haven
McDonald JN (1984) The recorded North American selection regime and late quaternary megafaunal extinctions. In: Martin PS, Klein RG (eds) Quaternary extinctions, a prehistoric revolution. University of Arizona Press, Tucson, pp 404–439
Persson J (1986) Trees, plants and a rural community in the southern Sudan. Unasylva 154:32–43
Polunin NVC (1985) Traditional marine practices in Indonesia and their bearing on conservation. In: McNeely JA, Pitt D (eds) Culture and conservation: the human dimension in environmental planning. Croom Helm, London, pp 155–179
Reeves BOK (1983) Six milleniums of buffalo kills. Sci Am 249(4):92–100
Richards P (1985) Indigenous agricultural revolution. Ecology and food production in West Africa. Hutchinson, London
Robbins WW, Harrington JP, Freire-Marreco B (1916) The ethnobotany of the Tewa Indians. Smithsonian Institution Bureau of American Ethnology, Government Printing Office, Washington, DC
Shepherd G (1992) Managing Africa's tropical dry forests: a review of indigenous methods. Overseas Development Institute, London
Silberbauer GB (1972) The G/wi Bushmen. In: Bicchieri MG (ed) Hunters and gatherers today. Holt, Rinehart and Winston, New York, pp 271–326
Smith KGV (ed) (1973) Insects and other arthropods of medical importance. British Museum (Natural History), London
Stringer C (2002) The morphological and behavioural origins of modern humans. Proc Br Acad 106:23–30
Thomas EM (1959) The harmless people. Secker & Warburg, London
Tindale NB (1933) Report on field trip. Oceania Sydney 4:101–105
Tindale NB (1972) The Pitjandjara. In: Bicchieri MG (ed) Hunters and gatherers today. Holt, Rinehard and Winston, New York, pp 217–268
Tishkoff SA, Varkonyi R, Cahinhinan N, Abbes S, Argyropoulos G, Destro-Bisol G, Drousiotou A, Dangerfield B, Lefranc G, Loiselet J, Piro A, Stoneking M, Tagarelli A, Tagarelli G, Touma EH, William SW, Clark AG (2001) Halotype diversity and linkage disequilibrium at human G6PD: recent origin of alleles that confer malarial resistance. Science 293:455–462
Turnbull C (1961) The forest people. Jonathon Cape, London
van der Post L (1958) The lost world of the Kalahari. Hogarth Press, London
van der Post L (1961) The heart of the hunter. Hogarth Press, London
Volkman SK, Barry AE, Lyons EJ, Nielsen KM, Thomas SM, Choi M, Thakore SS, Day KP, Wirth DF, Hartl DL (2001) Recent origin of *Plasmodium falciparum* from a single progenitor. Science 293:482–484
Wheat JB (1967) A Paleo-Indian bison kill. Sci Am 216(1):44–52

Chapter 2
Farming

2.1 Origins

Archaeological evidence suggests that hunter-gatherers started to settle down about 10,000 years ago. Just what factors prompted them to do so after such a long period of mobility is a matter for debate. One theory is that their populations grew beyond the point where they could still be supported by an itinerant collecting lifestyle; hence the need to find a fundamentally different way of obtaining food (Cohen 1977). An alternative theory assigns a central role to climate change (Fig. 2.1). Evidence for a cold spell at the critical time is offered by observed changes in the balance of oxygen isotopes at different horizons in polar ice cores. Decreases in the proportion of the heavy isotope (O^{18}) relative to the more common lighter one (O^{16}) provide a reliable indication of cooling (Atkinson et al. 1987; Dansgaard et al. 1989, 1993). The warming trend that followed the end of the last ice age would have been expected to have the opposite effect, but it is now postulated to have caused the melting of a large ice-dam in the region of what is now the St Lawrence seaway. The disappearance of this dam could have released large quantities of fresh water into the North Atlantic and to have brought the Gulf Stream to a halt (Broecker et al. 1989; Hostetler et al. 2000). The usual warming effect of this current would thus have been negated and because of the way the earth's ocean currents are linked, the cooling trend could have been felt world-wide (Mithen 2003).

A change such as this could well have provided the stimulus for the development of agriculture. A snapshot of the likely sequence of events can be gleaned from archaeological work carried out at Abu Hureyra, a site on the headwaters of the Euphrates in what is now Syria (Moore and Hillman 1992; Moore et al. 2000). As increasingly hostile conditions caused the forest to decline, plant remains show how the local community had first to abandon its use of tree fruits, then to turn to wild grasses and finally in desperation to try and survive on barely edible small-seeded legumes. At this point the site was abandoned, only to be re-occupied a few hundred years later by a people who had by now discovered how to cultivate cereals and

J. Edington, *Indigenous Environmental Knowledge*,
DOI 10.1007/978-3-319-62491-4_2

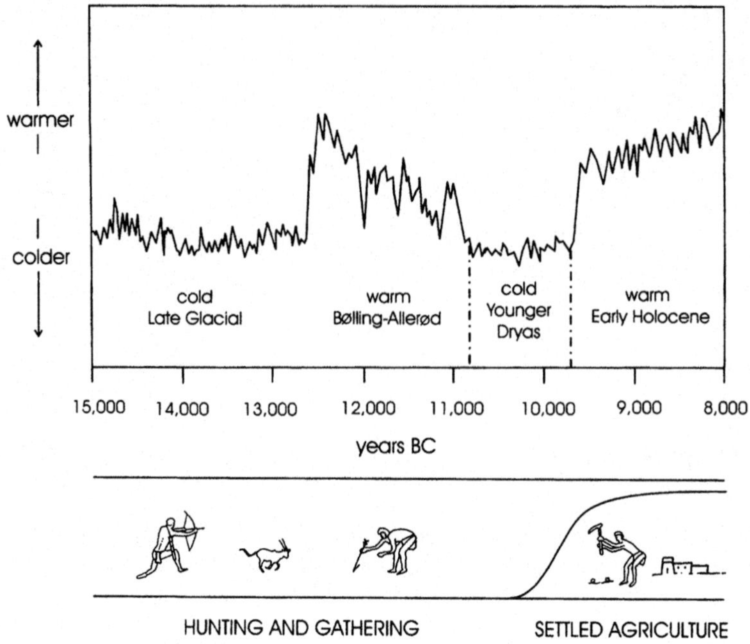

Fig. 2.1 Reconstruction of climatic changes occurring at the end of the last Ice Age (Developed from Dansgaard et al. 1993; Molleson 1994 and Sherratt 1997)

legumes (and subsequently how to domesticate sheep and goats). Furthermore, the newcomers had graduated from using their simple timber and reed shelters to living in villages with substantial mud-brick houses. The critical transformation from an itinerant existence to a settled agricultural one had evidently been achieved.

At one time it was thought that these pioneer attempts at domestication and settlement were limited to the Middle East. We now know this not to have been the case. Other centres were apparently also involved, including ones in North China, South-East Asia, Central Africa, South and Central America, the main distinction between the sites lying in the identity of the wild species selected. Thus maize and turkeys were first domesticated in Mexico, potatoes and llamas in South America, bananas and buffaloes in South East Asia and so on (Fig. 2.2). Wherever it happened, the process of domestication seems to have followed much the same course. It typically involved bringing selected wild plants and animals under human control, with the aim of modifying their attributes to meet human needs (Bar-Yosef and Meadow 1995; Sherratt 1997). With animals this usually meant developing strains with greater docility, increased milk yield or improved coat quality (Zeuner 1963) and with plants it usually meant selecting for the enlargement of edible structures (Fig. 2.3).

At first, farmers would have been restricted to using the crops available in their own immediate locality and this could have carried penalties. For example an initial over-concentration on maize (*Zea mays*) seems to have caused various kinds of

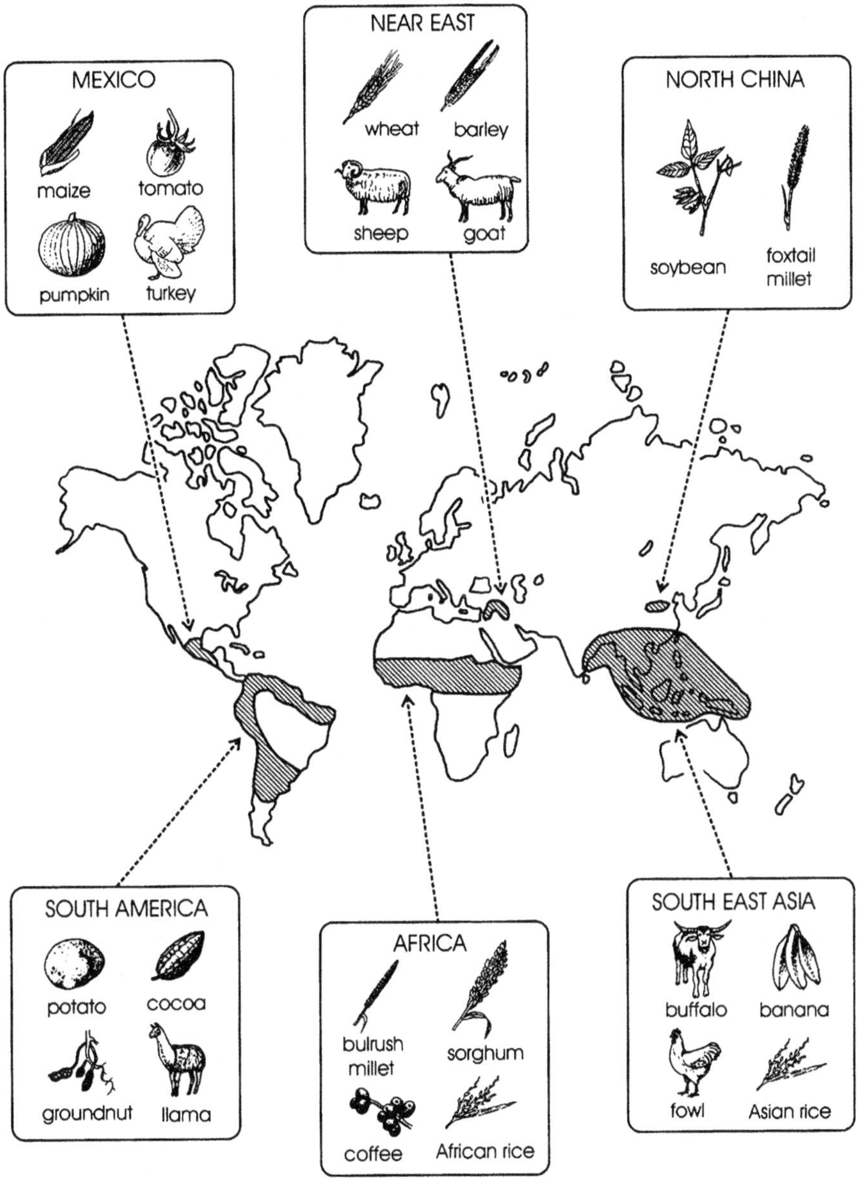

Fig. 2.2 Probable sites of domestication for selected plant and animal species (Based on Zeuner 1963; Harlan 1971; Smartt and Simmonds 1995 and Glover and Higham 1996)

skull malformations (Larsen 1995, 2000) and may also have precipitated the deficiency disease, *pellagra*. Eventually however the world-wide dispersion of other domesticated species would have allowed farmers to build up more varied crop repertoires. One can only speculate about what the dispersal agencies were, but they

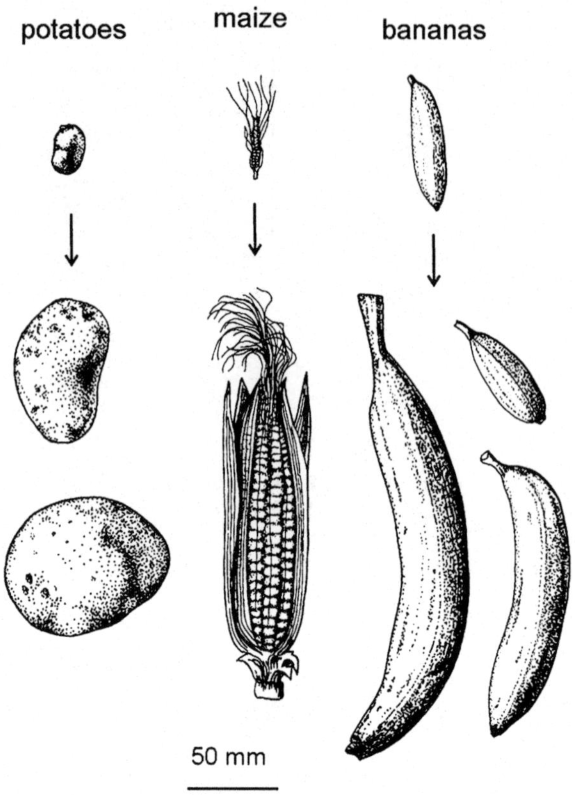

Fig. 2.3 Enlargement of the edible portions of food plants achieved through the processes of domestication and selection

would certainly have included travellers, explorers and probably also missionaries. The input from modern plant breeders came very much later. By way of illustration, Table 2.1 attempts to reconstruct the details of early crop dispersals in the West African forest zone. It can be seen that of the ten crops featured, only three are of local origin, the remainder coming from much further afield. There are some surprises in the list. Bananas, for example, which most people associate with Africa or the West Indies, seem actually to have originated in neither of those places but in South East Asia (Simmonds 1962), a region where wild bananas can still be found growing in the forest.

Whilst it eventually created a more reliable food-supply system, sacrificing the mobility enjoyed by hunter-gatherers would have generated challenges, including the risk of soil fertility decline and a greater vulnerability to pest attack (p. 5). Western farmers can call on outside assistance to address these and other difficulties. Agrochemicals are readily available to counter difficult soil conditions and to suppress weeds and pests. If they so choose, modern farmers can also commission specialists to survey the soils on their holdings and agricultural advisors are on hand to recommend the best crop mix for particular situations. Even severe obstacles such as insufficient rainfall and short day-lengths are susceptible to technical

Table 2.1 Origins and likely history of some West African crop plants

Crop plant	Probable domestication site	History in West Africa
○ Maize (*Zea mays*)	Central America	Sixteenth century Portuguese introduction
○ Rice (*Oryza sativa*)	Asia	Sixteenth century Portuguese introduction
● Rice (*Oryza glaberrima*)	West Africa	Local dissemination
○ Cowpea (*Vigna unguiculata*)	Southern Africa	Local diversification
○ Cassava (*Manihot esculenta*)	South America	Sixteenth century Portuguese introduction followed by local and modern diversification
● Yellow yam (*Dioscorea cayenensis*)	West Africa	Local dissemination and diversification
○ Greater yam (*Dioscorea alata*)	Asia	Sixteenth century Portuguese introduction
○ Cocoa (*Theobroma cacao*)	South America	Nineteenth century Spanish introduction, dissemination by missionaries
● Oil palm (*Elaeis guineensis*)	West Africa	Diversification by local farmers and modern plant breeders
○ Banana (*Musa sapientium*)	South East Asia	Fifth century introduction via Madagascar. Local diversification supplemented by later introductions from the West Indies

Based on Purseglove (1968, 1972), Irvine (1969) and Smartt and Simmonds (1995)
● = of indigenous origin, ○ = introduced

solutions, for instance in the form of sophisticated watering systems and artificially-illuminated greenhouses. Traditional farmers face the same range of obstacles, but are generally obliged to address them on their own terms. They are moreover frequently under strong pressure to increase food production to meet the demands of expanding human populations. How well do they cope?

2.2 Responding to Site and Soil Variations

2.2.1 Local Recognition of Soil Types

Bontok cultivators on the Philippine island of Luzon base their crop-choices on their own land classification systems (Prill-Brett 1986). Likewise the *Teduray* of Mindanao employ a system that incorporates both land-form and soil-type categories. Thus they know that *tamfur* or sandy-loam soils are suitable for growing bananas, whereas *batewan* stony soils should be reserved for undemanding crops such as squashes or egg plants (Schlegel 1979). The *Ketchi* of Central America are notable for employing a particularly comprehensive scheme of site classifications and one that takes account not only of the colour and texture of soils but also their

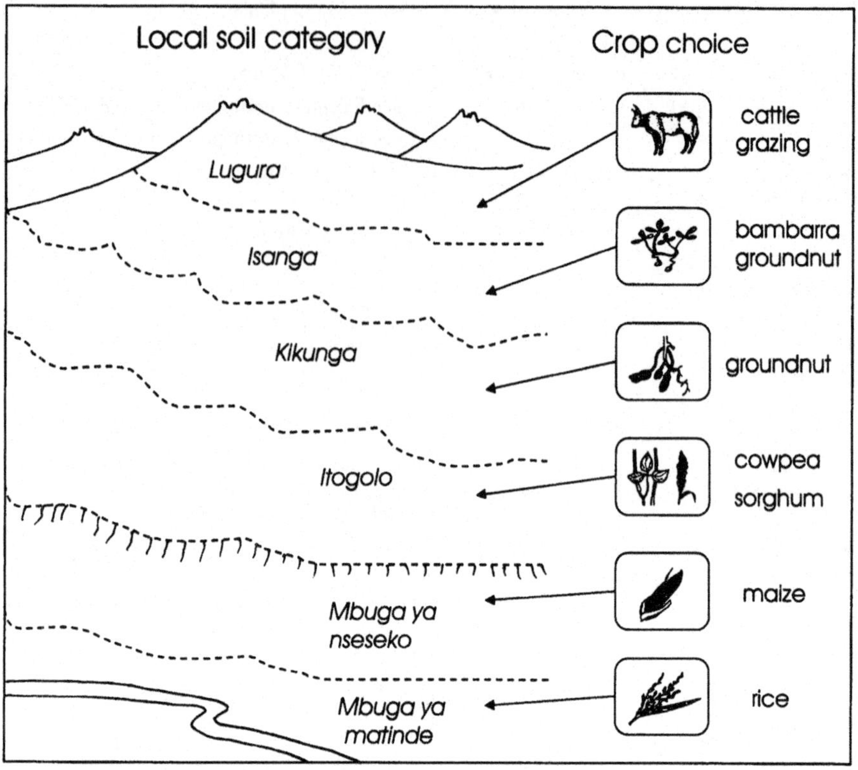

Fig. 2.4 The soil classification systems and crop choices of *Sukumaland* farmers (Developed from Malcolm 1953)

water-holding capacities (Carter 1969; Marten and Vityakon 1986). Although traditional soil classifications are less sophisticated than modern ones, in not taking vertical soil-profile features into account, modern scientific measurements of properties such as particle-size distribution and soil-nutrient status frequently confirm the local judgments (Ettema 1994).

In hilly terrain, erosive processes are usually the main determinants of soil type. The downstream passage of water removes fine particles from upper levels and also leaches out dissolved nutrients. Soils further down the slopes are the recipients of these displaced materials and become enriched in consequence. Traditional farmers show themselves to be generally well-acquainted with such processes and to be capable of turning this knowledge to their own advantage.

For instance, in *Sukumaland* (now part of Tanzania) where the landscape is spectacularly punctuated by granite-topped hills, local farmers appreciate that the hilltops themselves are unsuitable for any agricultural purpose other than grazing (Fig. 2.4). Further down the slope however they recognize various categories of useable soils. In an upper band are leached sandy soils of relatively poor quality. The farmers call these *isanga* soils, and generally use them for growing undemanding crops such as the *bambarra* groundnut (*Voandzeia subterranea*). At the next level down, where richer loams or *kikunga* soils develop, these are reserved for more challenging crops such as the common groundnut (*Arachis hypogaea*). At the foot of the slope, where the deposition of clay particles results in the formation of hardpans, the thinner crusts are usually planted with cowpeas (*Vigna unguiculata*) and the thicker ones with sorghum (*Sorghum vulgare*). Finally local farmers recognise that the land briefly inundated by river-flooding during the wet season is suitable for maize (*Zea mays*) cultivation, whereas that flooded for longer periods is ideal for rice (*Oryza sativa*) growing. As well as having their own set of names for the different soil types, the farmers also make use of wild tree species as soil indicators (Malcolm 1953). Underlying this latter practice is the realization that certain wild plant assemblages share their habitat requirements with particular crops.

These remarkably detailed local perceptions of soil types and agricultural potentials were first fully revealed in an exhaustive survey of *Sukuma* agriculture carried out by Donald Malcolm, a British colonial administrator (Malcolm 1953). Even earlier however, the *Sukumaland* farmers and their soil classification systems had earned their place in history by prompting Geoffrey Milne, a British soil surveyor, to develop his now widely used concept of the soil *catena* (Milne 1936a, b). The term derives from the Greek word for a chain and the concept predicts that on any particular geological formation, erosive and depositional processes will generate similar soil sequences on every slope; a finding with profound implications for agriculture on hilly terrain.

2.2.2 The Use of Multiple Seed Strains

Sometimes it is not so much a question of farmers using different species to cope with site variations, but different strains of the same species. When Peruvian potato growers plant their high-altitude fields, they select varieties they know to be frost resistant (although usually less tasty) and reserve their regular varieties for lower elevations. Such farmers are typically able to identify 35 different potato varieties and keep as many as 25 of these handily in stock (Brush 1980; Brush et al. 1981).

The *Mende* of Sierra Leone are similarly adept at matching strains with sites. They are familiar with 59 different rice varieties and typically make use of about eight of these at any one time (Richards 1985). They plant some strains on the hilltops where direct rainfall is the only source of water and others in a succession down the slope of the river valley; in each case with is a careful matching of known growth rates and the expected duration of flooding. Many of these judgements are based on genuine experimentation. The farmers routinely plant rice varieties in experimental plots to observe their performance; the varieties having been obtained either through exchange with other villages or by the collection of mutant strains from the fields. The experimental plots are carefully chosen to represent particular soil types or sites at specific points on the flood profile and the aim of the experiments is to assess performance in these various situations. Yields are quantified using homemade measuring devices such as calabashes or tobacco tins (Richards 1985).

Peruvian farmers similarly experiment with potato varieties at different elevations with the expressed intention of discovering the lower altitude limit for this crop (Rhoades and Bebbington 1995). In that they involve hypothesis framing followed by experimental testing; these explorations have most of the hallmarks of scientific procedures. The only missing element is a lack of the statistical techniques to detect small differences between experimental outcomes.

2.3 Coping with Seasonal Variations

2.3.1 The Farming Calendar

An obvious way for a farmer to react to seasonal variations is to devise a work calendar that fits the climatic progression of an average year. Factors taken into account include the time needed to prepare the land for individual crops, the maturation time of these crops and the desirability of doing the harvesting during dry weather. This last precaution avoids the problem of trying to store wet produce and denies crops such as groundnuts the opportunity to start sprouting whilst they are still in the ground. A typical work calendar from *Sukumaland* is shown in Fig. 2.5. Although the sequence varies slightly for the different crops, the general principle is to start planting once the rains have got under way and to do the harvesting during the drier months. When there are two rainy seasons, as there are elsewhere in East Africa, this fact is taken into the reckoning, with millet and maize usually being planted in anticipation of the first rains and beans, sorghum and groundnuts in advance of the second ones.

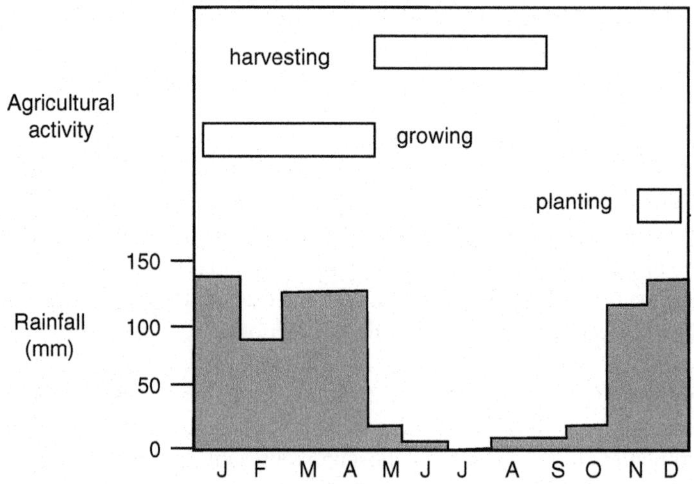

Fig. 2.5 A *Sukumaland* crop calendar (Developed from Malcolm 1953)

2.3.2 Multiple Crop Strains

Unfortunately rainfall sequences cannot be relied on to repeat themselves with perfect precision from 1 year to the next. This uncertainty has prompted the development of some interesting local insurance strategies, again involving multiple crop strains. In the drier parts of Africa, farmers are in the habit of planting multiple varieties of sorghum (*Sorghum bicolor*) and bulrush millet (*Pennisetum typhoides*) in the same field; often as many as three varieties of each. The millet strains tolerate drier conditions than the sorghums (200–800 mm compared with 550–1000 mm annual rainfall), but within each species strains exist with marginally different water requirements. By planting such mixtures together, the farmers are in effect covering all eventualities. If the rains turn out to be unusually light, then only the most drought-resistant of the millets will develop successfully. If on the other hand the rains are heavy, the most water-demanding of the sorghums will come into its own and outperform all the other varieties. These traditional insurance strategies are well documented from both West and Southern Africa (Hill 1972; Mushita 1993). A similar principle is also employed in the cooler highlands of Ethiopia, except there the mixture involves an assortment of wheat and barley strains; the varieties thriving in any particular year being largely dependent on whether the prevailing winds are dry ones blowing from the Red Sea or moister ones originating from the North West of the country. Local farmers claim that the orientation of weaver bird nests offers a reliable indication of overall wind direction; the nest openings being prudently positioned to face downwind.

2.3.3 Seasonal Indicators

Farmers regularly use natural indicators to chart the passage of the seasons. Prompts include the flowering and fruiting cycles of wild plants, the migratory movements of birds and the changing patterns of star constellations. For the *Desâna* shifting-cultivators living alongside the headwaters of the Rio Negro in Brazil, it is essential to have sections of the forest ready-felled in preparation for the burning operations that can be carried out only during dry periods. It is this burning process that generates the ash needed to fertilize new crop areas. The appearance in the sky of a constellation shaped like a pit viper (Fig. 2.6a) is the prompt for one round of felling (Ribeiro and Kenhíri 1989). This star pattern probably equates to the *Scorpius* constellation of conventional astronomy (Hugh-Jones 1982). In turn, the end of the fruiting season of leguminous *ingá* trees (Fig. 2.6b) serves to indicate that the short January dry season is about to start and that preparations for another bout of burning should be put in hand (Ribeiro and Kenhíri 1989). These trees are represented in the region by a number of species including *Inga edulis, Inga cinnamomea* and *Inga marginata*. All are notable for producing conspicuous large pods with fluffy edible seeds (Silva 2011).

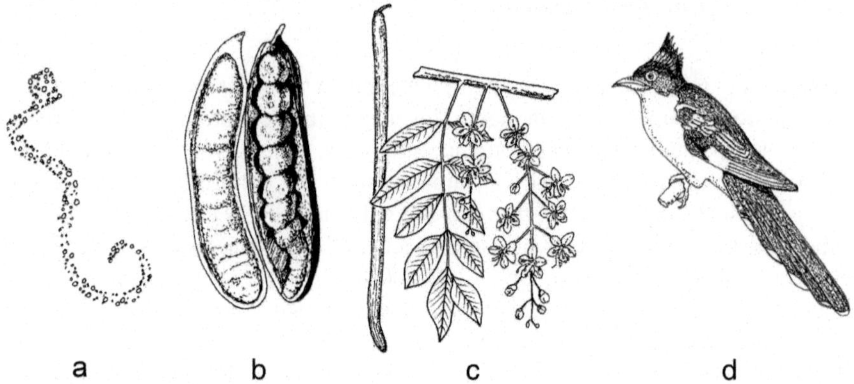

Fig. 2.6 Some traditional seasonal indicators. (**a**) Pit viper star constellation (after Ribeiro and Kenhiri 1989), (**b**) *ingá* fruit pod, (**c**) flowers of the Indian laburnum (*Cassia fistula*), (**d**) pied crested cuckoo (*Clamator jacobinus*)

For farmers living in parts of Africa and Asia influenced by the monsoons, speculation about their timing is an annual preoccupation. Whilst some of the signs suggested are clearly fanciful, others appear to have a sound biological and meteorological basis. In West Africa for example, the appearance of young reddish leaves on forest trees such as *eki* (*Lophira alata*) and *iroko* (*Chlorophora excelsa*) is locally interpreted as a firm indication that the rains are on their way. As leaf-break in many forest trees is known to be stimulated by moist oceanic airflows that bring the rains, these notions are probably well-founded (Oguntoyimbo and Richards 1978).

In India likewise, there is a folk belief that the monsoon will start 45 days after the blooming of the yellow-flowered, Indian laburnum tree (*Cassia fistula*) (Fig. 2.6c). Since 1996, Dr. Purshottambhai Kanani, an agronomist based at Gujarat Agricultural University has been examining this claim and, on testing it over a series of late and early monsoons, has found it to be a generally good guide (Anonymous 2001). Indian farmers also have their favourite monsoon indicator amongst the birds. This is the pied crested cuckoo (*Clamator jacobinus*), a striking and easily recognized species (Fig. 2.6d). Along the West coast its arrival is taken as a sure sign that the rains are only a few days away. Because this bird migrates from Africa and is generally carried to India on the monsoonal winds, the claim again has substance (Ali and Ripley 1984).

2.4 Maintaining Soil Fertility

2.4.1 Nomadic Pastoralists and Shifting Cultivators

Two groups have devised novel solutions to the threat to soil fertility posed by continued use of the same land. Nomadic pastoralists periodically move their herds from base areas to places where seasonal rains have produced a temporary flush of

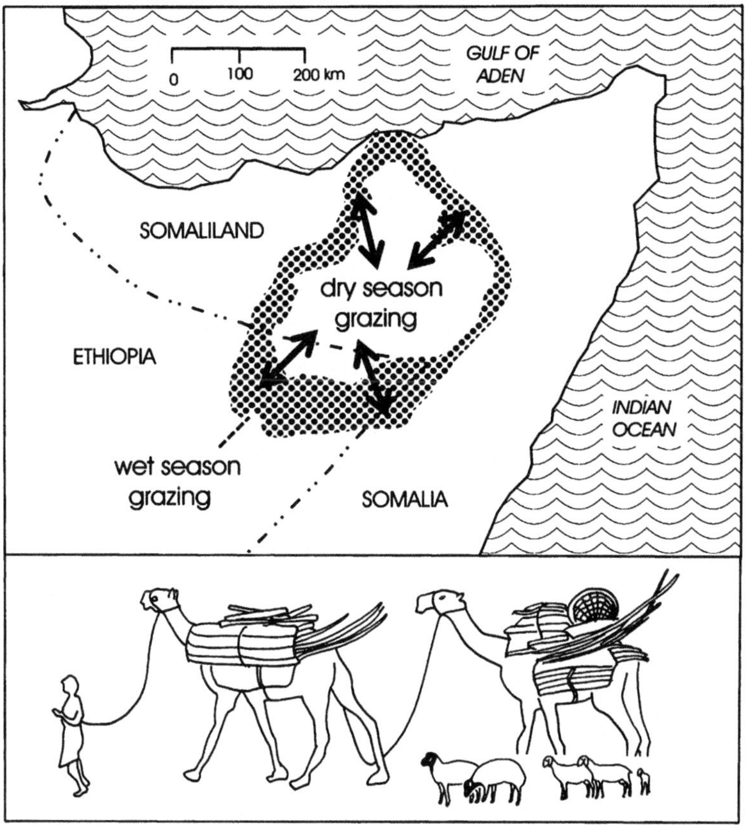

Fig. 2.7 Movement patterns of pastoral nomads in the Horn of Africa (Developed from Lewis 1965)

new grass. Figure 2.7 shows such a migration pattern as practiced by nomads operating in the dry areas in the horn of Africa. In this case the animals involved are camels, sheep and goats. Elsewhere in East Africa, cattle are the species of choice, the *Maasai* being well known exponents of this kind of pastoralism. Again the strategy is to rest core areas by moving stock to peripheral pastures that have recently received rain (Jacobs 1975). In Northern climates the general principle is the same but the herd animals are different, typically consisting of yaks or reindeer, and the stimulus for movement is not seasonal rain, but the exposure of pastureland by melting snow (Forde 1934; Ekvall 1968). The acute environmental awareness of pastoralists is exemplified by the fact that even *Maasai* boys of fifteen can to identify and gauge the grazing value of at least 15 species of grasses.

There are however certain blind spots. Because cattle numbers are equated with wealth and status, there is a temptation to keep larger herds than can be supported by the available pastures. In Africa this phenomenon is referred to as the "cattle complex" and is arguably an important cause of over-grazing and soil erosion

(Herskovits 1926). In a similar way the "camel complex" of the *Bedouin* in Arabia can result in the serious overstocking of grazing areas (Gallacher and Hill 2008).

Shifting cultivators counter the problem of soil-nutrient depletion by periodically moving the location of their farming operations. Farm plots cut from the tropical forest have a limited period of productive life because of the intrinsic infertility of the underlying soils (Gourou 1958). The remedial strategy is to abandon depleted plots to give them time to recover and to open up new areas elsewhere (Fig. 2.8a). The woody vegetation developing on the abandoned plots has the capacity to draw-up additional minerals from deeper strata. These nutrients then become released to the surface soil through the processes of leaf fall and the cutting and burning of the woody vegetation (Nye and Greenland 1960). This cycle of depletion and renewal can continue virtually indefinitely, as long as the recovery periods are kept sufficiently long (Fig. 2.8b); 10 years being an average length for each complete cycle (De Schlippé 1956; Denevan and Padoch 1988). Population pressure poses the main threat to such systems because it tempts farmers to try and increase production by reducing the length of the fallows.

Settled farmers are generally in a poor position to employ such strategies although they may use fallow periods in a limited way. One of the best-recorded historical examples is provided by the mediaeval three-field system of Western Europe. This worked by allowing each field a rest and recovery period every third year (Duby 1968, Baker 1973). Similar systems are still in use in many parts of the tropics, although with much more varied time allocations between the cultivation and resting phases (Ruthenberg 1976).

2.4.2 *Mulches and Manures*

Some traditional techniques depend, not so much on the *in situ* growth of natural vegetation, but rather on the burning or composting of materials brought in from elsewhere. In parts of Tanzania, the two techniques are combined, with gourds and cucumbers being grown in ash piles and peas and beans in compost beds (Mbegu 1996). An unusual variant practiced in Peru is for farmers to collect the fronds of a local fern (*Dennstaedtia glauca*) to use as a green manure. When this material is applied to terraces where potatoes and *ocas* (another traditional root vegetable, *Oxalis tuberosa*) are being grown, it can be shown to produce significant increases in levels of soil nitrogen, phosphorus and potassium (Camino and Johns 1988). A novel way of introducing organic material to a cultivation plot is that practiced by *Papago* Indians in the American South West (Nabhan 1987). They build wicker-work barriers across flood gullies so that, when the floods come, the plant detritus being carried along is deposited behind the barrier and creates a ready mulched seed bed.

A technique widely practiced in the Himalayan foothills is for farmers to collect grass and tree-leaf fodder to feed to animals being held overnight in byres or on

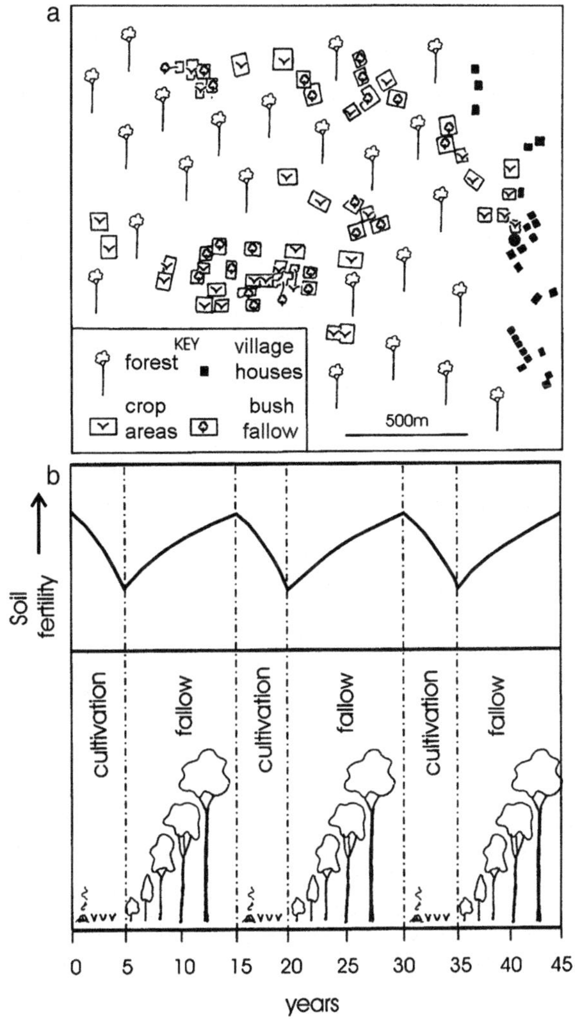

Fig. 2.8 Shifting cultivation. (**a**) Cultivation patterns amongst the *Bora* peoples of the Peruvian Amazon (based on Denevan and Treacy 1987), (**b**) the soil fertility renewal system employed by shifting cultivators

house forecourts (Fig. 2.9). These byres are sometimes free-standing structures, but often the entire ground floor of dwelling houses is given over to this purpose. After the animals have fed, the residual fodder, now mixed with manure and urine, is gathered up and carried out to the fields to be used as a fertilizer. An interesting feature of this cycle is the degree of discrimination exercised in the selection of fodder species (Joshi and Joshi 2003). Whilst any mix will eventually yield useful manure, some fodder plants are avoided because they are suspected of reducing milk-yields or of so altering the character of the milk as to render it unsuitable for cheese and butter production (Gaur et al. 1981–2).

Fig. 2.9 The leaf fodder-manure cycle as practiced in Northern India. (**a**) Fodder being carried into the village, (**b**) manure and animal bedding being transported back to the fields (**c**) buffalo being fed on house forecourts and in ground-floor byres

2.4.3 Biological Nitrogen Fixation

Even these examples do not quite exhaust the diversity of traditional soil-enrichment techniques. Some very ingenious ways are used to fix atmospheric nitrogen. Visitors to drier parts of the tropics are often surprised to see scatterings of trees retained in cereal fields. On the face of it this would seem to be an unhelpful practice, because of the risk that shading from the tree canopies would reduce crop growth. In reality however, this potential drawback is more than outweighed by the nitrogen-fixing capability of the trees. The species employed are all members of the pea family, Leguminosae. In Northern India *khejri* (*Prosopis cineraria*) is the species of choice, whilst in Ethiopia its place is taken by the white thorn (*Faidherbia albida*) (Fig. 2.10). The tree's nitrogen-fixing capability is located in its root nodules and some of the nitrogenous products formed there diffuse usefully into the soil. Further nitrogen additions result from the decomposition of fallen leaves. Through these various processes, the surrounding cereal crops are in effect receiving a free fertilizer. In Ethiopia, the potentially adverse shading effect of the thorn trees is lessened by their branches being periodically cut for firewood. Tree densities are typically in the region of 10 trees per hectare and the resultant increases in crop yields are estimated to amount to 36% for sorghum and 76% for maize (Poschen 1986). Using low-growing peas and beans for nitrogen-fixation is of course a much more familiar practice and one dating back at least to Roman times (Isely 1982; Beets 1982). However employing leguminous trees although less well known is evidently no less effective (Felker 1981).

Because they are periodically flooded with water, lowland rice fields do not lend themselves to such use of legumes. This does not mean however that there is no scope for deploying nitrogen-fixing techniques and rice farmers in South East Asia have devised an ingenious procedure based on the floating water fern *Azolla* (Fig. 2.11e). This plant grows on the water surface and its capacity to fix atmospheric nitrogen derives from the metabolic activities of symbiotic organisms present in its tissues (Ashton and Walmsley 1976). These are variously classified by biologists as blue-green algae or as cyanobacteria.

The farmer starts the process by scattering the ferns across his rice fields and then allows them to grow and spread over the water surface, renewing the coverage as necessary (Fig. 2.11). When the plants eventually die and disintegrate, the nitrogenous compounds contained in them are liberated into the water and are available for assimilation by the rice plants. Another option is to plough the living ferns directly into the soil. Estimated increases in rice production range from 14 to 40% (Lumpkin and Plucknett 1980). The technique was evidently invented by the farmers themselves, although who precisely first discovered it is a matter of conjecture. In Vietnam, the story is told of a lady farmer from the village of *La Van* in *Thai Ninh* province who developed the technique over a century ago and made a living by selling starter cultures to fellow farmers. Chinese historians are unimpressed by this tale and lay claim to a much earlier use of *Azolla* in their own country.

Fig. 2.10 Soil enhancement techniques based on nitrogen-fixation by leguminous trees. (**a**) *Khejri* trees (*Prosopis cineraria*) retained in a millet field in Rajasthan (the animals under the tree are desert gazelles, a species protected by the local *Bishnoi* tribe), (**b**) white thorn trees (*Faidherbia albida*) retained in an Ethiopian millet field

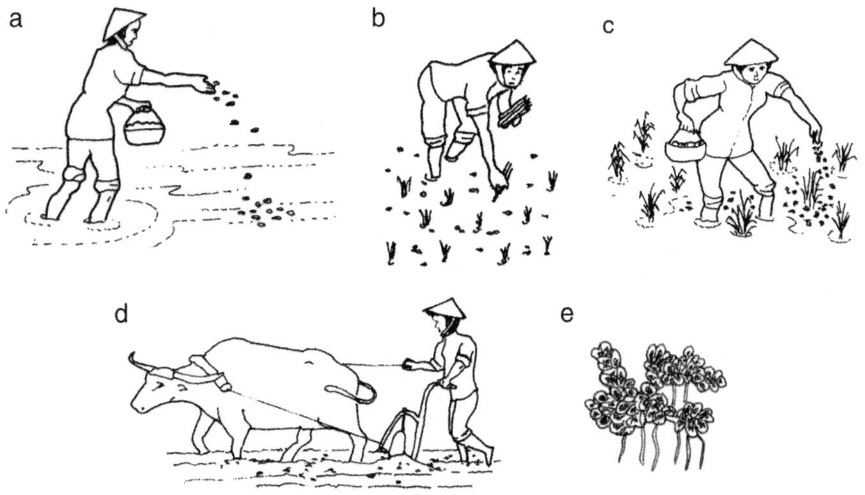

Fig. 2.11 Use of the water fern *Azolla* as a fertility-enhancing agent in rice cultivation. (**a**) Sprinkling ferns on a flooded field, (**b**) planting rice seedlings, (**c**) filling gaps with additional ferns, (**d**) ploughing-in ferns after the rice has been harvested and the field drained, (**e**) individual fern plant

2.5 Coping with Weeds and Pests

2.5.1 Weeds

Whilst Western farmers heartily disapprove of wild plants invading their crop fields and usually contrive to eliminate any species they suspect of interfering with crop production, the attitude of traditional farmers is generally more flexible. They certainly recognize that some wild plants compete so vigorously with their crops that they need to be removed, an exceedingly laborious process if done manually. For instance, to weed an acre of maize using a hand hoe can take as long as 3 weeks to accomplish, even working 8 h a day (Skjønsberg 1989).

However in traditional settings not all invasive wild plants are unwelcome. Some are recognized as having the potential to serve useful purposes, including acting as supplementary foodstuffs, animal fodder or sources of raw materials. Indian rice-fields are estimated to support as many as 158 useful species of this kind (Datta and Banerjee 1978). Moreover in Thailand some rice-field weeds are so highly prized that they are routinely transferred to gardens for growing-on and subsequent market sale as leaf vegetables (Price 1997). Some communities go so far as to make a formal distinction between useful and damaging wild species (Chacón and Gliessman 1982). Thus in the Tabasco region of Mexico, 20 species are classed as *buen monte* or "good" wild plants and are variously put to use as human foods, animal fodder, herbal medicines and composts. Placed in the opposite category are the *mal monte* or "bad" wild plants of which there are some 21 species. These are poorly regarded because they compete with crops, compact the soil or are poisonous to farm animals or humans.

2.5.2 Pests

There are many reasons why crop plants are susceptible to pest attack, amongst them the fact that crops are bred to be especially nutritious, are often grown in vulnerable monocultures and have commonly lost some of their natural defensive mechanisms. However, unlike their Western counterparts, traditional farmers are not necessarily unduly disconcerted by the presence of pests in their fields. Larger insects such as grasshoppers and locusts may simply be hand-picked from the crop and used as food. In South Africa for example, the *Pedi* routinely collect elegant grasshoppers (*Zonoceros elegans*) and brown locusts (*Locustana pardalina*) in this way, usually timing their collections for the early morning when the insects are still sluggish (Quin 1959). In West Africa the stinking grasshopper (*Zonoceros variegatus*) receives similar attention and, although the insect produces a repellent fluid, this can apparently be neutralized by boiling (Idowu and Modder 1996).

Crop invasions by vertebrate pests are similarly capable of being turned to advantage. In the forest zone of West Africa, palm-leaf barriers are constructed around farm plots in such a way as to leave gaps for the insertion of traps. The species most frequently caught in this manner are cane rats or "cutting grass" (*Thryonomys swinderianus*), pouched rats (*Cricetomys gambianus*) and brush-tailed porcupines (*Atherurus africanus*), all of them valuable sources of protein (p. 53). Likewise in the rice-fields of Southeast Asia, it is a common practice to set traps for seed-eating birds and to use the captures to supplement food supplies (Brown and Marten 1986).

2.5.3 Plant-Based Pest Repellents

When more general pest-control measures are thought necessary, village farmers often seek to harness the pest-repelling properties of wild plants. Some 672 species are reported to be used in this capacity (Secoy and Smith 1983) although, as is the case with medicinal plants, unambiguous evidence of effectiveness is often lacking (Prevett 1962). One undoubted success is the traditional Chinese use of root-bark from the thundergod vine (*Tripterygium wilfordii*). This contains toxic alkaloids (Beroza and Bottger 1954) and is employed to control caterpillars and aphids (Yang and Tang 1988).

There are also some notable instances of plant materials being used to protect stored produce. In India the leaves of the native *neem* tree (*Azadirachta indica*) have for many centuries functioned in this role (Fig. 2.12a–c). A common technique is to coat the interior of grain storage bins with a paste prepared from *neem* leaves and then to place further dried leaves on the grain surface (Ahmed and Grainge 1986). Modern studies confirm that *neem* preparations can be effective in controlling as many as a 100 species of pests, including insects, mites and plant nematode worms (Stoll 1988). A comparable practice in South America is where Andean potato farmers line their potato storage boxes with the leafy twigs of *muña* (*Minthostachys*

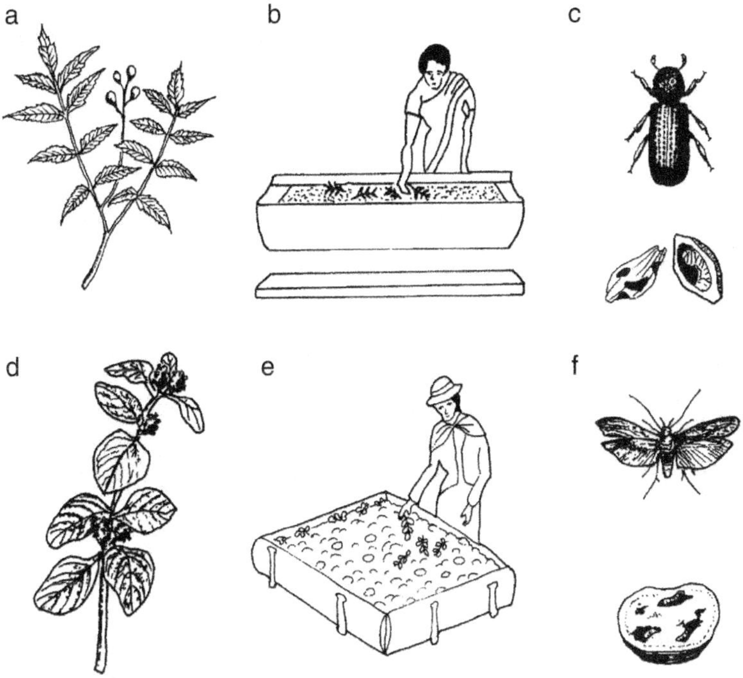

Fig. 2.12 Traditional use of pest-repellent plants. (**a**) *Neem* (*Azadirachta indica*), (**b**) leaf spray being placed in a traditional hollow log grain-storage bin in India, (**c**) a commonly targeted species – the lesser grain borer (*Rhizopertha dominica*) and the type of damage caused, (**d**) *muña* (*Minthostachys mollis*), (**e**) leaf spray being placed in a potato store in Peru, (**f**) targeted species – the potato tuber moth (*Phthorimaea operculella*) and the type of damage caused (Developed from Stoll 1988 and Alkire et al. 1994)

mollis) a shrubby species of mint (Fig. 2.12d–f). This discourages attacks by aphids and potato-tuber moths (*Phthorimaea operculella*), the repellent agent in this case being an oil contained in the plant's leaves.

2.6 Cultural and Religious Factors

2.6.1 *Pleas for Good Growing Weather*

Thus far we have been describing practical measures instituted by traditional farmers; measures that can be substantiated in scientific terms. Not all local actions fall into this logical framework. The notion that the gods control the weather, but are amenable to negotiation, forms the basis for the great variety of agricultural ceremonies that punctuate the farming calendar. The corn dances performed by the Pueblo Indians of New Mexico offer a typical example and illustrate the reliance often

Fig. 2.13 Pueblo Indian corn dance (Developed from Bahti 1982)

placed on sympathetic magic. Male dancers carry gourds containing pebbles which they rattle to simulate the sound of raindrops (Fig. 2.13). The women in their turn wear wooden headdresses carved to portray banks of clouds and drops of rain. All the dancers carry sprigs of evergreen trees such as juniper and Douglas-fir (*Juniperus osteosperma* and *Pseudotsuga menziesii*), two species associated with wetter conditions than those prevailing locally. Finally, the dance-troupe leader carries a tall staff topped with tail feathers from the thick-billed parrot (*Rhynchopsitta pachyrhyncha*). This is another allusion to rain because this particular species, although now mainly confined to Mexico, formerly arrived annually with the rain-bearing winds. On a rational basis, it is inconceivable that these rainmaking ceremonies could affect the weather; forecasting is hard enough without any further complications. They may however have the virtue of reassuring participants that everything has been done to forestall the possibility of crop failure. With ever-present threats from flooding, droughts and tropical storms, village communities in developing countries are arguably in particular need of comforting constructs of this kind.

2.6.2 *Heterodox Attitudes Towards Pests*

Religious factors also influence the way pest species are regarded in traditional communities. For example Buddhist rice farmers in Sri Lanka routinely set aside sections of their fields where birds can feed unmolested (Goldsmith and Tenakoon

Fig. 2.14 Village tolerance of potential pest species. (**a**) Weaver birds being permitted to nest in an oil palm tree in Nigeria, (**b**) humane trap used in a *Bishnoi* village to remove rats from the vicinity of grain stores (**c**) grain store roof being used as a bird-feeding station in the same village

1982). Likewise in Nigeria, smallholders with oil palm trees in their compounds are not greatly disconcerted if local weaver birds (*Ploceus cucullatus*) choose one of the trees as a communal nesting site (Fig. 2.14a). Although nest-construction involves the birds stripping the tree's leaves and effectively halting fruit production, householders not only tolerate these depredations but deem it an honour that their trees should have been chosen (Funmilayo 1975).

In dry areas of Rajasthan, agricultural advisors have had little success in persuading farmers to control rodents and marauding wild antelopes such as the *nilgai* (*Boselophus tragocamelus*) (Sankhala and Jackson 1985; Malhotra 1986). This is because the local *Bishnoi* community adheres strictly to a central tenet of Hinduism, which is that wild species come under the protection of the gods and should therefore not be molested. Thus when rats gain access to the communal grain store they are captured unharmed in specially-designed traps (Fig. 2.14b) and subsequently transported to the edge of the village for release (although probably returning rather rapidly to the grain stores). This particular instance of respect derives largely from the fact that all rodents are regarded as coming under the patronage of the god *Ganesa*.

Another regular practice in the same villages is for handfuls of millet seed to be taken from the grain store (which in years of poor rain is far from full) and spread on the flat roof for the benefit of local wild birds (Fig. 2.14c). Birds have the good fortune to be associated with a variety of Hindu deities including *Kartikeya, Durga and Vishnu*. In its turn the tolerance shown towards *nilgai* and other larger mammals such as desert gazelles (*Gazella gazella*) and blackbuck (*Antilope cervicapra*) has its origins in the association of cattle and similar cow-like animals with the god *Shiva*. These species all have the potential to compete with domesticated sheep and goats on village land but in practice none is ever molested. A subterfuge devised by the agriculture department was to try and persuade local people that *nilgai* were more akin to horses than to cows and could therefore be legitimately controlled. In the event this stratagem came to naught, because local people pointed out that horses have a supernatural patron too, in the form of the god *Kalki*. In fact so extensive is the range of deities in the Hindu pantheon that few species lack religious patronage of one sort or another.

2.7 Pressures for Change

2.7.1 The Need for Productivity Increases

Notwithstanding the ingenuity they bring to their various tasks, farmers latterly have come under strong pressure to increase their productivity. One motive for doing this is to supplement their incomes and to improve the quality of their lives; the prospects appearing especially favourable if some kind of international trading arrangement can be entered into. Secondly, governments are constantly instructing farmers to step-up food production to keep pace with population growth. Neither of these objectives is readily achieved. Attempts to increase trade involvement can fall foul of world market fluctuations and trade barriers, can incur trading losses to intermediaries and can sometimes trigger the unwelcome phenomenon of "cash-crop malnutrition" (p. 193).

Moreover attempts to increase productivity, for whatever reason, can sometimes prove incompatible with traditional farming techniques. The shortening of fallows is a case in point. If these are curtailed or eliminated with the intention of increasing yields, this strategy can backfire and lead to agricultural collapse. Hunter (1967) describes such a sequence from northern Ghana. There, efforts to raise food production to keep pace with population increases of 2.5–3.0% per annum disrupted agricultural cycles and by precipitating widespread soil erosion caused a dramatic fall in food production rather than the hoped for increase. Okafor (1987) describes a similar progression in south eastern Nigeria, where the shortening of fallow periods and the fragmentation of farm holdings caused soil erosion on such a large scale that farming operations were effectively brought to a halt.

This is part of the justification for boosting farm productivity by employing the package of inputs known as the "green revolution". This typically combines the use of industrially-manufactured fertilizers and pesticides with the deployment of high-yielding seed varieties (Borlaug and Dowswell 1995). This approach has produced some spectacular results, with notable successes including the doubling of cereal production in India and other Asian countries over the period 1970–1990. Such measures served to rescue these countries from the threat of serious food shortages (Hazell 2009). They work best on large well-capitalized farms or ones fortunate enough to receive government subsidies. The difficulty faced by small independent farmers is that all the prescribed inputs cost them money. Even the high-yield seeds, because they are specially-produced hybrids, need to be purchased anew for each planting season. In these circumstances, small farmers often prefer to upgrade their existing techniques, rather than discarding them altogether.

2.7.2 Fine-Tuning Traditional Soil-Improvement Techniques

Interesting examples of such fine-tuning are provided by the work of farmer-friendly scientists at the Los Baños campus of the University of the Philippines. They have discovered, for instance, that culturing *Azolla* in cool, high-elevation ponds so increases spore production that packets of spores can be made available to farmers for use as starter cultures (Fig. 2.15a, b) (Payawal and Paderon 1986). Another valuable innovation has been to develop the practice of inoculating rice straw with the fungus *Trichoderma harzianum* (Fig. 2.15c). This has the effect of shortening by some 3–4 weeks the time needed for the straw to decompose into a useable compost. Farmers with access to this material are reported to be able to reduce their fertilizer bills by as much as a half (Cuevas et al. 1988). The enhanced *Azolla* system likewise removes the need to apply nitrogen fertilizers to the rice crop, although conventional phosphate fertilizers may still be required (Yatazawa et al. 1980; Palm and Sandell 1989).

"Alley-cropping" is another traditional technique that has been fine-tuned. Instead of dispersing nitrogen-fixing trees throughout the cereal fields, they are instead planted in rows that alternate with a crop such as maize. Leguminous shrubs commonly used in this role include *Leucaena leucocephala* and *Desmanthus virgatus*. Leafy twigs cut from the shrubs and laid in the crop areas function as growth-stimulating mulch and nitrogenous compounds diffuse into the crop area from the tree roots (Kang et al. 1981). An integral part of the process is the periodic cutting of the shrub branches to serve as firewood. Apart from yielding fuel, this also prevents excessive shading from impeding growth of the cereal crop (Wijewardene and Waidyanatha 1984).

Fig. 2.15 Some examples from the Philippines of traditional soil-enrichment techniques being upgraded. (**a**) Upland culture ponds for the production of *Azolla* spores, (**b**) stockpiling the spores for distribution to farmers, (**c**) rice straw piles inoculated with the fungus *Trichoderma*

2.7.3 Improving Pest-Control Techniques

It is often assumed that the obvious way for traditional farmers to improve their pest control measures is to make greater use of synthetic pesticides. This is not however such a straightforward option as might first appear. Handling such materials in village situations, especially in the absence of suitable protective clothing, can be a hazardous business (Bhat and Krishnamachari 1977; Bull 1982; Jeyaratnam et al.

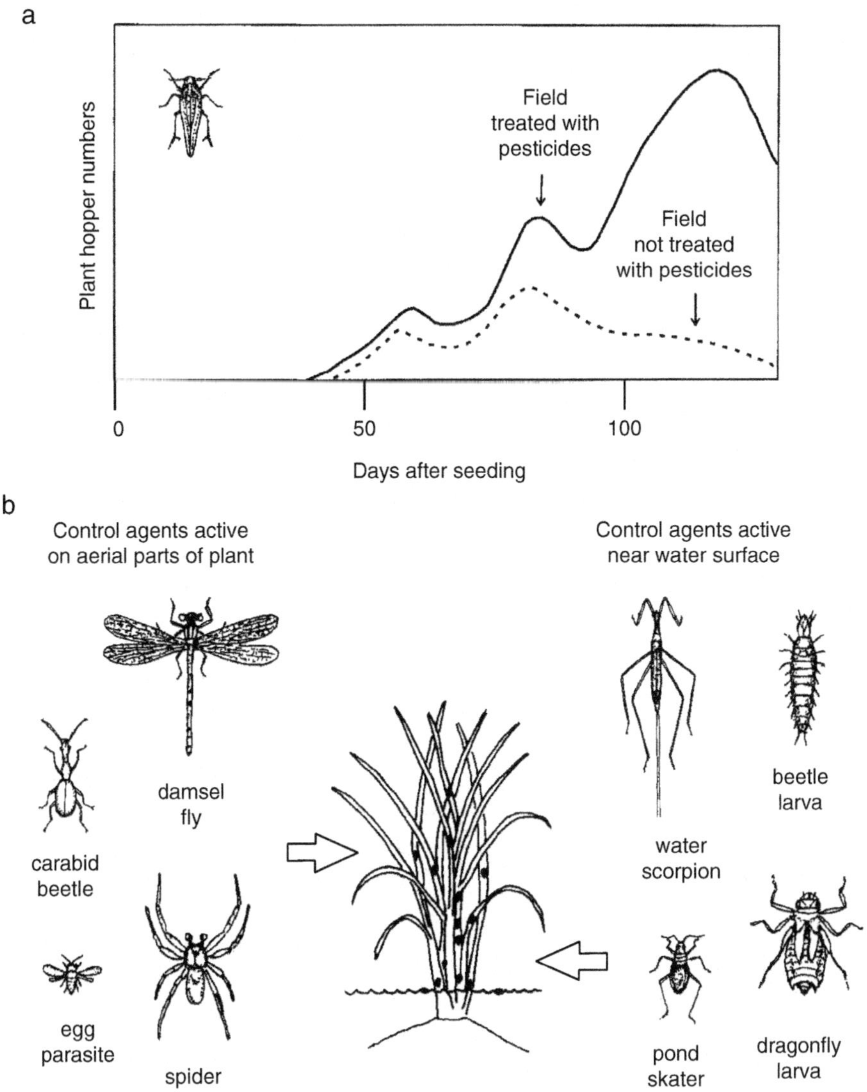

Fig. 2.16 (**a**) The anomalous situation in which a pesticide application increases a pest population (in this case of the brown plant hopper, *Nilaparvata lugens*) rather than reducing it, (**b**) natural control agents for this particular pest, indicating their sites of action (Developed from Reissig et al. 1986)

1987; Loevinsohn 1987; Maumbe and Swinton 2003). Not only this, there is evidence that pesticide applications can sometimes increase pest numbers rather than reducing them, an outcome precisely opposite to that intended (Fig. 2.16). This happens because the populations of insects (and other arthropods such as spiders and mites), normally acting as natural control agents, are more seriously affected by the pesticides than are the pests themselves. Consequently they are prevented from having their usual controlling effect. Unwanted "pest resurgences" of this kind are recorded from many major crops, including cocoa (Conway 1972), oil palms (Wood 1971), cotton (Bottrell and Adkisson 1977) soybeans (Shepard et al. 1977) and rice (Reissig et al. 1986). In such circumstances farmers are placed in the position, not only of having wasted their money in purchasing the pesticides, but also of being obliged to forego benefits they would otherwise have enjoyed at no cost.

A sounder approach is to promote the use of *biological control* techniques. These involve augmenting stocks of useful species (such as *chalcid* wasps), or else creating the conditions in which these can work more effectively. Although for centuries Chinese fruit farmers have been in the habit of controlling pest caterpillars on fruit trees by moving colonies of weaver ants (*Oecophylla smaragdina*) into the plantations (Needham et al. 1986), there are relatively few fully-documented cases of traditional farmers employing biological methods without outside prompting (Doutt 1964). The main obstacles seem to lie with identification problems and failures to understand the ecological relationships between the species involved. Historically in Western circles such an understanding had to await the development of microscopy and it is significant that Anthony van Leeuwenhoek was able to describe the life cycle of small pest insects such as aphids only after he had developed his special magnifying lenses (Magner 1979).

Lacking such tools, village observers often jump to mistaken conclusions. For instance, farmers in Honduras failed to distinguish between the parasites of pest weevils and the weevils themselves and consequently destroyed both of them indiscriminately (Bentley and Andrews 1991). Likewise farmers in the Sudan, noting that the arrival of ladybird beetles coincided with crop damage, concluded that these were the harmful pests and duly destroyed them. In fact, the real pests were small aphids and scale insects and, had the ladybirds been left unmolested, they would have helped to control the outbreaks (Sharland 1995). These are striking examples of local blind spots.

Sometimes, even with appropriate knowledge, remedying pest problems is beyond the scope of local farmers and requires international co-operation. This was the case for example when a minute parasitoid "wasp" (*Epidinocarsis lopezi*) was transported from South America to Africa to control outbreaks of the cassava mealy bug (*Phenacoccus manihot*) (Neuenschwander and Herren 1988, Nweke 2009). In effect this was a matter of re-uniting the crop, the pest and the parasite – all original inhabitants of South America – and facilitating their coexistence in a more balanced state.

There is however some scope for simpler local initiatives. In China for example, museums have been organized to produce mounted sets of key harmful and beneficial insects and to distribute them to farmers (Williams 1979). An interesting scheme in Indonesia has been to encourage farmers to grow potted rice plants in their houses. This allows them to observe at first hand the interactions between rice pest caterpillars and their natural enemies (Brader 1979). In India much reliance is placed on the distribution of instructional booklets printed in local languages. Not only do these help farmers to distinguish between harmful and useful species, they also offer guidance on how to combine biological control techniques with the use of natural repellents (Vijayalalaksmi et al. 1998).

The best plant-based repellents combine the advantages of being locally available and being free of damaging side-effects on humans and non-target species (Jacobson 1989). They usually also have the merit of containing such a wide mix of active ingredients that they are unlikely to trigger pest resistance. *Neem* extracts have all these desirable properties and, whilst at one time they were thought to be damaging to honey bees, the latest evidence is that they may actually benefit these insects by controlling their parasites (Melathopoulus et al. 2000).

There may also be some scope for extending the Western technique of growing non-crop species in farm plots, with the intention that these should serve as refuges for useful predatory and parasitic insects (Zandstra and Motooka 1978; Altieri and Whitcomb 1979; Wratten 1992). Currently, traditional farmers rarely use such methods on their own initiative largely because the underlying concept is unfamiliar to them. They do however sometimes plant sacrificial crops to divert pests from the valuable ones (Matteson et al. 1984; Brown and Marten 1986).

2.7.4 The Continuing Importance of Seed Diversity

Agriculture departments make strenuous efforts to persuade traditional farmers to abandon their traditional seed varieties in favour of the new high-yielding ones. Sometimes these efforts go as far as confiscating or destroying traditional seed stocks (Salazar 1992; Pretty 1995). Such draconian measures are usually quite inappropriate. Farmers often need to draw on traditional seed varieties to help them cope with difficult environmental conditions. Moreover, as well as being valuable to the farmers themselves, these stocks continue to serve as a repository of genetic material that modern plant breeders can draw on. This custodial function is rendered all the more significant by the fact that some of the seeds in question do not remain viable in modern refrigerated cold stores. Consequently they will remain in existence only if farmers continue to use them (Altieri and Merrick 1987).

It has often been left to non-governmental organizations (NGOs) to protect the farmer's long-term interests in this matter, although a few national seed banks contrive to act in a genuinely supportive capacity (Cooper et al. 1992; de Boef et al. 1993). The Plant Genetic Resource Centre at Addis Ababa in Ethiopia is one of these. This Centre stores and classifies seeds contributed by local farmers, multiplies them in special plots and makes batches freely available on request (Fig. 2.17). It also co-ordinates the farm testing of seeds and offers subsidies to farmers who actively preserve traditional varieties (Worede and Mekbib 1993). Ethiopia incidentally has a special significance in the history of plant breeding. *Arabica* coffee (*Coffea arabica*) was almost certainly first domesticated there and local farmers have contributed significantly to the diversification of important crops such as sorghum (*Sorghum bicolor*) and pearl millet (*Pennisetum glaucum*) (Smartt and Simmonds 1995).

Fig. 2.17 A farmer-friendly seed bank, the Plant Genetic Resources Centre at Addis Ababa in Ethiopia. (**a**) Display of local sorghum varieties, (**b**) compiling an inventory of available seed strains, (**c**) testing and multiplying seeds in trial plots

In short, a few deficiencies aside, most traditional farming practices continue to have relevance in their original settings. Not only this, attention now needs to be given to the possibility of them offering useful pointers to Western farmers. Most scope exists in situations where it is hoped to reduce agricultural energy-consumption and to offset the effects of climate change.

References

Ahmed S, Grainge M (1986) Potential of the neem tree *(Azadirachta indica)* for pest control and rural development. Econ Bot 40:201–209

Ali S, Ripley SD (1984) Handbook of the birds of India and Pakistan. Oxford University Press, New Delhi

Alkire BH, Tucker AO, Maciarello MJ (1994) Tipo, *Minthostachys mollis* (Lamiaceae) an Ecuadorian mint. Econ Bot 48:60–64

Altieri MA, Whitcomb WH (1979) The potential use of weeds in the manipulation of beneficial insects. Hortscience 14:12–18

Altieri MA, Merrick LC (1987) In situ conservation of crop genetic resources through maintenance of traditional farming systems. Econ Bot 41:86–96

Anonymous (2001) Folk wisdom. Traditional weather forecasting on trial. *The Economist*, November 24, 2001, 102–104

Ashton PJ, Walmsley RD (1976) The aquatic fern *Azolla* and its *Anabaena* symbiont. Endeavour 35:39–43

Atkinson TC, Briffa KR, Coope GR (1987) Seasonal temperatures in Britain during the past 22,000 years, reconstructed using beetle remains. Nature 375:587–592

Bahti T (1982) Southwestern Indian ceremonials. KC Publications, Las Vegas

Baker ARH (1973) Field systems in medieval England. In: Baker RH, Harley JB (eds) Man made the land, essays in English historical geography. David and Charles, Newton Abbot, pp 59–68

Bar-Yosef O, Meadow RH (1995) The origins of agriculture in the Near East. In: Price TD, Gebauer B (eds) *Last hunters –first farmers. new perspectives on the prehistoric transition to agriculture*. School of American Research Press, Santa Fé, pp 39–94

Beets WC (1982) Multiple cropping and tropical farming systems. Westview Press, Boulder

Bentley JW, Andrews KL (1991) Pests, peasants, and publications: anthropological and entomological views of an integrated pest management program for small-scale Honduran farmers. Hum Organ 50:113–270

Beroza M, Bottger GT (1954) The insecticidal value of *Tripterygium wilfordii*. J Econ Entomol 47(1):188–189

Bhat RV, Krishnamachari KAVR (1977) Endemic familial arthritis of Malnad – an epidemiological study. Indian J Med Res 66:777–786

Borlaug NE, Dowswell CR (1995) Mobilising science and technology to get agriculture moving in Africa. Dev Policy Rev 13:115–129

Bottrell DG, Adkisson PL (1977) Cotton insect pest management. Annu Rev Entomol 22:451–481

Brader L (1979) Integrated pest control in the developing world. Annu Rev Entomol 24:225–254

Broecker WS, Kennett JP, Flower BP, Teller JP, Trumbore S, Bonani G, Wolfli W (1989) Routeing of meltwater from Lawrentide ice sheet during the Yonger Dryas cold episode. Nature 341:318–321

Brown BJ, Marten GG (1986) The ecology of traditional pest management in Southeast Asia. In: Marten GG (ed) Traditional agriculture in Southeast Asia. A human ecology perspective. Westview Press, Boulder, pp 241–272

Brush SB (1980) Potato taxonomies in Andean agriculture. In: Brokensha D, Warren M, Werner O (eds) Indigenous knowledge systems and development. University Press of America, New York, pp 37–47

Brush SB, Carney HJ, Humán Z (1981) Dynamics of Andean potato agriculture. Econ Bot 35:70–88

Bull D (1982) A growing problem. Pesticides and the third world poor. OXFAM, Oxford

Camino A, Johns T (1988) Laki-laki *(Dennstaedtia glauca,* Polypodiaceae): a green manure used in traditional Andean agriculture. Econ Bot 42:45–53

Carter WE (1969) New lands and old traditions. Kekchi cultivators in the Guatemalan lowlands. University of Florida Press, Gainesville

Chacón JC, Gliessman SR (1982) Use of the "non-weed" concept in traditional tropical agroecosystems of south-eastern Mexico. Agro-Ecosyst 8:1–11
Cohen MND (1977) The food crisis in prehistory: overpopulation and the origins of agriculture. Yale University Press, Newhaven
Conway GR (1972) Ecological aspects of pest control in Malaysia. In: Farvar MT, Milton JP (eds) The careless technology: ecology and international development. The Natural History Press, New York, pp 467–488
Cooper D, Vellvé R, Hobbelink H (1992) Growing diversity, genetic resources and local food security. Intermediate Technology Publications, London
Cuevas VC, Samulde SN, Pajaro PG (1988) *Trichoderma harzianum* Rifai as activator for rapid composting of agricultural wastes. Philipp Agric 71(4):461–469
Dansgaard W, White JWC, Johnsen SJ (1989) The abrupt termination of the Younger Dryas event. Nature 339:532–534
Dansgaard W, Johnsen SJ, Clausen HB, Dahl-Jensen D, Gundestrup NS, Hammer CU, Hvidberg CS, Steffensen JP, Sveinbjörnsdottir AE, Jouzel J, Bond G (1993) Evidence for general instability of past climate from a 250-kyr ice-core record. Nature 364:218–220
Datta SC, Banerjee AK (1978) Useful weeds of Bengal rice fields. Econ Bot 32:297–310
de Boef W, Amanor K, Wellard K, Bebbington A (1993) Cultivating knowledge, genetic diversity, farmer experimentation and crop research. Intermediate Technology Publications, London
De Schlippé P (1956) Shifting cultivation in Africa. The Zande system of agriculture. Routledge/Kegan Paul, London
Denevan WM, Padoch C (1988) Introduction: the bora agroforestry project. In: Denevan WM, Padoch C (eds) Swidden-fallow agroforestry in the Peruvian Amazon. Advances in economic botany 5. New York Botanical Garden, New York, pp 1–7
Denevan WM, Treacy JM (1987) Young managed fallows at Brillo Nuevo. In: Denevan WM, Padoch C (eds) Swidden-fallow agroforestry in the Peruvian Amazo. Adavances in economic botany 5. New York Botanical Garden, New York, pp 8–46
Doutt RL (1964) The historical development of biological control. In: DeBach P (ed) Biological control of insect pests and weeds. Chapman and Hall, London, pp 21–42
Duby G (1968) Rural economy and country life in the Mediaeval West. University of South Carolina Press, Columbia
Ekvall RB (1968) Fields on the hoof, nexus of Tibetan nomadic pastoralism. Holt, Rinehart and Winston, New York
Ettema CH (1994) Indigenous soil classification systems. What is their structure and function and how do they compare to scientific soil classifications? University of Georgia, Athens
Felker P (1981) Uses of tree legumes in semiarid regions. Econ Bot 35:174–186
Forde CD (1934) Habitat, economy and society. A geographical introduction to ethnology. Methuen, London
Funmilayo O (1975) The village weaverbird and the villagers: a protected pest. Nigerian Field 40:183–186
Gallacher DJ, Hill JP (2008) Effects of camel grazing on density and species diversity of seedling emergence in the Dubai (UAE) inland desert. J Arid Env 72:853–860
Gaur RD, Semwel JK, Negi KS (1981–2) Fodder plants of Garwhal hills and their impact on the quality and quantity of milk. J Himalayan Stud Region Dev, 56, 87–91
Glover IC, Higham CFW (1996) New evidence for early rice cultivation in south, Southeast and East Asia. In: Harris DR (ed) The origins and spread of agriculture and pastoralism in Eurasia. University College London Press, London, pp 413–441
Goldsmith E, Tenakoon M (1982) Traditional agriculture in Sri Lanka. Ecologist 12(5):209–215
Gourou P (1958) The tropical world. Its social and economic conditions and its future status. Longmans & Green, London
Harlan JR (1971) Agricultural origins; centers and non-centers. Science 174:465–473
Hazell PBR (2009) Transforming agriculture: the green revolution in Asia. In: Spielman DJ, Pandya-Lorch R (eds) Millons fed, proven successes in agricultural development. International Food Policy Research Institute, Washington, DC, pp 25–32

Herskovits MJ (1926) The cattle complex in East Africa. 28:230-272, 361-380, 494-528, 630-664
Hill P (1972) Rural Hausa, a village and a setting. Cambridge University Press, Cambridge
Hostetler SW, Bartlein PJ, Clark PU, Small EE, Solomon AM (2000) Simulated influences of Lake Agassiz on the climate of central North America 11,000 years ago. Nature 405:334–337
Hugh-Jones S (1982) The Pleiades and Scorpius in Barasana cosmology. In: Aveni AF, Urton G (eds) Ethnoastronomy and archaeoastronomy in the American tropics. New York Academy of Sciences, New York, pp 183–201
Hunter JM (1967) Population pressure in a part of the West African Savanna: a study of Nangodi, Northeast Ghana. Ann Assoc Am Geogr 57:101–114
Idowu AB, Modder WWD (1996) Possible control of the stinking grasshopper, *Zonocerus variegatus* (L) (Orthoptera, Pyrgomorphidae) in Ondo state through human consumption. Nigerian Field 61(1–2):7–14
Irvine FR (1969) West African crops. Oxford University Press, London
Isely D (1982) Leguminosae and Homo sapiens. Econ Bot 36:46–70
Jacobs AH (1975) Maasai pastoralism in historical perspective. In: Monod T (ed) Pastoralism in Tropical Africa. International African Institute and Oxford University Press, Oxford, pp 406–425
Jacobson M (1989) Botanical pesticides, present and future. In: Arnason JT, Philogene JR, Morand P (eds) Pesticides of plant origin. American Chemical Society, Washington, DC, pp 1–10
Jeyaratnam J, Lun KC, Phoon WO (1987) Survey of acute poisoning among agricultural workers in four Asian countries. Bull World Health Organ 65:521–527
Joshi AR, Joshi K (2003) Fodder plants of hilly areas of Bagmati and Langtang watershed, Nepal: ethnobotany and future conservation strategy. Ethnobotany 15:107–114
Kang BT, Wilson GF, Spikens L (1981) Alley cropping maize *(Zea mays L.)* and leucaena *(Leucaena leucocephala* Lam) in southern Nigeria. Plant Soil 63:165–179
Larsen CS (1995) Biological changes in human populations with agriculture. Annu Rev Anthropol 24:185–213
Larsen CS (2000) Skeletons in our closet. Revealing our past through bioarchaeology. Princeton University Press, Princeton
Lewis IM (1965) The Northern pastoral Somali of the horn. In: Gibbs JL (ed) Peoples of Africa. Holt, Rinehart and Winston, New York, pp 321–360
Loevinsohn ME (1987) Insecticide use and increased mortality in rural Central Luzon, Philippines. Lancet i:1359–1362
Lumpkin TA, Plucknett DL (1980) Azolla; botany, physiology and use as a green manure. Econ Bot 34:111–153
Magner LN (1979) A history of the life sciences. Marcel Dekker, New York
Malcolm DW (1953) Sukumaland. An African people and their country. A study of land use in Tanganyika. International African Institute and Oxford University Press, London
Malhotra SP (1986) Bishnois – their role in conservation of desert ecosystem. In: Shankarharyan KA, Shankar V (eds) Desert environment, conservation and management. Central Arid Zone Research Institute, Jodhpur, pp 23–24
Marten GG, Vitakon P (1986) Soil management in traditional agriculture. In: Marten GC (ed) Traditional agriculture in southeast Asia. A human ecology perspective. Westview Press, Boulder, pp 199–225
Matteson PC, Altieri MA, Gagne WC (1984) Modification of small farmer practices for better pest management. Annu Rev Entomol 29:383–402
Maumbe BM, Swinton SM (2003) Hidden health costs of pesticide use in Zimbabwe's smallholder cotton growers. Soc Sci Med 57:1559–1571
Mbegu AC (1996) Making the most of compost: a look at *wafipa* mounds in Tanzania. In: Reij C, Scoones I, Toulmin C (eds) Sustaining the soil. Indigenous soil and water conservation in Africa. Earthscan Publications, London, pp 134–138
Melathopoulus AP, Winston ML, Whittington R, Smith R, Lindberg C, Mukai A, Moore M (2000) Comparative laboratory toxicity of neem pesticides to honey bees (Hymenoptera: Apidae), their mite parasites *Varroa jacobsoni* (Acari: Varroidae) and *Acarapis woodi* (Acari: Tarsonemidae), and brood pathogens *Paenibacillus larvae* and *Ascophaera apis*. J Econ Entomol 93(2):199–209

Milne G (1936a) Normal erosion as a factor in soil profile development. Nature 138:548–549
Milne G (1936b) A provisional soil map of East Africa. Kenya, Uganda, Tanganyika and Zanzibar with explanatory memoir. Crown Agents, London
Mithen S (2003) After the ice. A global human history, 20,000–5000 BC. Harvard University Press, Cambridge, MA
Molleson T (1994) The eloquent bones of Abu Hureya. Sci Am 270(2):60–65
Moore AMT, Hillman GC (1992) The Pleistocene to Holocene transition and human economy in Southwest Asia: the impact of the Younger Dryas. Am Antiq 5(3):482–494
Moore AMT, Hillman GC, Legge AJ (2000) Village on the Euphrates: from foraging to farming at Abu Hureyra. Oxford University Press, Oxford
Mushita AT (1993) Strengthening the informal seed system in communal areas of Zimbabwe. In: De Boef W, Amanor K, Wellard K, Bebbington A (eds) Cultivating knowledge, genetic diversity, farmer experimentation and crop research. Intermediate Technology Publications, London, pp 85–88
Nabhan GP (1987) The desert smells like rain. A naturalist in Papago Indian Country. North Point Press, San Francisco
Needham J, Lu GD, Huang HT (1986) Science and civilisation in China. Biology and biological technology. Cambridge University Press, Cambridge
Neuenschwander P, Herren HR (1988) Biological control of the cassava mealybug, *Phenacoccus manihoti*, by the exotic parasitoid *Epidinocarsis lopezi* in Africa. Philos Trans R Soc Lond B 318:319–333
Nweke F (2009) Resisting viruses and bugs: cassava in Sub-Saharan Africa. In: Spielman DJ, Pandya-Lorch R (eds) Millions fed, proven successes in agricultural development. International Food Policy Research Institute, Washington, DC, pp 41–46
Nye PH, Greenland DJ (1960) The soil under shifting cultivation. Commonwealth Agricultural Bureaux, Farnham Royal
Oguntoyimbo JS, Richards P (1978) Drought and the Nigerian farmer. J Arid Environ 1:165–194
Okafor FC (1987) Population pressure and land resource depletion in southeastern Nigeria. Appl Geogr 7:243–256
Palm O, Sandell K (1989) Sustainable agriculture and nitrogen supply: farmers' and scientists' perspective. Ambio 18(8):442–448
Payawal PC, Paderon EM (1986) Sporocarp formation and spore germination of *Azolla* strains in the Philippines. Philipp Agric 69:633–643
Poschen P (1986) An evaluation of the *Acacia albida*-based agroforestry practices in the Haranghe Highlands of Ethiopia. Agrofor Syst 4(2):129–143
Pretty JN (1995) Regenerating agriculture. Policies and practice for sustainability and self-reliance. Earthscan Publications, London
Prevett PF (1962) The reduction of Bruchid infestation of Cowpeas by postharvest storage methods. Reports of the West African Stored Products Research Unit, Nigeria
Price LL (1997) Wild plant food in agricultural environments: a study of occurrence, management, and gathering rights in Northeast Thailand. Hum Organ 56:209–221
Prill-Brett J (1986) The Bontok: traditional wet-rice and swidden cultivators of the Philippines. In: Marten GG (ed) Traditional agriculture in southeast asia. A human ecology perspective. Westview Press, Boulder, pp 54–84
Purseglove JW (1968) Tropical crops, dicotyledons. Longman, London
Purseglove JW (1972) Tropical crops, monocotyledons. Longman, London
Quin PJ (1959) Food and feeding habits of the Pedi. Witwatersrand University Press, Johannesburg
Reissig WH, Heinrichs EA, Litsinger JA, Moody K, Fiedler L, Mew TW, Barrion AT (1986) Illustrated guide to integrated pest management in rice in Tropical Asia. International Rice Research Institute, Los Baños
Rhoades R, Bebbington A (1995) Farmers who experiment: an untapped resource for agricultural research and development. In: Warren DM, Slikkerveer LJ, Brokensha D (eds) The cultural dimension of development. Indigenous knowledge systems. Intermediate Technology Publications, London, pp 296–307

Ribeiro BG, Kenhíri T (1989) Rainy seasons and constellations: the Desâna economic calendar. In: Posey DA, Balée W (eds) Resource management in Amazonia: indigenous and folk strategies, advances in economic botany 7. New York Botanical Gardens, New York, pp 97–114
Richards P (1985) Indigenous agricultural revolution. Ecology and food production in West Africa. Hutchinson, London
Ruthenberg H (1976) Farming systems in the tropics. Clarendon Press, Oxford
Salazar R (1992) Community plant genetic resources management: experiences in Southeast Asia. In: Cooper D, Vellvé R, Hobbelink H (eds) Growing diversity. genetic resources and local food security. Intermediate Technology Publications, London, pp 17–29
Sankhala KS, Jackson P (1985) People, trees and antelope in the Indian Desert. In: McNeely JA, Pitt D (eds) Culture and conservation: the human dimension in environmental planning. Croom Helm, London, pp 205–210
Schlegel SA (1979) Tiruray subsistence: from shifting cultivation to plow agriculture. Ateneo de Manila University Press, Quezon City
Secoy DM, Smith AE (1983) Use of plants in control of agricultural and domestic pests. Econ Bot 37:28–57
Sharland RW (1995) Using indigenous knowledge in a subsistence society of Sudan. In: Warren DM, Slikkerveer LJ, Brokensha D (eds) The cultural dimension of development. Indigenous knowledge systems. Intermediate Technology Publications, London, pp 385–395
Shepard M, Carner GR, Turnipseed SG (1977) Colonization and resurgence of insect pests of soybean in response to insecticides and field isolation. Environ Entomol 6:501–506
Sherratt A (1997) Climatic cycles and behavioural revolutions: the emergence of modern humans and the beginning of farming. Antiquity 71:271–287
Silva S (2011) Frutas da Amazônia Brasileira. Metalivros, São Paulo
Simmonds NW (1962) The evolution of bananas. Longman, London
Skjønsberg E (1989) Change in an African Village. Kefa speaks. Kumarian Press, West Hartford
Smartt J, Simmonds NW (1995) Evolution of crop plants. Longman Scientific and Technical, Harlow
Stoll G (1988) Natural crop protection, based on local farm resources in the tropics and subtropics. AGRECOL & Margraf Publishers, Weikersheim
Vijayalalaksmi K, Sridhar S, Damodharan E (1998) Rice, non-chemical pest control. Centre for Indian Knowledge Systems, Chennai
Wijewardene R, Waidyanatha P (1984) Conservation farming for small farmers in the humid Tropics. Department of Agriculture, Sri Lanka and the Commonwealth Consultative Group for the Asia Pacific Region, Colombo
Williams PH (1979) Vegetable crop protection in the People's Republic of China. Annu Rev Phytopathol 17:311–324
Wood BJ (1971) Development of integrated control programmes for pests of tropical perennial crops in Malaysia. In: Huffaker CB (ed) Biological control. Plenum Press, New York, pp 422–457
Worede M, Mekbib H (1993) Linking genetic resource conservation to farmers in Ethiopia. In: Boef W d, Amanor K, Wellard K, Bebbington A (eds) Cultivating knowledge: genetic diversity, farmer experimentation and crop research. Intermediate Technology Publications, London, pp 78–84
Wratten S (1992, August 22) Farmers weed out the cereal killers. *New Scientist*, 31–35
Yang RZ, Tang CS (1988) Plants used for pest control in China: a literature review. Econ Bot 42:373–400
Yatazawa M, Tomomatsu N, Hosoda N, Nunome K (1980) Nitrogen fixation in *Azolla-Anabaena* symbiosis as affected by mineral nutrient status. Soil Sci Plant Nutr 26:415–426
Zandstra BH, Motooka PS (1978) Beneficial effects of weeds in pest management – a review. PANS 24(3):333–338
Zeuner FE (1963) A history of domesticted animals. Hutchinson, London

Chapter 3
Food Supplies and Nutrition

3.1 Traditional Diets

3.1.1 The Continuing Use of Wild Foods

Whilst it might seem strange to discuss food supplies separately from agriculture, there is a good reason for doing this. In many traditional villages, a substantial proportion of the food consumed derives, neither from cultivated crops nor domesticated animals, but from local wild species most of them collected from local village land. The bulk of this contribution comes from plants but, in the absence of taboos against killing animals, these also feature. The wild plant materials include leaf vegetables, flowers, fruits, seeds, root tubers and fungi and for any particular community this input can be extensive, ranging in number from a few tens of species to over a hundred (Table 3.1). The animal foods may be equally varied and include not only vertebrates such as mammals, birds, reptiles, amphibians and fish, but also more lowly creatures such as snails and insects. For instance, amongst the *Tswana* of Botswana, this tally of wild animal foods features 31 mammal species, 36 birds, 4 amphibians, reptiles and fish and 16 insects (Grivetti 1979). An obvious question to ask is whether these wild foods serve some particular dietary purpose or whether they merely bulk up the nutrients already obtained from agricultural sources?

3.1.2 Staples and Relishes

Pursuing this line of thought, it is instructive to examine the overall composition of traditional meals. They usually have two distinguishable components. There is first a major element or *staple,* consisting of a large serving of a cereal, a starchy fruit (such as cooking bananas or "plantain") or a starchy root (such as potatoes or

J. Edington, *Indigenous Environmental Knowledge*,
DOI 10.1007/978-3-319-62491-4_3

Table 3.1 Use of wild food plants by village communities from different vegetation zones

Vegetation zone	Staple foods	Number of species used				
		Leaf vegetables	Fruits & seeds	Roots & tubers	Fungi	Total
Tropical rain forest, Malaysia (Laderman 1982, 1984)	Fish rice	38	29	0	3	70
Temperate ever-green forest, Nepal (Joshi 1987)	Maize potatoes	15	27	1	3	46
Tropical savanna woodland, Tanzania (Newman 1980)	Maize sorghum	5	31	0	1	37
Tropical thorn-bush grassland, Botswana (Grivetti 1978, 1979)	Sorghum maize millet	22	47	31	3	103
Tropical bushveld, South Africa sorghum (Quin 1959)	Maize millet	16	11	0	0	27

cassava). Then alongside this is a smaller component or *relish* made up of wild food items, dairy products or material derived from minor crops.

For a specific illustration of dietary composition one can scarcely do better than look at P. J. Quin's comprehensive study of the South African *Pedi* (Quin 1959). Before this tribal group succumbed to modern influences, the staple component of their meals consisted of a porridge or a whole-grain stew, in either case made from cereals such as maize (*Zea mays*), sorghum (*Sorghum bicolor*) or millet (*Pennisetum typhoides*). This was accompanied by a smaller relish component, which could take any of 55 different forms. Of these, 32 variants derived from wild plant or animal materials; typical items being the leaves of wild vegetables, seasonal fruits, seeds or flesh from wild melons, portions of meat from game animals, or a serving of a seasonally-available insect. The remaining 23 options came from minor cultivated crops such as peas, beans, melons, pumpkins, or the immature heads of maize and sorghum. Beer (home-brewed from cereals), milk, and small quantities of meat from domesticated animals completed the dietary mix. Overall, the staples contributed about 2000 g to total daily food intake and the relishes about 250 g. Although the details vary, such combinations of major and minor food components are typical of traditional village diets.

The balance of wild and cultivated elements may however change with the seasons. Intake of wild food tends to increase during the so-called "hungry season" when cultivated foods are in short supply (Falconer 1990). In Zimbabwe for example, wild fruit collection becomes most intensive when inputs from cultivated crops are at their lowest ebb (Campbell 1987). In Bangladesh most use is made of wild fruits, roots and tubers during the two periods of food scarcity that precede the rice harvests (Hassan et al. 1985). Likewise in parts of India, the use of wild mushrooms, fruits and seeds increases significantly during the four-month "hunger" gap in agricultural production. At this time, an unusual food supplement makes its appearance in the form of a concoction produced by boiling *sal* tree leaves (*Shorea robusta*) with *mahua* tree flowers (*Madhuca indica*) (Basu 1987).

Wild-food use also increases in years when cultivated crops are affected by drought. For example, when the rains fail in the Thar Desert of Rajasthan, local people draw on a selection of 25 wild plants to supplement their diets (Bhandari 1974, 1978). These include fleshy fruits, roots, tree gums and the seeds of legumes and grasses. Many are dried, powdered and used to enrich dwindling supplies of millet flour. In these arid zone situations, wild species, because they have not had their hardiness undermined by selection for other qualities, generally have the merit of being able to survive drought conditions better than cultivated varieties. Of course if the drought is very severe even these stratagems are insufficient and external assistance becomes necessary.

3.2 Insights from Nutritional Science

3.2.1 Daily Dietary Requirements

To interpret these patterns further we need to examine what modern nutritional science has to say about dietary needs. Arriving at this modern understanding has been a lengthy process. As early as 1840, the German chemist Justus von Liebig drew the important distinction between foodstuffs that contain carbon, hydrogen and oxygen and serve mainly to meet the body's energy needs, and those that also contain nitrogen and are required for growth and tissue repair. The first group comprises carbohydrates and fats, the second, proteins. He mistakenly believed all proteins to be of similar nutritional value and it took some pioneer animal-feeding experiments by the American team of Osborne and Mendel to demonstrate that the value of any particular protein depends on the amino acids it contains; 8 being "essential" in an adult's diet and 10 in a child's (Osborne and Mendel 1911). These amino acids cannot be synthesized by the body and consequently have to be obtained directly from the food eaten.

At around the same time, another class of foodstuffs was being shown to be indispensable. Experiments by Professor Gowland Hopkins in Britain revealed that rats fed on a balanced diet of pure protein, carbohydrate, fat and minerals still failed to thrive and eventually died. They could however be saved by adding a small quantity of milk to their food, amounting to as little as 4% of the total (Hopkins 1912). The significance of the milk was that it contained substances (what we now call "vitamins"), which although needed in only small amounts are nonetheless essential for good health.

Some of the breakthroughs in understanding the role of vitamins came about in roundabout ways. For example, the potential of cereal seed-coats to cure the deficiency disease *beriberi* was revealed when Christian Eijkman working in the Dutch

East Indies noticed that his chickens suffered from the same kind of leg paralysis that afflicted local people (Drummond and Wilbraham 1958). The common factor was that both had been eating rice from which the vitamin-rich seed coats had been removed by the process of "polishing", a procedure designed to make the grain look whiter and more attractive. Likewise the discovery by mariners (most famously Captain James Cook) that shipboard scurvy could be prevented by the inclusion of citrus fruit (such as limes) in the diet was eventually fully explained with the isolation of the compound we now call vitamin C.

Other breakthroughs followed. In 1929 workers at the Minnesota Medical School showed that linoleic acid, a fatty acid, is essential for good health in a similar way to vitamins and certain amino acids (Burr and Burr 1929, 1930), albeit to serve a very particular function. The compound is necessary to maintain the integrity and fluidity of cell membranes including those of the skin (Elias et al. 1980); a dietary deficiency of this component being one of the main causes of infant eczema.

And so the story continued. In 1933 Cicely Williams a young doctor working in the Gold Coast (now Ghana), identified a common nutritional disorder amongst young children and called it *kwashiorkor* (Williams 1935). This is a local word, meaning literally "the deposed one" and is applied to a child displaced from its mother's breast by the arrival of a sibling. Typical symptoms include swelling of the body, wasting of the muscles and a change in hair-colour to a reddish hue. It results from an inadequate intake of protein, a conclusion that was initially resisted (along with the exotic name) by the medical establishment of the day, but eventually fully accepted. Another related disorder, caused by deficiencies in both protein and calorie supply, is known as *marasmus* (from the Greek word for wasting) and results in the child's body becoming thin and its face wizened.

The most recent advance – dating from the 1960s – has been the realization that certain fibrous plant materials contribute to good health (Burkitt 1969; Mendeloff 1977; Trowell et al. 1985). Whilst such matter is indigestible and has no direct nutritional value, it appears to be necessary for the proper mechanical functioning of the large intestine. The results of these and other discoveries can be summarized in a list of *daily dietary requirements* (Table 3.2). It will be noticed that large inputs, such as those for protein and fibre, are expressed in grams whereas smaller inputs, such as those for vitamins and minerals are expressed in milligrams. The values indicated are *average* ones for adults. To some degree, low inputs over a short period can be compensated for by increased inputs later on. The table also identifies the deficiency disorders likely to develop if the appropriate nutritional inputs are not forthcoming.

Some of the problems take on a different complexion in the tropics. *Marasmus* and *kwashiorkor* have already been noted, but the situation with *goitre* also deserves mention. This condition can result, not only from a lack of iodine in soils and therefore in foodstuffs, but also from the fact that people who eat large amounts of cassava (*Manihot esculenta*) may accidentally be ingesting thiocyanate-like substances capable of interfering with the thyroid gland's capacity to concentrate iodine (Delange and Ahluwalia 1983). *Rickets* too takes on a somewhat different guise in the tropics. This deformity of the leg bones, long-associated with diets lacking

Table 3.2 Daily dietary requirements, also indicating the deficiency disorders associated with shortfalls

Dietary component	Required daily intake	Deficiency disorder	Symptoms
Energy	10,600 kJ (= 2550 kcal)	Marasmus	Thin body, "old man" face
Protein	55 g	Kwashiorkor	Swollen face, feet and hands
			Hair turns reddish
Linoleic acid	14.5 g	Eczema	Inflammation and scaling of the skin
Vitamin A	0.7 mg	Xerophthalmia	Night blindness
			Cornea of eye damaged
Vitamin B_1 (*thiamin*)	1.0 mg	Beriberi	Swollen legs or wasting and paralysis of legs
Vitamin B_2 (*riboflavin*)	1.3 mg	Cheilosis	Cracks at mouth corners
			Sore tongue and skin
Vitamin B_3 (*niacin*)	17 mg	Pellagra	Dry cracked peeling skin
			Sore mouth and tongue
Vitamin C skin	40 mg	Scurvy	Bleeding gums and bruised skin
Vitamin D	0.01 mg or sunlight	Rickets	Deformity of leg bones
Iron	10 mg	Anaemia	Pale gums and inside of eyelids
Calcium	500 mg	Osteomalacia rickets	Weakening of bones and teeth
			Deformity of leg bones
Zinc	9 mg	Impaired immunity	Vulnerability to infection
Iodine	0.1 mg	Goitre	Growth retardation
Dietary fibre	33 g	Diverticulitis	Abdominal pain

Intake values based on Department of Health (1991)

vitamin D and life in sunless industrial city streets, now seems in Africa and parts of Asia (and possibly in retrospect in Europe too) to have calcium deficiency as a subsidiary cause (Pettifor 2004).

3.2.2 *Food Analysis Tables*

Knowing what the human body needs to stay healthy is clearly only half the story; the other concerns the nutrient contents of individual foods. This approach was pioneered by Robert McCance and Elsie Widdowson, two nutritionists working in wartime Britain. With the country obliged to be more self-reliant in food, it became vital to understand how individual items could contribute to an adequate diet. This was the era of Dr. Carrot and Potato Pete, two cartoon characters introduced by the Ministry of Food and designed respectively to encourage the consumption of carrots (rich in *carotene* convertible by the body into vitamin A) and of potatoes (a good source of both energy and vitamins). The resultant *food composition tables* were first produced in 1940 and continue to be updated (Food Standards Agency 2002). Chemical analyses of wild foods are harder to come by. They have generally been carried out piecemeal, either on the plants from particular regions or on plants belonging to particular botanical groups (Zennie and Ogzewalla 1977; Felker and Bandurski 1977). Nonetheless the tally of species tested has been growing steadily and has now reached the point where useful conclusions can be drawn about the dietary relationships between wild and cultivated foods. Table 3.3 provides some typical entries of both types.

3.2.3 Staples and Relishes Revisited

These two kinds of analysis allow us to view the traditional practice of distinguishing between staples and relishes in a somewhat different light. Staples (such as millet, bananas, potatoes or cassava) can evidently provide a large proportion of the nutrients the body requires (Table 3.3). These foodstuffs are however bulky and include a great deal of indigestible material. Were they the only items consumed, the stomach would become filled before all nutritional needs had been met.

Relishes on the other hand, whilst containing the same range of nutrients (i.e. energy, protein, vitamins and minerals), do so in a much more compact form. They are therefore ideal complements to the bulky staples and, if substituted for a small proportion of them, allow nutritional sufficiency to be achieved within the physical capacity of the digestive system. One commentator has usefully likened the situation to a three-legged stool (Werner and Bower 1982). The seat represents the input from the staples, but cannot function without the support of the three legs, which in this analogy represent respectively additional inputs from compact sources of energy, protein, and vitamins plus minerals.

To take a specific example, in his nutritional study of a Nepalese hill village Ananda Joshi showed how the two staple foods (potatoes and maize porridge) were accompanied by cultivated relishes (including mung beans, radishes, pumpkins and leaf vegetables) and combinations of 46 wild food plants whose identity varied with the seasons (Joshi 1987; Joshi et al. 2003). He calculated that had the staples and cultivated items stood alone, this would have led to significant dietary deficiencies in B_1 and B_2 vitamins. Although hampered by difficulties in measuring wild food intakes and a lack of food-composition data, he concluded that the potential shortfalls were substantially reduced and perhaps wholly remedied by the inclusion of wild plant relishes. Locally-used species such as chestnuts (*Castanopsis indica, Castanopsis tribuloides*), mushrooms (*Agaricus campestris*) and various leafy weeds (*Plantago major, Chenopodium album*) are all potential sources of the missing B vitamins (Duke 1985; Tan et al. 1985).

Table 3.3 Composition tables for selected cultivated and wild food materials, expressed per 100 g of edible portion

	Protein	Energy	Vitamin A (*as carotene*)	Vitamin B_1 *thiamin*	Vitamin B_2 *riboflavin*	Vitamin B_3 *niacin*	Vitamin C	Iron	Fibre
Food type	g	kJ	mg	mg	mg	mg	mg	mg	g
STAPLES									
Millet	5.8	1481	0	**0.68**	0.19	**2.80**	0	NA	NA
Bananas	1.2	379	0.7	0.03	0.02	0.40	14	0.20	2.2
Potatoes	1.8	306	Tr	0.18	0.01	0.50	6	0.40	1.4
Cassava	1.0	565	0	0.05	0.04	0.60	0	0.90	NA
PROTEIN AND ENERGY SOURCES									
Pine seeds	***14.0***	***2840***	*Tr*	***0.73***	*0.19*	***3.80***	*Tr*	***5.60***	*1.9*
Melon seeds	**25.0**	**2431**	0	0.10	0.15	1.50	Tr	**8.00**	NA
Beef	**27.3**	912	Tr	0.09	**0.40**	**6.30**	0	**3.40**	0
Beans	**7.8**	379	Tr	0.11	0.09	0.50	Tr	2.00	**5.8**
VITAMIN AND MINERAL SOURCES									
Fat-hen	*4.2*	*183*	***6.9***	*0.16*	***0.44***	*1.20*	***80***	***3.50***	*2.1*
Amaranth	*3.6*	*109*	***6.5***	*0.04*	*0.22*	*0.70*	***23***	***2.90***	*1.3*
Carrots	0.6	100	**7.5**	0.09	Tr	Tr	2	0.40	**2.7**
Oranges	1.1	158	Tr	0.11	0.04	0.40	**54**	0.10	1.8
Brown bread	**8.5**	927	0	**0.27**	0.09	**2.50**	0	2.20	**5.9**

Wild species are indicated in italics and noteworthy values of all kinds in bold type (Based on Duke (1985), Tan et al. (1985) and Holland et al. (1991)
NA = no analysis figures available; *Tr* = trace only present

3.3 Input Patterns for Different Classes of Nutrients

3.3.1 Energy Intake

Although in Western countries sugar is popularly regarded as the most readily available energy source, sugars feature only minimally in traditional village diets. Daily intakes range from 10 to 20 g, which is dramatically lower than the average value of 150 g for Western diets. The situation seems to arise, not so much from any conscious village decision to avoid sugar, but rather from difficulties encountered in obtaining significant amounts of it. Many traditional villages fall outside the orbit of commercial sugar-trading based on processed cane sugar or sugar beet and are therefore obliged, if they want supplies, to obtain them from local sources. One variety of sorghum (*Sorgum vulgare* var. *saccaratum*), the so-called "sweet-reed" of Southern Africa, contains appreciable amounts of sugar (Quin 1959). So too do some fruits; with the starch in bananas for example converting substantially to sugar as the fruit ripens. Also fruits such as dates, figs and grapes (especially when the last items are converted to raisins) are all sugar-rich, but not all of them are widely distributed in the tropics.

More typically, sugary foods are hard to obtain, with the daunting options including the theft of honey from the nests of wild bees or the climbing of palm trees to tap sweet sap from inflorescence stalks, neither of them procedures likely to lead to extravagant consumption. Even in Europe, sugar did not come into general use until the seventeenth century when imports started to arrive from the Caribbean sugar plantations. Interestingly, another period of limited sugar availability occurred in Britain during the Second World War, a time when nutritional health is widely considered to have been better than it is currently (Patten 1985).

In traditional situations, fatty fruits, seeds and nuts rather than sugary materials, commonly meet the need for compact and convenient energy sources. One should bear in mind that 1 gm of fat yields 37 kilojoules of energy, more than twice the 16 kilojoules produced from 1 gm of sugar. In this role, fatty fruits and nuts are weight for weight far superior to the staples that typically accompany them. For example, the pine-seed cakes featured in the diets of some North American Indians contain twice as much energy as the accompanying maize and millet breads (Table 3.3). Likewise the seeds of melons (*egusi*) used in West Africa as a snack or to thicken soups, contain four times as much energy as associated staples such as cassava or yam (Table 3.3). Most village cultures use at least one concentrated foodstuff of this kind. Even the inhabitants of the Thar Desert of Rajasthan laboriously extract fatty seeds from the wild gourd, *Citrullus colocynthis* (known locally as *toos* or *tumba*) (Fig. 3.1a). These they wash with salt water to remove any adhering bitter pith and then grind into a powder that is added to millet flour to enrich it (Bhandari 1974). These same gourd seeds are incidentally an excellent source of linoleic acid (Sawaya et al. 1983).

In this habit of obtaining energy from plant fats rather than from sugar, our village counterparts can count themselves fortunate. The Western situation where

Fig. 3.1 Examples of wild food-plant use in the Thar Desert of India. (**a**) Wild gourds (*Citrullus colocynthis*) being harvested for their seeds, (**b**) collection of *kumut* (*Acacia senegal*) seed pods, (**c**) *khejri* (*Prosopis cineraria*) pods being offered for sale in a local market (*insets* show the key edible components all pictured on the same scale)

sugary foods are endlessly available is not one to aspire to; there now being ample evidence to show that sugar-rich diets carry with them an increased risk of obesity (with its attendant problems) and a high vulnerability to dental decay (Sreeby 1982, WHO 2003).

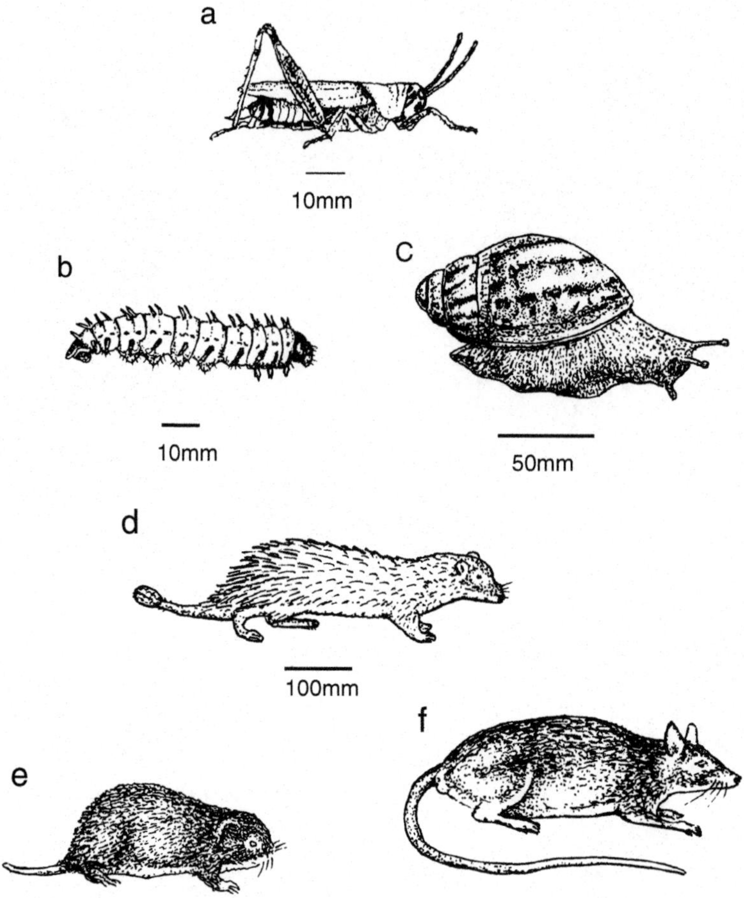

Fig. 3.2 Examples of wild animal protein sources from Africa. (**a**) Grasshopper (*Zonoceros variegatus*), (**b**) caterpillar (*Gonimbrasia belina*), the larva of the Emperor Moth and popularly known as the *mopane* "worm", (**c**) giant land snail (*Archachatina marginata*), (**d**) brush-tailed porcupine (*Atherurus africanus*), (**e**) cane rat (*Thryonomys swinderianus*), (**f**) pouched rat (*Cricetomys gambianus*)

3.3.2 Protein Intake

Protein in village diets can come from both animal and plant sources. Although in theory the species used could be wild or domesticated, in practice there is considerable village reluctance to kill farm animals except on special ceremonial occasions. This is because of their usefulness as draught animals or as producers of milk, butter or eggs. Opportunities to use wild species as food are therefore readily taken and this so-called *bushmeat* includes not only vertebrates such as mammals, reptiles and fish, but also invertebrate animals such as snails and insects (Fig. 3.2).

As noted earlier, the nutritional value of proteins depends on the diversity of amino acids they contain. On this score, the quality of protein obtained from favoured African bushmeat species, such as cane rats (*Thryonomys swinderianus*) and pouched rats (*Cricetomys gambianus*) is very high and equal to that derived from beef or mutton (Asibey 1974; Kyle 1987). Although snails and insects generally contain a less useful range of amino acids, they can nevertheless make a useful dietary contribution (Orraca-Tetteh 1963; Oliveira et al. 1976).

Another advantage of using wild mammals as protein sources is that their meat has a far lower saturated-fat content than that derived from modern domesticated stock. In the meat of wild animals the ratio of saturated to unsaturated fats is in the region of 2:1, whereas in domesticated ones it can be as high as 50:1 (Crawford 1968). This is no accident. Domesticated animals have been intentionally bred to have fatty carcasses because this increases the flavour of the meat and makes it easier to cook. There is however a penalty incurred, with strong evidence to connect a high intake of saturated fat with an increased risk of coronary heart disease and an increased propensity towards obesity and its attendant complications (WHO 1990, 2003). Such problems have previously been rare in traditional communities, although there is now evidence to show that the situation is changing there too and very much in the wrong direction (Reddy 2002).

Some communities, because of their religious convictions, avoid killing animals for food, or indeed any other purpose. In India an estimated 30% of the population comes into this category, including a large proportion of the followers of Hinduism, Janism, Sikhism and Buddhism (Chakravarti 1974). These groups can nonetheless obtain the amino acids they need by eating leguminous plants. Most of the species used are cultivated ones. The famous *dals* in Indian cuisine come into this category and are prepared from a range of cultivated peas, beans and lentils. However, wild legumes may also contribute; notable examples from the drier parts of India include pods or seeds from *kumut* and *khejri* trees (Fig. 3.1b, c). Both items feature prominently in local dry curries (Saxena 1986). Strict vegetarians and especially *vegans* in Western countries (who avoid all animal food materials including dairy products) run the risk of incurring a dietary deficiency of vitamin B_{12} (*cyanocobalamin*), which can result in anaemia. This is rarely a problem in traditional village diets because of the widespread consumption of dairy products. As little as 100 ml of milk or its equivalent in the form of buttermilk is sufficient to provide an adequate safeguard (Rajalakshmi 1981).

3.3.3 Sources of Vitamins and Minerals

Vitamins play a vital role in facilitating the body's enzymatic reactions, and minerals such as iron and calcium are indispensable for blood and bone formation. The vegetable matter contained in village relishes is a good source of these substances, with wild plants being just as important as cultivated ones (Table 3.3). In the diet of Nepalese hill villagers, many of the wild species used are temperate-zone weeds,

Fig. 3.3 Soup being prepared from stinging nettles in Nepal (Note the prudent use of bamboo forceps)

and include such familiar species as shepherd's purse, fat-hen and common nettle (Figs. 3.3 and 3.4a–c). The inclusion of fat-hen is particularly interesting. In Anglo-Saxon England not only were the leaves of this plant used as a green vegetable, its seeds were ground into flour (Mabey 1972). A similar pattern prevails in Southern Africa, with field-edge weeds such as spider flower, pigweed and common nightshade featuring prominently as relishes (Fig. 3.4d–f) (Fleuret 1979; Tredgold 1986). It might be noted with this last-named species that although the leaves are perfectly edible the unripe berries are poisonous, a fact that children usually learn through painful experimentation. The three species are often cooked together, with peanut butter being added to improve the flavour. They are also commonly dried for subsequent winter use. The main dietary significance of all these wild species is as sources of *carotene*, the plant pigment convertible by the body into vitamin A. In neither situation would the staple foods (cereals and potatoes) by themselves have provided sufficient *carotene* to meet daily vitamin A needs.

Vitamin A is especially important in protecting against *xerophthalmia*, an eye disorder that starts with a loss of night vision and, if left unchecked, leads to such serious damage to the conjunctiva and cornea as eventually to cause blindness. Most wild leaf vegetables are good sources of *carotene*, with concentrations often exceeding those of the familiar cultivated sources. For instance, the carotene content of nettles – rated at 20.2 mg/100 g – considerably exceeds that of the proverbial carrot, which contains only 7.5 mg/100 g. Chemical analysis of wild fruits from Amazonia likewise reveals that some of them have high *carotene* contents. One of these is the *tucaman* fruit derived from the palm, *Astrocaryum aculeatum* (Fig. 3.5). This contains three times as much *carotene* as a carrot. It needs to be noted in passing that recent research, whilst confirming the importance of these sources, indicates that

Fig. 3.4 Wild leaf-vegetables serving as useful vitamin A sources. (**a**–**c**) In Nepalese hill villages, (**d**–**f**) in the African highlands. (**a**) Shepherd's purse (*Capsella bursa-pastoris*) (**b**) fat-hen (*Chenopodium album*), (**c**) common stinging nettle (*Urtica dioica*), (**d**) spider flower (*Gynandropsis pentaphylla*), (**e**) pigweed (*Amaranthus thunbergii*), (**f**) common nightshade (*Solanum nigrum*)

only a proportion of the *carotene* ingested is converted by the body into vitamin A. The precise amount depends on the overall food mix and individual physiology and ranges from 3.6 to 28% (Tang 2010). Leaf vegetables are also useful sources of iron, calcium and folic acid; the last named compound offering important protection against anaemia during pregnancy (Oomen and Grubben 1978).

Paradoxically, in certain circumstances difficulties can arise, not from a shortage of vitamins, but from a surfeit of them. For centuries, the *Inuit* of the north Polar Regions had been aware that eating the liver from animals such as polar bears (*Thalarctos maritimus*) and bearded seals (*Phoca barbata*) can make them seriously ill. Subsequently, arctic explorers made the same discovery to their considerable cost. The problem substance turned out to be vitamin A, which is present in relatively small amounts in the bodies of fishes, but builds up to very high concentrations in predators such as bears (Rodahl and Moore 1943; Rodahl 1949). The consumption of excessive amounts of this vitamin causes skin damage, internal

Fig. 3.5 *Tucaman* fruits about to be sliced for use in a breakfast sandwich

haemorrhages and abnormal bone growth. There has been an interesting re-emergence of such problems in recent times. This is associated with the growing Western enthusiasm for consuming vitamin supplements. With a normal diet this is rarely necessary and can have unwelcome side effects. There is for example evidence that excess vitamin A intake can increase the risk of osteoporosis; also that the overuse of vitamin C can lead to digestive upsets (Kmietowicz 2003).

In village situations, although wild leaf vegetables also contain B and C vitamins, they are less important in this latter role. The main reason being that vitamin C is water-soluble and, because wild vegetables are typically cooked before eating, much of the vitamin is destroyed in the process (Gomez 1982). For instance, cooking 100 g of amaranth leaves reduces their vitamin C content from 169 to 25 mg. This focuses attention on soft fruits as potential vitamin C sources. These are typically eaten raw and the wild species often have vitamin C contents that rival, or exceed, those found in cultivated citrus fruits (Fig. 3.6). For instance *marula* fruits have a vitamin C concentration of 68 mg/100 g, those of kei apple 83 mg/100 g and those of baobab an astonishing 310 mg/100 g. These values compare very favourably with the 54.0 mg/100 g achieved by cultivated oranges (Wehmeyer 1966).

In fact, *marula* fruits can contribute to village nutrition in a variety of ways (Quin 1959). A single large tree may produce thousands of fruits in a single season and prize trees are often protected with thorn-hedge barriers to prevent the fallen fruit being eaten by livestock. Not only are the fruits consumed raw, their juice can be used to flavour porridge and, when fermented, to produce a potent alcoholic drink. Even the fruit stone is not wasted and when cracked open yields plant embryos rich in oil and calories.

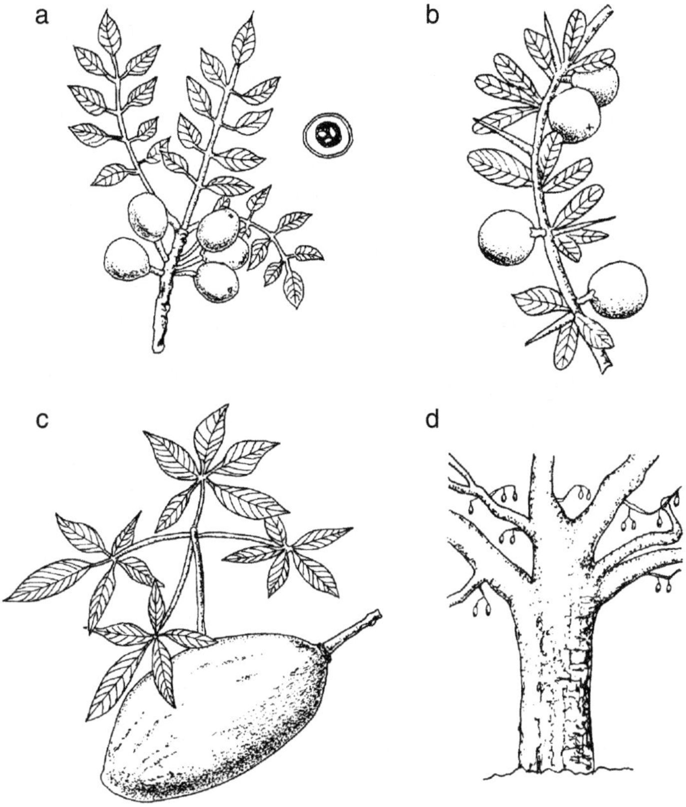

Fig. 3.6 Wild fruits serving as vitamin C sources in Southern Africa. (**a**) *Marula* (*Sclerocarya birrea*), also showing fruit kernel and embryos in cross-section, (**b**) kei apple (*Dovyalis caffra*), (**c**) *baobab* (*Adansonia digitata*), (**d**) *baobab* fruits in situ

As far as B vitamins are concerned, substantial amounts can be derived from cultivated crops, as long as these are processed in traditional ways. For example village techniques for brewing beer from millet or sorghum grain (Fig. 3.7) raise the *riboflavin* (vitamin B_2) content of such materials by as much as 105% (Platt 1964). It is ironic that missionaries, who often attempted to stop the preparation of local beers on supposed moral grounds, were frequently endangering the health of their charges by reducing their vitamin B intakes and increasing their vulnerability to *cheilosis* and *pellagra*.

Likewise traditional stone-grinding of flour, unlike modern flour-milling techniques, has the merit of retaining the germ of the grain together with its outer seed coats and aleurone layer (Fig. 3.8). All these components are rich in B vitamins and in iron. Modern milling processes tend to discard these fractions on the grounds that the baking and keeping properties of the bread are thereby improved and also on the assumption that the public at large prefers white bread to brown. This incidentally was another dietary issue to arise in Britain during the Second World War. With

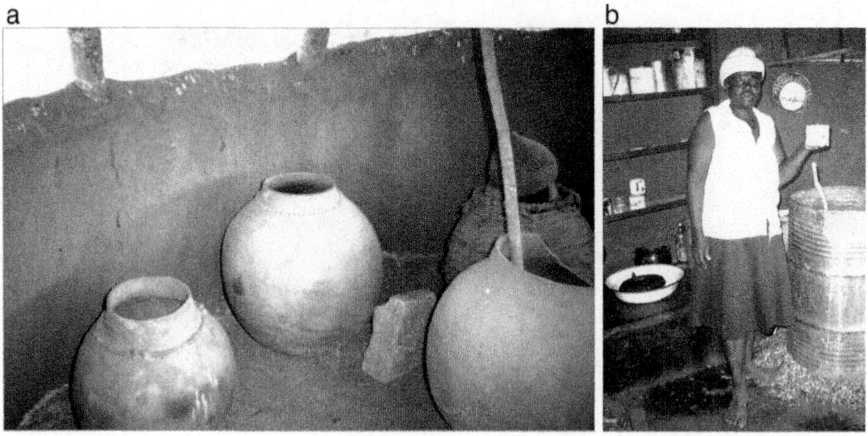

Fig. 3.7 Local beer brewing. (**a**) Traditional beer-brewing equipment in a *MaShona* household in Zimbabwe, (**b**) a more recent technique involving the use of an old oil drum; notice the millet heads strewn around the base of the drum

Fig. 3.8 Longitudinal cross-section of a wheat grain (Based on McCance 1946)

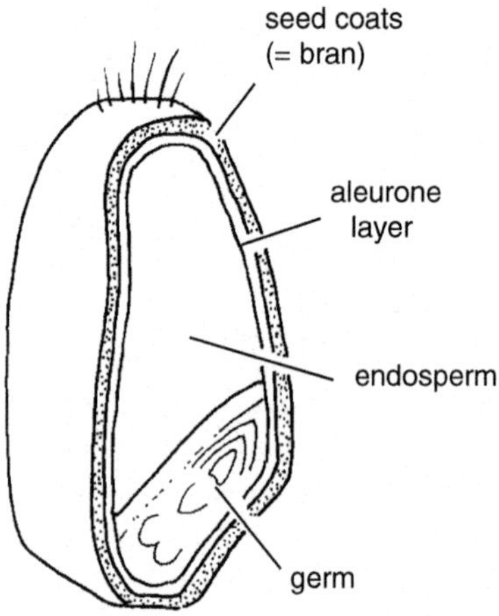

limited wheat supplies and concern being felt about the sufficiency of vitamin intakes, flour manufacturers were instructed in 1942 to include 85% of the original grain fractions in the bread they produced (Drummond and Wilbraham 1958). This was in contrast to the 70–80% that had previously become the norm, again largely following consumer preferences. The revised strategy significantly raised vitamin and mineral contents of loaves, but such was the public demand for white bread that this beneficial practice was eventually discontinued. Now, paradoxically, many modern milling processes continue to remove the useful fractions, but subsequently to replace them by a process of "fortification".

Amongst wild plants, leaf vegetables often feature prominently as sources of vitamin B_2, as do wild fungi. Fat-hen (Fig. 3.4b) for example contains 0.44 mg/100 g and the widely used ear-shaped fungus, *Aricularia auricula-judea,* 0.48 mg/100 g. Both values thus marginally exceed the value of 0.40 mg/100 g recorded for beef.

3.3.4 Dietary Fibre

Another important consequence of consuming vegetables and fruits together with stone-ground flour is that they all provide dietary fibre. It has come to be realised that this indigestible material performs the important function of keeping the lumen of the large intestine distended. This is believed to prevent pockets forming in the intestinal wall, thus avoiding the unpleasant condition known as *diverticulitis* (Bogardus 2006). Consistent with this interpretation is the fact that, compared with an incidence of 5–10% in the West, *diverticulitis* is a very rare condition in rural areas of the developing world (Painter and Burkitt 1971).

It has also been argued in the past that, because it makes waste evacuation more efficient, a large amount of fibrous material in the colon has the merit of reducing the retention time of toxins believed to trigger colonic cancer, a problem that is at least eight times more common in Western countries than in developing ones (Armstrong and Doll 1975; Mendeloff 1977; Willet and MacMahon 1984). Modern researchers however, whilst still postulating a dietary cause for this disorder, are less inclined to see a lack of fibre as the aggravating factor in Western diets. Current theories postulate a dietary excess of fatty red meat and a deficiency in fruit and vegetable intake (Giovannucci and Willet 1994).

3.3.5 Dietary Taboos

Notwithstanding these many instances of sound practice, dysfunctional dietary habits are not entirely absent from traditional villages. Just as diseases and herbal medicines are often locally assigned to "hot" or "cold" categories, so too may these dubious principles be applied to foods (Fieldhouse 1986; Helman 1990). This can result in a withdrawal of particular foods in precisely the circumstances they would have proved most useful. For example, village children in Malaysia suffering from the eye disorder *xerophthalmia* – classed by them as a "cold" disease – are often denied the green vegetables capable of curing the condition. This is on the spurious grounds that these are "cold" foods and as such are likely to aggravate an already "cold" condition (McKay 1980). Likewise in Bengal, children suffering from diarrhoea – classed as a "hot" illness – are denied much needed milk on the grounds that this is a "hot" food (Jelliffe 1957).

Notions of sympathetic magic can also influence local food choices. For example, pregnant women in West Africa often shun certain animal foods because they fear that some unwelcome characteristic of the animal eaten will be visited upon the forthcoming child. Thus to eat the flesh of a porcupine during pregnancy is to run the risk of a difficult birth (porcupines being spiky), to eat giant snails is to run the risk of the newborn child salivating excessively (snails being slimy) and to eat pork is to run the risk of the child developing a pig-like face (Ogbeide 1974; Chiwuzie et al. 1987). Such fancies can cause valuable protein foods to be avoided at precisely the time they are most needed. There also exists in parts of East Africa a curious cultural prohibition on eating fish (Simoons 1974). Its origins are obscure, but it has the unfortunate effect of causing a valuable protein source to be neglected.

3.4 Starvation and Malnutrition

Praise for traditional food systems might seem to sit uneasily alongside frequent media and aid-agency portrayals of starving adults and children. Statistics from the United Nations Food and Agriculture Organization confirm that such problems are real and that some countries (particularly ones in Africa) experience malnourishment affecting in excess of 35% of the population. In a great many of these cases however, the problem seems to have been generated by external factors, rather than arising from within the communities themselves. Excessive population growth (itself originally stimulated by Western medical inputs) has been a prime factor causing food demands to outstrip production. Such population expansions also tempt communities to settle in regions (such as the Sahel zone of Africa) that are unsuitable for permanent occupation (Dalby et al. 1977). Other severe problems can arise as a consequence of civil wars, tribal conflicts and political maladministration (D'Souza 1988).

Starvation represents a much more fundamental nutritional disturbance than those caused by deficiencies of individual nutrients, although these too can prove fatal if they become prolonged. Unlike a lack of water that can cause people to die in a matter of days, starvation takes a course that can extend over a period as long as three months, often with temporary respites being gained from the consumption of wild "famine foods". When supplies run out altogether, the body first uses glycogen stores contained in the muscles and liver, then fat stores and finally as a last resort protein in the tissues, including fatally that contained in the heart muscles. When food crises reach this point, traditional dietary practices are of little help and there has to be recourse to food inputs provided by external aid agencies. These usually consist of pre-cooked blends of carbohydrates and fats, supplemented with vitamins and minerals.

It is rare however for traditional communities left to their own devices to approach anything like this state of affairs. They usually react well to minor or moderate climatic disturbances. For example in his study of the *Tswana* of Botswana – a group that combines cereal farming and pastoralism with the use of wild foods – Grivetti (1978) found no evidence of *kwashiorkor* or *marasmus*, even after 7 years of low rainfall. Likewise in an examination of 1047 children from the *Karamajong* – a farming, pastoral and food-collecting tribe living in a drought-prone region of Uganda – Jelliffe and his co-workers found no obvious symptoms of vitamin deficiencies and only four cases of *marasmus* and one of *kwashiorkor* (Jelliffe et al. 1964).

3.5 The Paradoxical Efficacy of Traditional Diets

It might be imagined that, because village food providers make so many sound dietary decisions, they must in some way have anticipated the body of nutritional knowledge that took Western scientists over a 100 years to accumulate. In fact, this is true in only a very limited sense. Native North Americans apparently discovered quite independently that *scurvy* could be prevented by consuming certain plant and animal materials, including cranberries, spruce tea and moose adrenal glands, all good sources of vitamin C (Vogel 1970). They also discovered that *pemmican* produced from a mixture of shredded dried meats and animal fats could provide an extremely compact food source capable of sustaining them on long journeys. There is a similar traditional recognition that some seeds can serve the same purpose. Hence the *Pedi* know that as little as two ounces (56 g) of meal prepared from melon seeds will sustain them for a whole day's hunting (Quin 1959). Otherwise however, it seems that traditional decisions about the make-up of meals stem largely from a culinary desire to enjoy contrasts in food tastes and textures, an understandable sentiment when trying to relieve the monotony of starchy staples. On this basis, the beneficial nutritional outcomes have to be viewed as largely fortuitous and not the result of some form of comprehensive knowledge system. Natural selection may however have played a part by favouring groups whose dietary practices were adaptive, thereby offering better prospects for good health and longevity.

Evolutionary history seems also to have been a major factor. Mankind's nutritional physiology appears to have become fixed in the hunter-gatherer phase, was adapted to the circumstances then prevailing and has retained its basic character ever since. The range of foods originally available would have been large, but the abundance of each rather limited except during the fruiting season of a particularly prolific species or when a large animal had been killed. Sugar and salt would have been in short supply; the carcasses of wild animals used for food would have contained only low levels of saturated fatty acids, and the nutrient content of most plant foods would have been complemented with useful fibrous materials (Eaton and Konner 1985; Eaton et al. 1988). Early agriculture would have added starchy staples to the mix, but as long as the subsidiary items continued to be included, there need have been no serious mismatch between dietary intakes and physiological needs. In this sense, traditional communities have the good fortune of still operating in food environments that match their underlying physiology. Western communities on the other hand, partly from a lack of dietary self control and partly through the blandishments of a commercially-oriented food industry, seem frequently to have adopted diets that render them vulnerable to a variety of chronic disorders (p.224).

References

Armstrong B, Doll R (1975) Enviromental factors and cancer incidence and mortality in different countries with special reference to dietary practices. Int J Cancer 15:617–631

Asibey EOA (1974) Wildlife as a source of protein south of the Sahara. Biol Conserv 6:32–39

Basu NG (1987) Forests and tribals. Manisha Granthalaya, Calcutta

Bhandari MM (1974) Famine foods in the Rajasthan desert. Econ Bot 28:73–81

Bhandari MM (1978) Flora of the Indian desert. Scientific Publishers, Jodhpur

Bogardus ST (2006) What do we know about diverticular disease? A brief overview. J Clin Gastroenterol 40(Suppl 3):108–111

Burkitt DP (1969) Epidemiology of cancer of the colon and rectum. Cancer 28:1–13

Burr GO, Burr MM (1929) A new deficiency disease produced by the rigid exclusion of fat from the diet. J Biol Chem 82:345–367

Burr GO, Burr MM (1930) The nature and role of the fatty acids essential in nutrition. J Biol Chem 86:587–621

Campbell BM (1987) The use of wild fruits in Zimbabwe. Econ Bot 41(3):375–385

Chakravarti AK (1974) Regional preference for food: some aspects of food habit patterns in India. Can Geogr 17:395–410

Chiwuzie J, Ukoli F, Okojie O, Isah E, Eriator I (1987) Traditional practitioners are here to stay. World Health Forum 8:240–244

Crawford MA (1968) Fatty-acid ratios in free-living and domestic animals. Possible implications for atheroma. Lancet i:1329–1333

Dalby D, Harrison Church RJ, Bezzaz F (1977) Drought in Africa. International African Institute, London

Delange F, Ahluwalia R (eds) (1983) Cassava toxicity and thyroid: research and public health issues. International Development Research Centre, Ottawa

Department of Health (1991) Dietary reference values for food, energy and nutrients for the United Kingdom. HMSO, London

Drummond JC, Wilbraham A (1958) The Englishman's food. A history of five centuries of English diet. Jonathan Cape, London

D'Souza F (1988) Famine: social security and an analysis of vulnerability. In: Harrison GA (ed) Famine. Oxford University Press, Oxford, pp 1–56

Duke JA (1985) CRC handbook of medicinal herbs. CRC Press, Boca Raton

Eaton SB, Konner M (1985) Palaeolithic nutrition. A consideration of its nature and current implications. N Engl J Med 312(5):283–289

Eaton SB, Konner M, Shostak M (1988) Stone agers in the fast lane: chronic degenerative diseases in evolutionary perpective. Am J Med 84:739–749

Elias PM, Brown BE, Ziboh VA (1980) The permeability barrier in essential fatty acid deficiency: evidence for a direct role for linoleic acid in barrier function. J Investig Dermatol 74:230–233

Falconer J (1990) "Hungry season" food from the forest. Unasylva 160:14–19

Felker P, Bandurski RS (1977) Protein and amino acid composition of tree legume seeds. J Sci Food Agric 28:791–797

Fieldhouse P (1986) Food and nutrition. Customs and culture. Croom Helm, London

Fleuret A (1979) The role of wild foliage plants in the diet, a case study from Lushoto, Tanzania. Ecol Food Nutr 8:87–93

Food Standards Agency (2002) McCance and Widdowson's the composition of foods. Sixth Summary Edition. Royal Society of Chemistry, Cambridge

Giovannucci E, Willet WC (1994) Dietary factors and risk of colon cancer. Ann Med 26(6):443–452

Gomez MI (1982) Sources of vitamin C in the Kenya diet and their stability to cooking and processing. Ecol Food Nutr 12:179–184

Grivetti LE (1978) Nutritional success in a semi-arid land: examination of Tswana agro-pastoralists of the eastern Kalahari, Botswana. Am J Clin Nutr 31:1204–1220

Grivetti LE (1979) Kalahari agro-pastoral-hunter-gatherers: the Tswana example. Ecol Food Nutr 7:235–256

Hassan N, Huda N, Ahmad K (1985) Seasonal patterns of food intake in rural Bangladesh: its impact on nutritional status. Ecol Food Nutr 17:175–186

Helman CG (1990) Culture, health and illness. An introduction for health professionals. Wright, London

Holland B, Welch AA, Unwin ID, Buss DH, Paul AA, Southgate AAT (1991) McCance and Widdowson's the comparison of foods, 5th edn. Royal Society of Chemistry and Ministry of Agriculture, Fisheries and Food, Cambridge

Hopkins FG (1912) Feeding experiments illustrating the importance of accessory factors in normal dietaries. J Physiol 44:425–460

Jelliffe DB (1957) Social culture and nutrition. Cultural blocks and protein malnutrition in early childhood in rural West Bengal. J Pediatr 20:128–138

Jelliffe DB, Bennett FJ, Jelliffe EFP, White RHR (1964) Ecology of childhood disease in the Karamajong of Uganda. Arch Environ Health 9:25–36

Joshi AR (1987) A study of the environmental relationships of certain village communities in the central development region of Nepal. Doctoral Thesis, University of Wales, Cardiff

Joshi AR, Shrestha SL, Joshi K (2003) Environmental management & sustainable development at the crossroad. AnKuS, Kathmandu

Kmietowicz Z (2003) Food watchdog warns against high doses of vitamins and minerals. Br Med J 326:1001

Kyle R (1987) A feast in the wild. Kudu Publishing, Oxford

Laderman C (1982) Wild vegetable consumption on the east coast of Peninsular Malaysia. Malay Nat J 35:165–171

Laderman C (1984) Food ideology and eating behaviour: contributions from Malay studies. Soc Sci Med 19:547–559

Mabey R (1972) Food for free. A guide to the edible wild plants of Britain. Collins, Glasgow

McCance RA (1946) Bread. Lancet 6386:77–81

McKay DA (1980) Foods, illness and folk medicine: insights from Ulu Trengganu, West Malaysia. In: Robson JRK (ed) Food, ecology and culture. Readings in the anthropology of dietary practices. Gordon and Breach Science Publishers, New York, pp 61–66

Mendeloff AI (1977) Dietary fiber and human health. N Engl J Med 297(15):811–814

Newman JL (1980) Dimensions of Sandawe diet. In: Robson JRK (ed) Food, ecology and culture. Readings in the anthropology of dietary practice. Gordon and Breach Science Publishers, New York, pp 27–33

Ogbeide O (1974) Nutritional hazards of food taboos and preferences in Mid-West Nigeria. Am J Clin Nutr 27:213–216

Oliveira JFS, de Çarvalho JP, de Sousa RFXB, Simao MM (1976) The nutritional value of four species of insects consumed in Angola. Ecol Food Nutr 5:91–97

Oomen HAPC, Grubben GJH (1978) Tropical leaf vegetables in human nutrition. Royal Tropical Institute, Amsterdam

Orraca-Tetteh R (1963) The giant African snail as a source of food. In: Ovington JD (ed) The better use of the world's fauna for food. Institute of Biology, London, pp 53–61

Osborne T, Mendel LB (1911) Feeding experiments with isolated food substances. Carnegie Institute of Washington, Publication No 156, Washington

Painter NS, Burkitt DP (1971) Diverticular disease of the colon: a deficiency disease of western civilisation. Br Med J ii:450–454

Patten M (1985) We'll eat again. A collection of recipes from the war years selected by Marguerite Patten. Imperial War Museum and Hamlyn Publishing, London

Pettifor JM (2004) Nutritional rickets: deficiency of vitamin D, calcium or both? Am J Clin Nutr 80(6):1725–1729

Platt BS (1964) Biological ennoblement: improvement of the nutritive value of foods and dietary regimes by biological agencies. Food Technol 18:662–670

Quin PJ (1959) Food and feeding habits of the Pedi. Witwatersrand University Press, Johannesburg
Rajalakshmi R (1981) Applied nutrition. Oxford and IBH Publishing, New Delhi
Reddy KS (2002) Cardiovascular diseases in the developing countries: dimensions, determinants, dynamics and directions for public action. Public Health Nutr 5:231–237
Rodahl K (1949) Toxicity of polar bear liver. Nature 164(4169):530–531
Rodahl K, Moore T (1943) The vitamin A content and toxicity of bear and seal liver. Biochem J 37:166–168
Sawaya WN, Daghir NJ, Khan P (1983) Chemical characterization and edibility of the oil extracted from *Citrullus colocynthis* seeds. J Food Sci 48(1):104–106
Saxena SK (1986) Desert plants used as human food during scarcity and famines. In: Shankarnarayan KA, Shankar V (eds) Desert environment – conservation and management. Central Arid Zone Research Institute, Jodhpur, pp 43–47
Simoons FJ (1974) Rejection of fish as human food in Africa: a problem in history and ecology. Ecol Food Nutr 3:89–105
Sreeby LM (1982) Sugar availability, sugar consumption and dental caries. Community Dent Oral Epidemiol 10(6):1–7
Tan SP, Wenlock RW, Buss DH (1985) Immigrant foods, second supplement to McCance and Widddowson's the composition of foods. HMSO, London
Tang G (2010) Bioconversion of dietary provitamin A carotenoids to vitamin A in humans. Am J Clin Nutr 91(5):1468–1473
Tredgold MH (1986) Food plants of Zimbabwe. Mambo Press, Gweru
Trowell H, Burkitt D, Heaton K (1985) Dietary fibre, fibre-depleted foods and disease. Academic, London
Vogel VJ (1970) American Indian medicine. University of Oklahoma Press, Norman
Wehmeyer AS (1966) The nutrient composition of some edible wild fruits found in the Transvaal. S Afr Med J 40:1102–1104
Werner DB, Bower BL (1982) Helping health workers learn. The Hesperian Foundation, Palo Alto
WHO (1990) Diet, nutrition, and the prevention of chronic diseases. Technical Report 797, World Health Organization, Geneva
WHO (2003) Diet, nutrition and the prevention of chronic diseases. Technical Report 916, World Health Organization, Geneva
Willet WC, MacMahon B (1984) Diet and cancer – an overview. N Engl J Med 310(10–11):633–638
Williams CD (1935) Kwashiorkor: a nutritional disease of children associated with a maize diet. Lancet 226:1151–1152
Zennie TM, Ogzewalla CD (1977) Ascorbic acid and vitamin a content of edible wild plants of Ohio and Kentucky. Econ Bot 31:76–79

Chapter 4
House Design and Construction

4.1 Design Criteria

Two useful yardsticks are available for gauging the achievements of traditional house builders. Firstly, one can explore how effectively the structures they create protect house occupants from climatic extremes and secondly, one can try to assess how efficiently the builders make use of local constructional materials. On the first score, the idea of the "climatic comfort zone" is a useful one. When human volunteers placed in experimental climate chambers are asked to express their environmental preferences, whilst their reactions to humidity conditions prove fairly flexible, their idea of what constitutes a comfortable temperature range is more circumscribed, with preferred values typically falling within the relatively narrow range of 21–30 °C (Ellis 1953; Olgyay 1992; Karyono 1996; Nicol and Humphreys 2001). When the preferred temperature and humidity values are plotted on a graph, a tapering shape is produced, with a narrow tail reflecting the fact that high temperatures are harder to tolerate if humidity levels are also high (Fig. 4.1a). When average monthly outdoor temperature and humidity values recorded from various parts of the world are added to the same graph (Fig. 4.1b), it becomes clear that most regions experience periods of the year when outdoor conditions fall outside the comfort zone, being variously too hot, too cold, too humid or too dry. This is a measure of the climatic challenge faced by local house builders.

As regards the availability of constructional materials, especially those of plant origin, the first point to make is that this availability is also climate related (Fig. 4.2). Amongst tropical climates, the wet ones are the most prolific, with their output including not just durable timbers and bamboos but also palm leaves suitable for roof coverings and lianas suitable for use as bindings. Drier savannah-type habitats are much poorer timber sources because tree cover there is relatively sparse, but make up for this by supporting grasses suitable for thatching. Desert fringe habitats are the least productive of all. Of the few trees present, most are disqualified for constructional use either because they are too small or have poor load-bearing

J. Edington, *Indigenous Environmental Knowledge*,
DOI 10.1007/978-3-319-62491-4_4

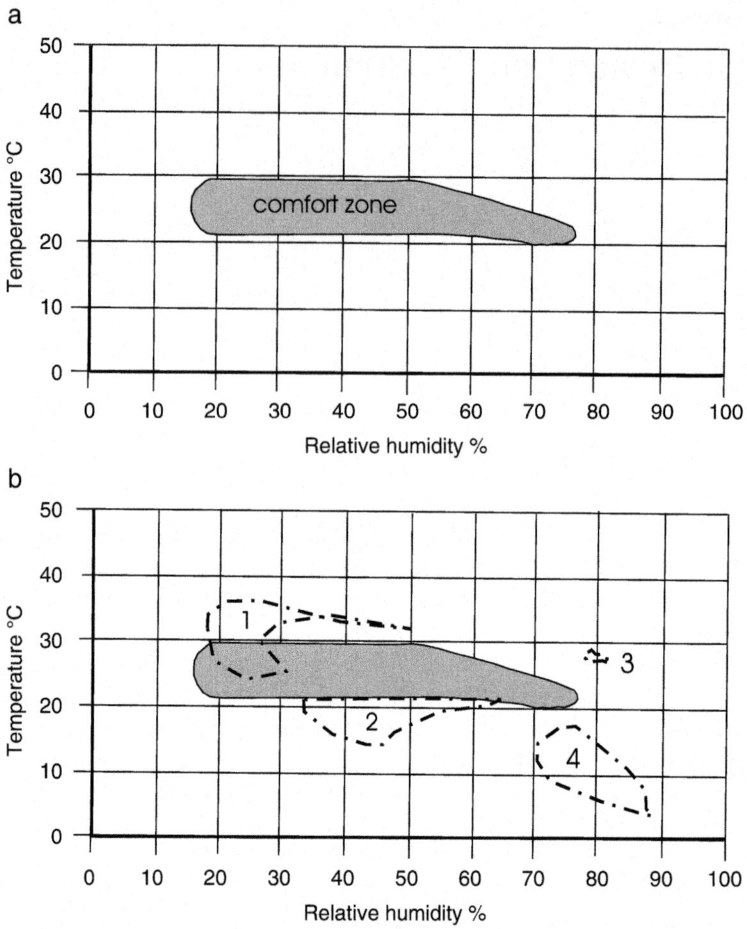

Fig. 4.1 (**a**) The "comfort zone", a representation of human climatic preferences (Based on Koenigsberger et al. 1974 and Olgyay 1992), (**b**) average monthly temperature and humidity regimes for a variety of climates (1) hot-dry Khartoum, (2) upland tropical Harare, (3) hot-humid Kuala Lumpur, (4) temperate York

properties. Temperate habitats fall somewhere between the two extremes. The supply of forest trees has usually been much reduced by agriculture, but the cereals often grown in their stead can make suitable roof coverings (Letts 1999). The ability to select materials that are "fit for purpose" thus offers a second criterion for assessing the work of local house builders.

In addition to these strictly practical considerations, houses also commonly display features whose attributed functions are less tangible. Most decorative motifs come into this category, as so too do various expressions of religious symbolism (Oliver 1987; Waterson 1990). Occasionally a cultural design concept is so pervasive as to virtually over-ride all other considerations. For instance the *Batammiliba*

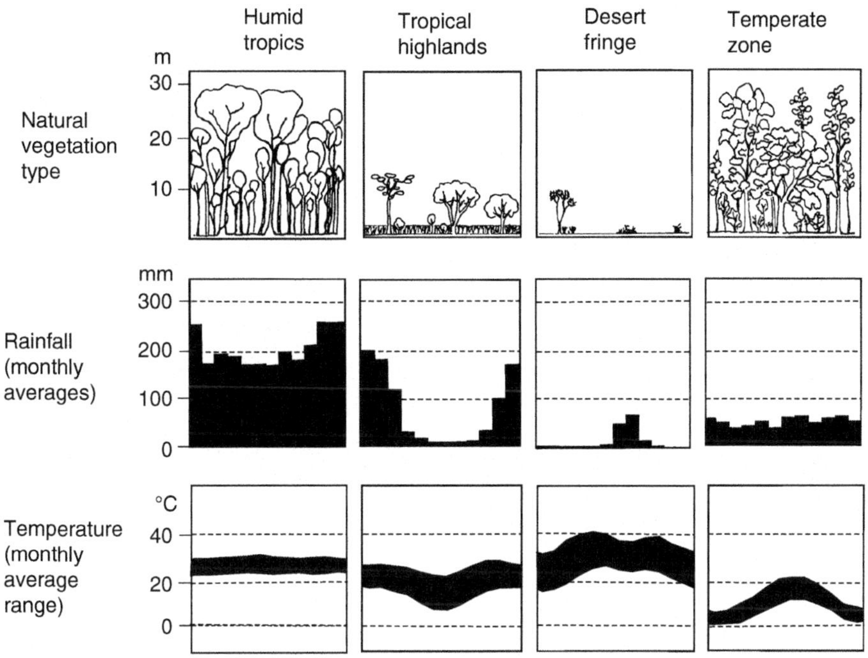

Fig. 4.2 The availability of biological raw materials in the main climatic zones (also indicating average rainfall and temperature values)

of Dahomey and Togo equate the various components of the house with different parts of the human body (Blier 1987). Thus the entrance portal is regarded as a mouth and accordingly has grinding stones to simulate teeth and overlooking windows to signify eyes (Fig. 4.3a). The stomach is represented by granaries, the womb by a central sleeping hut and the walls are pierced by apertures meant to represent excretory outlets. Another similarly all-embracing cultural prescription is the "house as a ship" concept pursued by the *Toraja* people of Sulawesi (Fig. 4.3b) (Lewcock and Brans 1975). Whilst the resulting structure is certainly spectacular, it is something of an indulgence in practical terms because of the additional materials and extra work needed to create it. Moreover it offers only very cramped living quarters (Fig. 4.3b).

Cultural expressions of a more modest kind are to be found in many South East Asian houses in the form of roof finials variously carved to portray sacred buffaloes or mythical half-dragon, half-snake like creatures known as *nagas* (Fig. 4.4a). The rationale is that by paying homage to these supernatural creatures, the well-being of the house and its occupants will be assured. In Thailand, actions to placate *nagas* start early in the building process. Because these creatures are perceived to be constantly moving about in the soil, it is considered prudent to seek advice from a soothsayer about their current position. In this way it is hoped to avoid striking and enraging them when house posts are being driven into the ground (Conklin 1991).

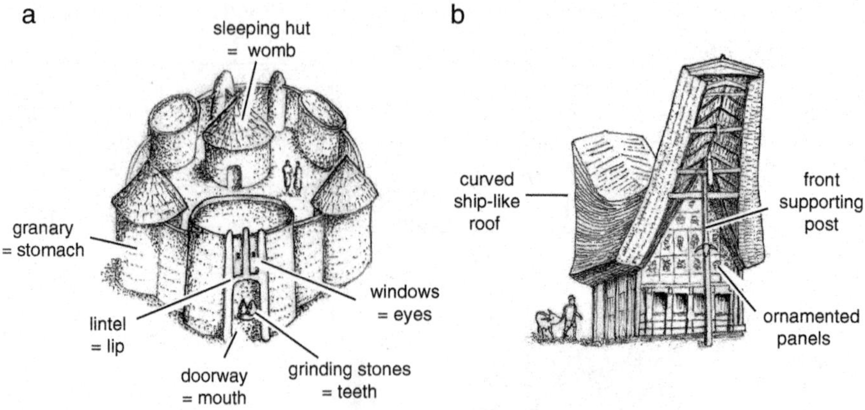

Fig. 4.3 Examples of culturally determined house designs. (**a**) House in Dahomey designed to simulate parts of the human body (Developed from Blier 1987), (**b**) house in Sulawesi designed to simulate a giant boat (Developed from Waterson 1990)

Fig. 4.4 Thai house features intended to solicit spiritual protection. (**a**) Roof finials portraying *nagas*, (**b**) a spirit house

Another placatory measure practiced by Thai house builders is to erect a miniature facsimile house in the compound of each new building (Fig. 4.4b). This is intended to provide a refuge for spirits displaced by building operations and offerings of food and drink are regularly placed alongside it.

Modern architects are well positioned to meet the practical (and for that matter the cultural) requirements of building construction. They can source raw materials on a world-wide scale; they have access to test data on individual building components and can, if they so choose, experiment with models of prospective designs in wind tunnels and light chambers (Olgyay 1992). They can moreover adjust indoor climatic regimes using externally-powered devices such as air conditioners and room heaters. Village house builders lack these advantages and are obliged to do the best they can using a mixture of passive energy sources, locally-available materials and their own ingenuity. How well do they fare?

4.2 The Humid Tropics

4.2.1 Climatic Challenges

Beginning with the humid tropics, one simple way for house builders to respond to the local environment is to construct their houses with steeply-pitched roofs. This allows rain water to be shed without causing structural damage. Constantly high temperatures and humidity levels are more difficult to cope with, but their effects can still be mitigated by arranging for air to circulate freely through and around the house. Sometimes this is achieved simply by leaving gaps in slatted walls; an approach adopted by Malaysian shifting cultivators such as the *Semai* (Dentan 1979), but there are also more sophisticated options. In Malay village or *kampung* houses, window openings are intentionally positioned to allow a free flow of air through the house; a process also helped by minimizing the number of internal room partitions (Fig. 4.5). Raising the house on stilts serves a similar purpose by elevating the structure above the height zone where wind speeds are slowed by frictional resistance with the ground. As long as the wind is actually blowing and the atmosphere is not fully saturated with moisture, such measures increase comfort levels by speeding air-flow over the skin and hence the efficiency of the sweating process. Even with humidity values of 70%, air moving over the skin at 0.1 metres per second will still produce a cooling sensation equivalent to a 2 °C reduction in temperature. Clearly this cooling process cannot function on windless days and in these circumstances, rooms in Malay houses can become uncomfortably hot in the afternoon (Rahman 1995). This is a good example of perfect solutions not being wholly achievable within a strictly traditional framework.

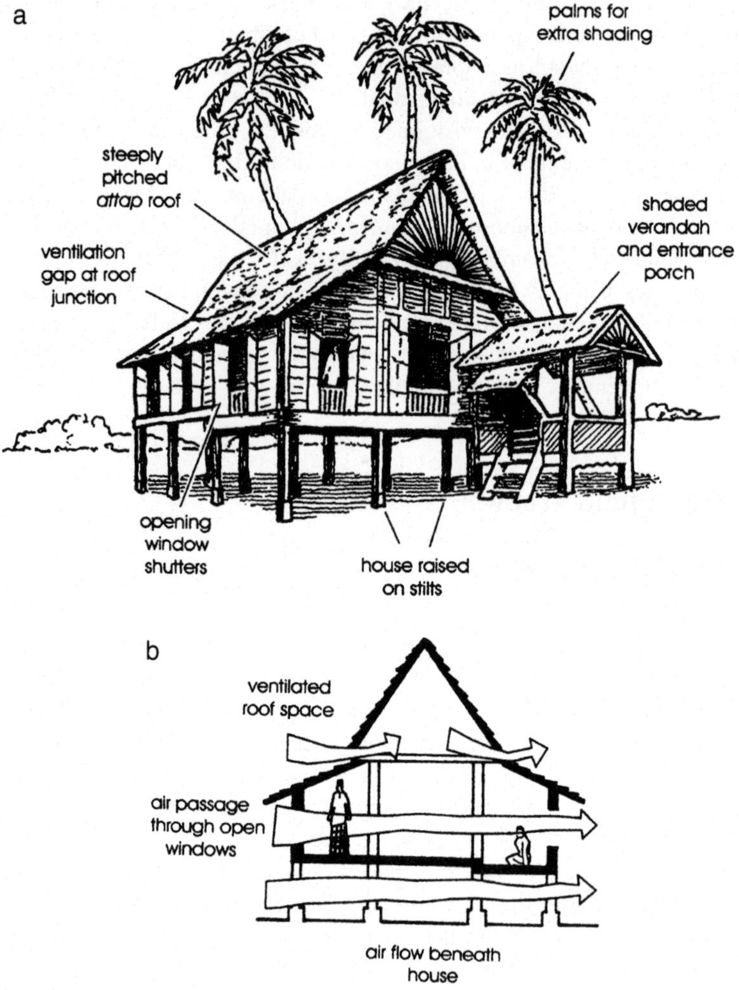

Fig. 4.5 A Malay *kampung* house, (**a**) General features, (**b**) cross section to show air movements (Developed from Lim 1987)

4.2.2 Durability of Constructional Materials

Any house in the tropics faces the threat of having its timbers damaged by wood-boring insects and rot-inducing fungi. Amongst the insects, termites or "white ants" (Fig. 4.6a) are notorious for their destructive powers and it is for this reason that Malay *kampung* house builders try to build with timbers they know to be termite-resistant. For this reason the main house posts and cross beams are usually made from trunk sections of *chengal* (*Neobalanocarpus heimii*) and *resak* (*Vatica cuspidata*) (Gibbs 1987). On the same grounds, floor slats and wall panels are typically

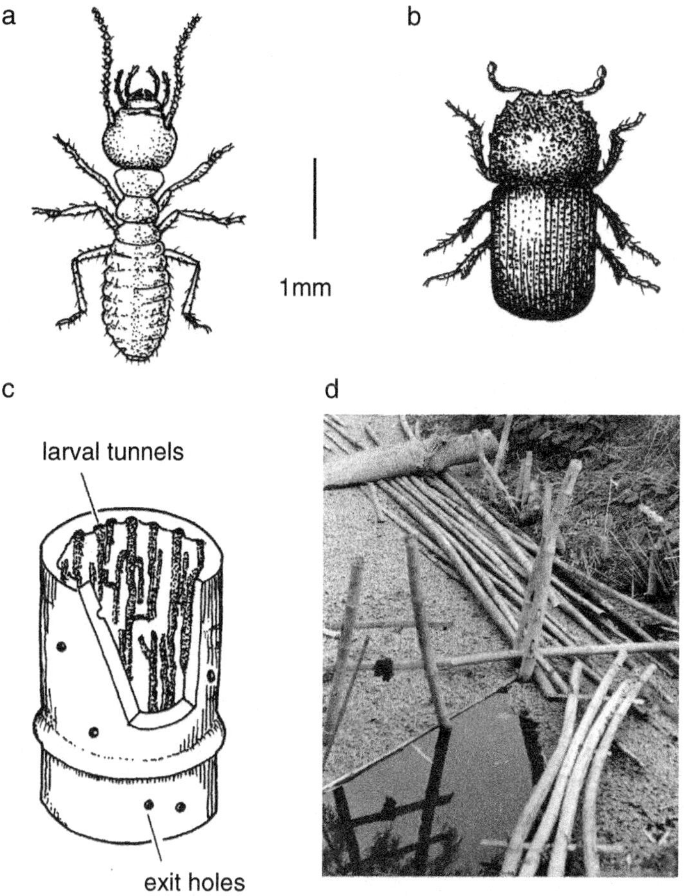

Fig. 4.6 Insects capable of damaging constructional timbers in the tropics. (**a**) Termite (**b**) bamboo borer, (**c**) typical bamboo borer damage, (**d**) technique of soaking bamboo stems to lessen the risk of borer attack (Vietnam)

made from split lengths of nibong palm (*Oncosperma tigillarium*) or bamboo (*Bambusa* species). Reliable species identification is important in making such selections and frequently the distinguishing features used by local builders are precisely the same as those employed by professional foresters. With trees they relate to such features as the shape of fruits and leaves and the surface characteristics of bark (Symington 1974). Indeed it is quite usual for local species-classification systems to run parallel with those devised by professional taxonomists (Berlin et al. 1966; Berlin 1992).

Although bamboo is resistant to attack by termites, it is at risk from the destructive activities of another insect, a small beetle known as the bamboo borer (*Dinoderus minutus*) (Fig. 4.6b, c). Fortunately, house builders in South-east Asia have developed their own solution to this particular threat. Soaking bamboo stems in water for

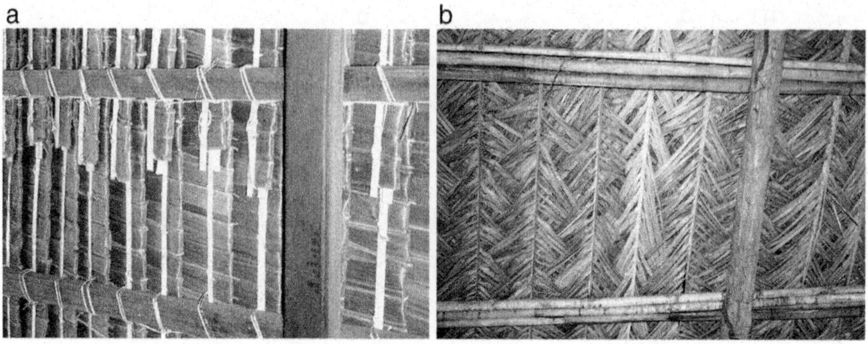

Fig. 4.7 Construction details of palm leaf roofs, viewed from the room interior. (**a**) Leaflets of sago palm (*Metroxylon sagu*) wound around bamboo laths in Malaysia, (**b**) interwoven leaflets of *acuri* palm (*Scheelea phalerata*) in a Brazilian house roof

some months prior to use (often in the village fishpond), has the effect of leaching out starches and sugars from the stems, thereby making them much less attractive to the borers (Fig. 4.6d) (Roonwall et al. 1966).

Palm leaves frequently feature in the roof coverings of rain forest houses. In Malaysia the leaves are wrapped around laths of bamboo or *nibong* palm to create *attap* panels that are laid in an overlapping sequence on the roof (Fig. 4.7a). In the highlands the *bertam* palm (*Eugeissona tristis*) is the main source of such leaves; in the lowlands its place is taken by the *rumbia* or sago palm (*Metroxylon sagu*) and in coastal areas by the *nipa* palm (*Nypa fructicans*). Roof panels made from any of these materials can be relied on to last between 5 and 15 years before needing replacement. House builders rash enough to use less durable leaves such as those from the coconut palm (*Cocos nucifera*) are likely to suffer the mortification of seeing them disintegrate very rapidly, even during the course of a single season.

A common enemy of palm-leaf roofs in Asia is the caterpillar of the pyralid moth, *Herculia nigrivitta*. Biological agents such as tiny chalcid wasps usually keep this insect in partial check but, if this natural control mechanism breaks down the results can be catastrophic. This was dramatically demonstrated when a pesticide spraying programme to control mosquitoes inside houses in Malaysia caused unexpected collapses of their palm-leaf roofs. The explanation was that the pesticide spray had inadvertently destroyed the natural control agents, but had left the offending caterpillars virtually unaffected.

Parallel choices of materials are made by house builders in other tropical forest areas. Those operating in the Peruvian Amazon region have their own preferred durable house timbers in the form of trunk sections of *huacapú* (*Minquartia guinensis*) and *palisangre* (*Brosimum guianense*). In this case the palm species favoured for roof coverings is *Lepidocaryum tessmannii* (Parodi 1988). The house walls are made from split staves of the *açai* palm (*Euterpe oleracea*), a material that is only marginally durable but, being quite abundant is readily renewable. Amongst the *Kekchi Maya* of Belize, *cohune* palm (*Orbigyna cohune*) leaves usually serve as the rot-resistant roof coverings and trunk sections of durable timbers such as rosewood (*Dalbergia stevensonii*) and *sabodilla* (*Achras zapota*), as the house posts. In much of South and Central America where palm-leaves are used as roof coverings, the panels are held together not by winding the leaflets around laths but by intertwining the individual leaflets (Fig. 4.7b).

4.2.3 Ritual Aspects

As elsewhere, various ritual elements feature in the building practices of this zone. In Trinidad for example, there is a local belief that cutting bamboo shortly after the full moon will make it less vulnerable to borer attack. This particular notion is actually one that has been experimentally tested and shown to have no scientific basis (Kirkpatrick and Simmonds 1958). An interesting contention amongst Malay house builders is that when the main house uprights are formed (as they usually are) by splitting sections from a single tree trunk, it is prudent to pacify the disturbed tree-spirit by arranging for the posts to occupy the same relative positions in the construction as they enjoyed in the growing tree (Lim 1987). Another postulated way of avoiding offence is to ensure that all timber sections are aligned according to their original direction of growth. Thus uprights should never be inserted upside down in the ground with their root ends pointing ignominiously skywards and horizontal members should always be butted on to one another, base to crown, to simulate the direction of natural growth (Waterson 1990, Vellinga et al. 2007). The practical advantages of applying these rules may be questionable but, like the comparable agricultural ceremonies, they at least have the merit of providing reassurance in uncertain circumstances and of fostering an attitude of respect towards the plant species providing raw-materials.

4.3 The Tropical Highlands

4.3.1 Climatic Challenges

Seasonal variability, rather than continuously hot and humid conditions, is the main challenge faced by house builders in the tropical highlands. Typically, hot-wet summers alternate with cold-dry winters (Fig. 4.2) and there is therefore a premium on creating a house structure that can cope with both extremes. Whether the iconic structure of this zone, the African round hut (Fig. 4.8a), qualifies is a matter of debate. Hut owners generally describe their dwellings as pleasantly cool in summer and even those who have built new houses with modern materials often retreat back to their old round huts during the hottest weather. This summer-coolness derives

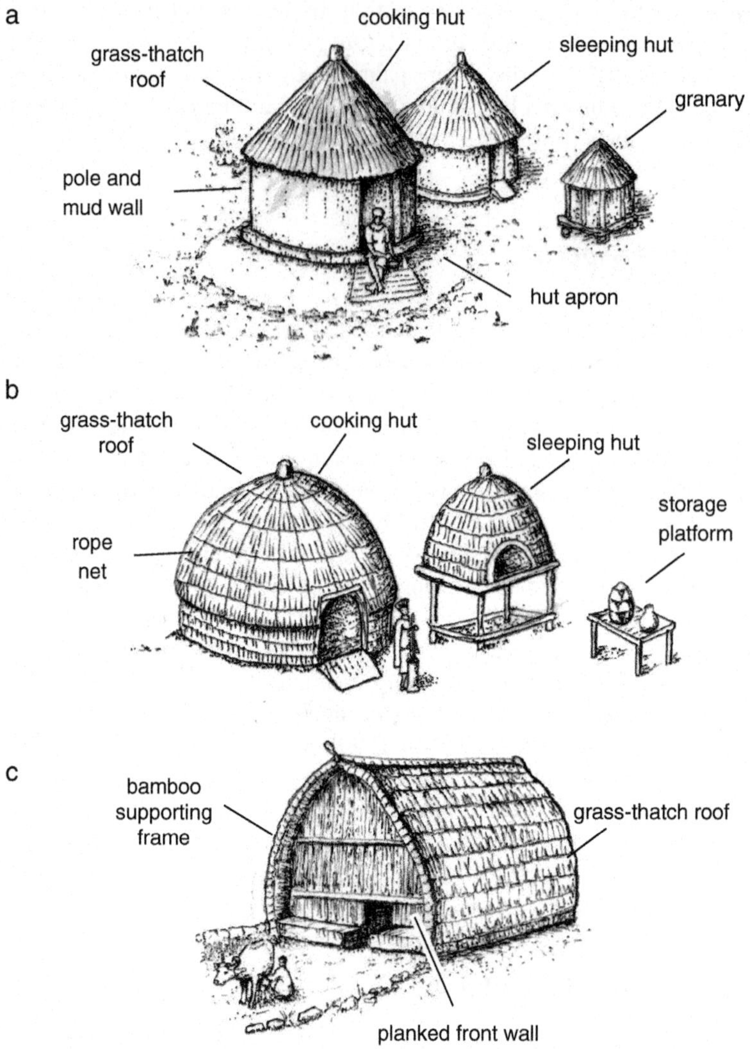

Fig. 4.8 Examples of houses from the tropical highlands. (**a**) Cone and cylinder hut in Mashonaland, (**b**) "beehive hut" in Swaziland, (**c**) arched *Toda* house in India

principally from the good insulating properties of the thick mud wall and the substantial thatched roof.

On the face of it, winter conditions look distinctly less satisfactory. Indoor measurements show that during this period temperatures can fall as low as 10°C at night, well below the climatic comfort zone (Malama and Sharples 1997; Sharples and Malama 1997). In cooking huts, as opposed to those used for sleeping, hearth fires will usually raise temperatures to acceptable levels, but if the house lacks a chimney, this is achieved only at the cost of unacceptable levels of smoke pollution. The hut apron rarely features in such discussions but perhaps it should. On cool but sunny winter days when the hut interior is unpleasantly cool, a useful option is to sit out on the surrounding apron either to do domestic chores or simply to relax in the sun (Fig. 4.8a). Although air temperatures are low at such times, the heating effect of direct insolation can more than compensate for this and winter conditions may not therefore be quite as taxing as they first appear.

4.3.2 Constructional Materials

In the matter of the availability of constructional materials, savannah habitats are by definition a rich source of grass species and at least some of these can serve as thatching materials. The properties looked for are long stems and a resistance to fungal and insect attack; caterpillars again posing the main insect threat. In Southern Africa the grass species of choice are *Hyparrhenia filipendula* and *Hyperthelia dissoluta*. Both have hard insect-resistant stems that grow over two metres long. Whilst in the African round hut, thatching is restricted to the conical roof (Fig. 4.6a), in other regional designs most of the house-surface is covered with grassy materials. Such is the case for example with the "beehive" huts of Swaziland and Zululand and the barrel-shaped houses of the *Toda* peoples of the Nilgiri hills of India (Fig. 4.8b, c). With the beehive huts, the underlying frame consists of a latticework of flexible saplings (nowadays usually derived from stems of introduced *Acacia mearnsii* trees), with *Hyparrhenia hirta* providing the outer grass covering. In the *Toda* house, the roof framework is constructed from over-arching stems of the bamboo *Pseudoxytenanthera monadelpha*, with the outer covering formed from bundles of the local swamp grass *Eriochrysis rangachari*. Both are endemic species and their declining availability is currently giving cause for concern (Gunawardene et al. 2007).

All these house structures have timber frames and there is the usual preoccupation with obtaining timbers resistant to termite attack. In the highveld regions of Southern Africa, this is the reason for preference being given to species such as *mobola* plum (*Parinari curatellifolia*), *mushava* (*Monotes glaber*), bloodwood (*Pterocarpus angolensis*) and silver terminalia (*Terminalia sericea*). At lower altitudes attention focuses on leadwood (*Combretum imberbe*), ironwood (*Androstachys johnsonii*) and *mopane* (*Colophospermum mopane*) (Palgrave 1983). House components made from any of these timbers can be relied on to last for at least 40 years. The Kruger National Park offers some striking testaments to the longevity of such woods. Ironwood chalet supports dating from 1919 are still in excellent condition, and the landscape is dotted with the trunks of leadwood trees which, although long since dead, still resist termite attack. Scientific tests on the durability of timbers frequently confirm the judgments made by traditional builders.

Spiritual allusions are rather rare in the houses of this zone, but the conical apex of the thatched roof often gets special attention. It may be adorned with a carved post, an upturned pot, a gourd or even an ostrich egg. These devices serve the practical purpose of preventing water ingress and, because they fracture under stress, of providing an early warning of imminent roof collapse (Denyer 1978). However they may also be locally regarded as serving less tangible functions such as warding off ill-fortune or advertising the high social status of the hut owner.

4.4 The Desert Fringe

4.4.1 Climatic Adaptation

In arid areas where humidity levels regularly fall below 55% there are excellent opportunities for harnessing the capacity of dry air to absorb moisture and be cooled in the process. A simple way for householders to do this is by inserting wettable screens into window apertures. By periodically wetting these screens, the incoming air becomes cooled by evaporation. In the dry areas of Rajasthan, interwoven twigs of the spiny desert shrub *Zizyphus nummularia* are much favoured for this purpose (Fig. 4.9). Elsewhere in India, wettable screens are made from woven root-strips of the grass *Vetiveria zizanioides*. In colonial times such screens were known as *khus-khus tatties* and were recognized as having the merit, not only of cooling incoming air, but also of investing it with a pleasant scent. A variant employed in some Middle Eastern houses (although mainly in larger urban dwellings), is to place wettable

Fig. 4.9 Householder in the Thar Desert of India wetting a window screen to cool incoming air

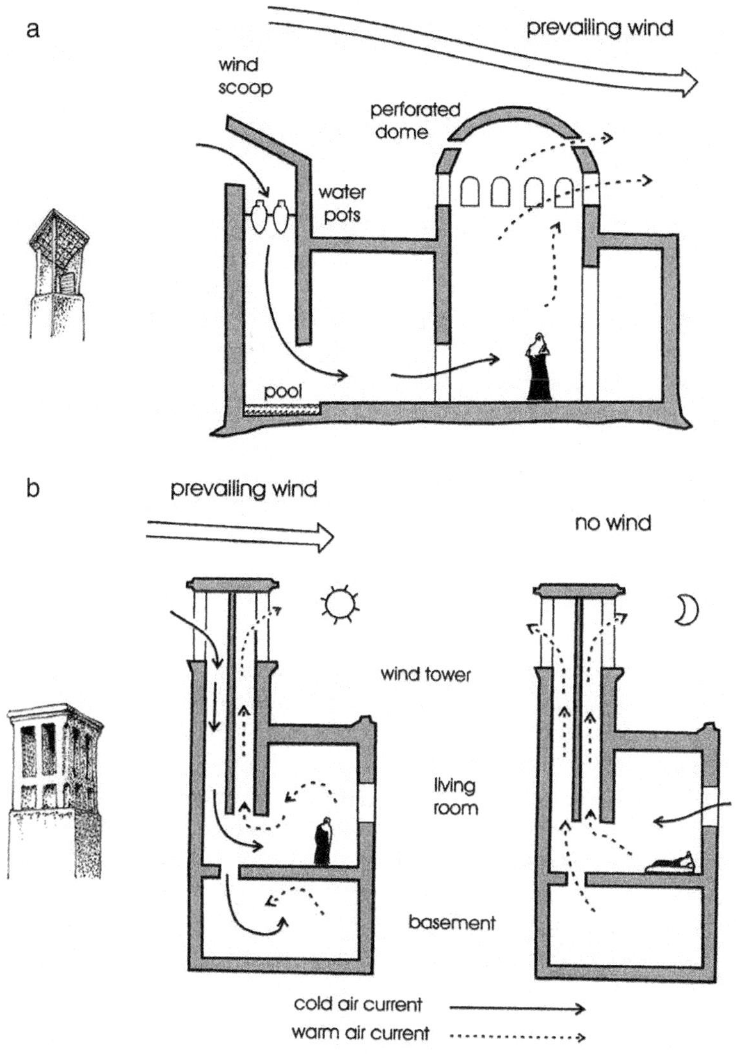

Fig. 4.10 Traditional ventilation systems from the Middle East (**a**) Wind scoop or *malqaf*, (**b**) wind tower or *bādgīr*, in day time and night time operation (Developed from Cain et al. 1976 and Bahadori 1978)

straw screens and porous water-filled pots in the path of the air captured by *wind scoops* or *wind towers*.

Wind scoops (or wind catchers) are designed to operate in regions where the wind blows predominantly from one direction (Fig. 4.10a). For instance in the Sind region of Southern Pakistan, the scoops are angled to catch the prevailing south-westerly winds and to provide relief during a hot season that extends from April to June (Bourgeois 1980). A trap-door incorporated into the structure allows the inlet conduit to be closed during sand storms and also in winter when the incoming air would be too cold for use. Houses with a wind scoop may also feature a domed roof perforated with outlet vents. The significance of the dome is that the air passing over it creates an area of reduced pressure on the lee side, thereby helping to draw out stale air from the house interior. Such a dome is calculated to generate a through flow of air that is 70% greater than that achieved in a conventional flat-roofed building (Bahadori and Haghighat 1985).

Wind towers are more flexible than wind scoops because they are designed to tap air currents from any point of the compass (Bahadori 1978). In the mornings when the walls of the tower are still cold from the previous night, air passing through it is cooled *en route* to the house interior (Fig. 4.10b). One only has to sit in the room beneath a wind tower to appreciate how comfortable conditions can be. Air currents in these living areas can reach speeds as high as 3 m a second and can generate as many as 30 air changes an hour (Roaf 1983). Later in the day as the tower heats up, the temperature differential disappears but the fact that the incoming air is moving still provides some relief from overheating. Even when there is no wind, as is often the case towards the end of the day, the tower acts as a heat chimney and in this mode, stale air rising up the shaft serves to draw-in fresh air through window openings.

Desert fringe houses are commonly constructed with thick mud-brick walls (Fig. 4.11 a-c); the crucial property of these being their high thermal capacity. Whilst during the day this causes them to absorb heat and to limit its penetration into the house, at night as temperatures fall they perform the equally valuable function of releasing heat into room interiors. This keeps living areas pleasantly warm at a time when outside temperatures are plummeting. Over the 24 h period, the difference between indoor and outdoor conditions can thus become very marked and very much to the advantage of house occupants. Whilst outdoor temperatures show large fluctuations, those indoors remain stable and comfortable (Fig. 4.12a). These intrinsic benefits are lost if mud is replaced by concrete (Fig. 4.12b). Then only the installation of an air conditioner will retrieve the situation. Arid zone houses usually include additional features designed to reduce temperature fluctuations (Rodger 1974; Kidd 1982). For example, with the Sudanese house shown in Fig. 4.11a, the long axis of the structure is purposely arranged to have an east-west orientation and to lack window openings in its end walls. This layout reduces the risk of overheating when the sun is low in the sky during the early morning and late evening. Moreover windows with moveable shutters are arranged along the north and south walls and on hot summer nights these can be opened to encourage a through-flow of cooler air from the prevailing wind, which at this time of the year blows mainly from the north.

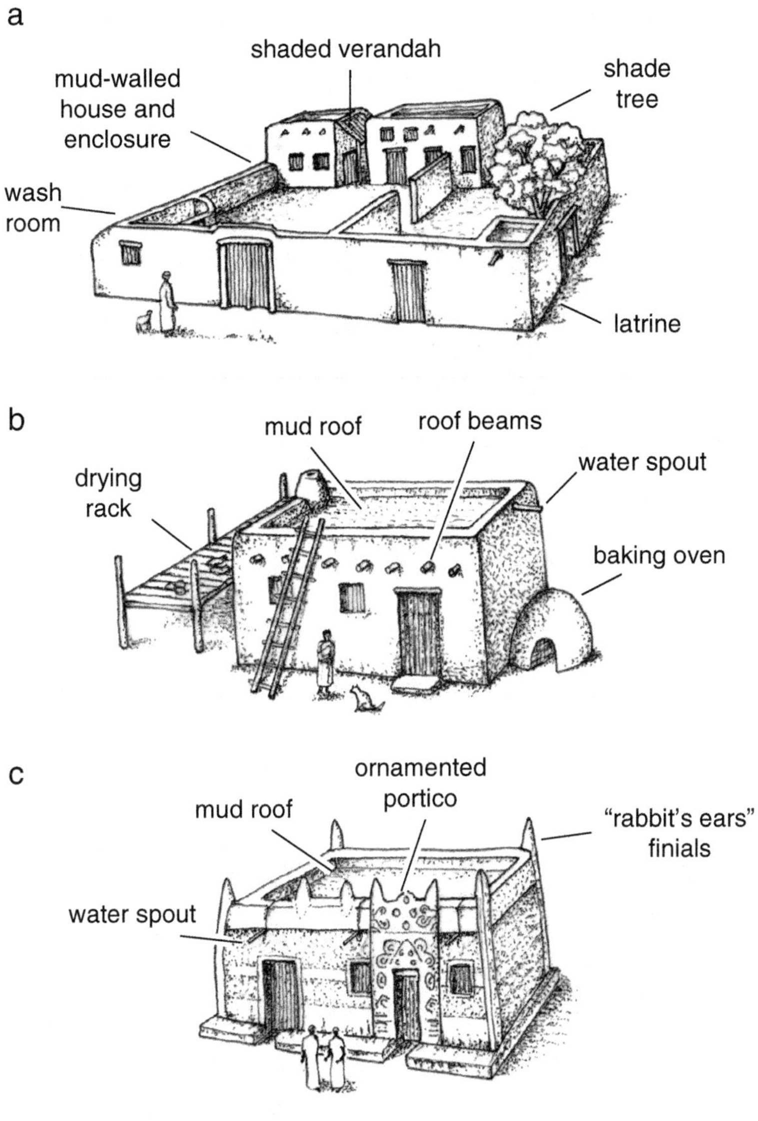

Fig. 4.11 Examples of desert fringe houses. (**a**) Courtyard house in Northern Sudan (Based on Kidd 1982), (**b**) Pueblo house in New Mexico, (**c**) *Hausa* house in Northern Nigeria

4.4.2 Constructional Materials

On the desert fringe there is usually no shortage of earthy materials suitable for making mud-bricks. Local people are well aware that a good mix combines coarse with fine particles, and that extra strength can be gained by adding binding materials such as straw or animal hair. Sometimes pith from the seed pod of the locust bean

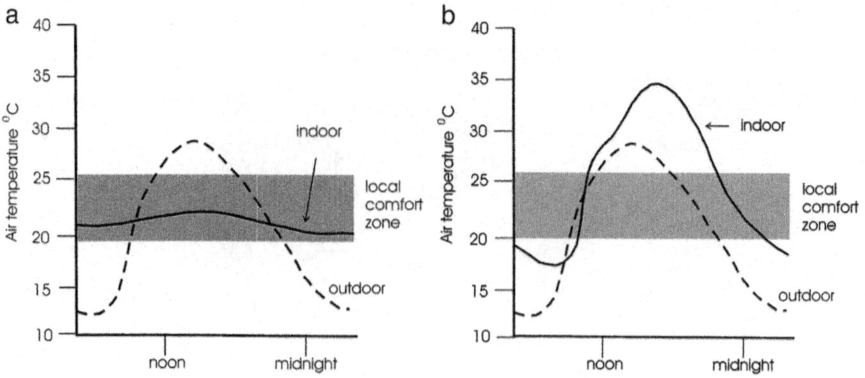

Fig. 4.12 Comparisons of indoor and outdoor temperature cycles. (**a**) In a traditional mud-walled house, (**b**) in a house constructed with modern materials (Based on Fathy 1986 and Norton 1986)

(*Parkia biglobosa*) is applied to the wall surface to water-proof it, a treatment reputed to remain effective for at least a year.

Acquiring sufficient building timber can be more of a problem. Wood is needed for doors, window frames and shutters, and most critically to provide the thick beams required to support heavy mud roofs. Unfortunately these dry environments have only a limited capacity to yield timbers of the necessary length, size and strength. One solution is to collect the required materials from further afield. This is the strategy followed by the Pueblo Indians of New Mexico when they make forays into wetter mountain regions to collect stout trunks of ponderosa pine (*Pinus ponderosa*).

The other approach is to take whatever woody material comes to hand and, if this is only marginally suitable, to modify the house design accordingly. This strategy is strikingly demonstrated in traditional Jordanian houses. Because the local woody vegetation consists mainly of stubby, multi-stemmed juniper bushes (*Juniperus phoenicea*), the practice has developed of compensating for the shortness of the trunk sections by adding an extra stone supporting-arch to the roof (Fig. 4.13a).

Another elegant traditional solution is found in the dry regions of North and West Africa where stems of *doum* palms (*Hyphaene thebaica*) are employed as roof supports. Although these stems have the merit of being termite-resistant, they have poor load-bearing properties and consequently can be used only in short lengths. Local builders have responded to this challenge by constructing corbels and beams from a mixture of mud and stem fragments (Moughtin 1982; Aradeon 1987). These then serve as supports for a criss-cross arrangement of short stem sections, all cut and positioned to fall within the prescribed length limit (Fig. 4.13b).

Dry-region houses in West Africa are often ornamented with embossed patterns (Fig. 4.11c). Because these patterns are most commonly placed on structural members such as door jambs and lintels, it has been suggested that their main function is to give an early warning of impending structural collapse; the patterns becoming

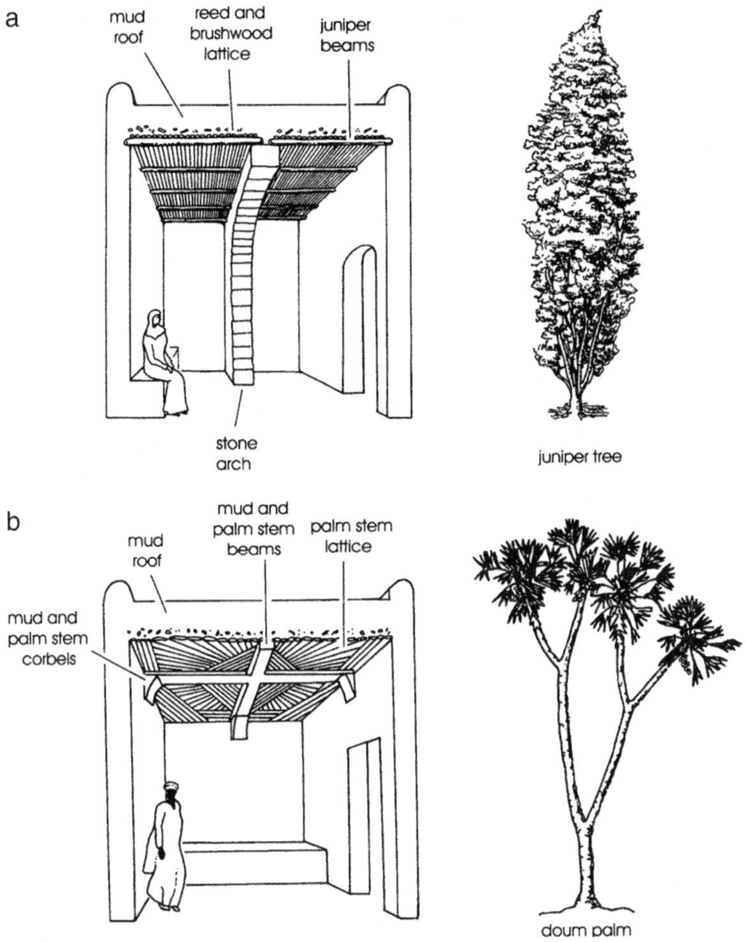

Fig. 4.13 House designs adapted to the limitations of available raw materials. (**a**) A Jordanian house with an extra stone arch to compensate for the shortness of the juniper stems used for the roof, (**b**) Northern Nigerian house incorporating a criss-cross arrangement of *doum* palm lengths designed to compensate for their limited mechanical strength

visibly distorted when put under pressure (Denyer 1978). Various interpretations have been placed on the "rabbits-ears" roof embellishments of Northern Nigerian houses (Fig. 4.11c). One of the more charming is that they honour the Togo hare (*Lepus capensis*), an actual species but also one whose fictional exploits feature prominently in local folk tales. This same animal is the likely inspiration for the "brer rabbit" stories brought to North America by African slaves.

4.5 The Temperate Zone

4.5.1 Climatic Adaptations

There is again some debate about whether traditional houses in the temperate zone can be regarded as climatically well-adapted. Certainly, the flimsy paper wall-panels of many Japanese houses are not effective in resisting the passage of either heat or cold (Sdei 2006). Some protection from the summer sun is gained from wide roof overhangs, but the entry of winter cold is usually scarcely offset by the limited heat generated from a small hearth fire. Wearing more clothes then seems to be the only other option.

Elsewhere, houses with stone or earth walls perform well in situations where hot days alternate with cold nights; the heat accumulated in the walls during the day being usefully liberated into room interiors at night in much the same way as happens in desert houses. In winter however, additional indoor heat sources are required to offset the substantial heat losses occurring through the walls. That this is a significant problem is reflected in the modern recommendation that, for stone-walled houses to be truly energy-efficient, they need to have insulating materials applied to their inner wall surfaces.

Climatic adaptation is not necessarily limited to the house structure itself; it may also dictate the positioning of the house in the landscape. Striking examples can to be found in both alpine and Himalayan regions of houses being intentionally sited some distance up hill-slopes rather than on the valley floors themselves. The intention is to place them out of reach of the pools of cold air that typically accumulate in valley bottoms at the night (Geiger 1965). Sometimes the prescriptions include distinctly mystical elements. For instance the famous *Feng Shui* house-placement system of China, whilst including some perfectly rational recommendations such as positioning houses where mountains ranges offer protection from cold northerly winds, also obscurely recommends that a site should be sought where the hills on the eastern side are shaped like a tiger and those on west like a dragon (Eitel 1984).

4.5.2 Use of Raw Materials

Traditional house builders in temperate regions usually have access both to a wide range of timber trees and also to cereal stalks suitable for thatching. At least this is the case where the wheat being grown is still of the traditional long-stemmed type (Letts 1999). Modern short-stemmed wheats are generally unsuitable and in Europe this has prompted a switch either to wild reeds (*Phragmites australis*) – a considerably less durable material – or to straw from long-stemmed varieties grown specifically for the purpose (Fig. 4.14a). In Britain the John Innes Centre has been active in reviving the old strains and making them available to thatchers.

Where suitable stone is accessible this is widely used for house walls; the final structures often reflecting local geology in interesting ways (Wright 1985). A tradi-

Fig. 4.14 Some traditional building techniques from Wales. (**a**) Thatching with long-stemmed wheat straw in the Vale of Glamorgan, (**b**) renewing a wattle and daub wall panel, (**c**) "cruck" roof supports in a mediaeval house, note also the hazel lattice supporting the thatch (Photographs **b** and **c** taken at the National History Museum of Wales, St. Fagans)

tional alternative has been to erect stout timber house frames and then to fill the intervening spaces with "wattle and daub" panels. These consist of a framework of woody stems onto which a mixture of mud, cow dung and animal hair has been plastered (Fig. 4.14b). Usually the panel receives a final protective coat of lime mortar.

As termites are only patchily distributed in temperate regions (Harris 1971), there is generally no need to take them into account when selecting timbers. Other wood-boring insects certainly do occur in temperate regions, most notably the "woodworm" (*Anobium punctatum*) and the death watch beetle (*Xestobium rufovillosum*); but neither of these has sufficiently well-defined preferences to have an over-riding influence on the selection of timbers (Hickin 1975). Moisture content is a much more critical factor and when wood moisture levels rise above 25%, the risk of infestation by both insects and rot fungi increases significantly.

In temperate settings, traditional timber choices are much more determined by the shape, size and strength of the components. With mediaeval houses in Britain, the roof supports typically consisted of pairs of angled struts or "crucks", cut from combined trunk and branch sections of large oaks (*Quercus robur*) (Fig. 4.14c). Unlike their Asian counterparts, house builders in Britain apparently felt no compunction about turning the crucks upside down if this served their purpose better. As well as being employed as wattle wall lattices, split and woven stems of coppiced hazel (*Corylus avellana*) are often used to provide the underlying support for roof thatch (Fig. 4.14c).

4.6 Pressures for Change

4.6.1 Neglect of Health Issues

Traditional house builders are certainly not incapable of omissions and errors of judgment. Various blind spots are apparent in matters of health. For example whilst the cooling air-flow entering *kampung* houses undoubtedly increases comfort levels, it also facilitates the entry of mosquitoes. Some mosquitoes (those belonging to the genus *Anopheles*) are of course notorious for having evolved the habit of entering houses at night to feed on room occupants, often infecting them with malarial parasites in the process. People living alongside rice-fields are likely to be in double jeopardy. Not only are their airy houses readily penetrated by the winged adult insects, the open waters of adjacent rice fields provide a virtually endless supply of suitable habitats for larval development (IRRI 1988).

Formerly local people gave little attention to this issue, regarding the biting mosquitoes merely as a nuisance rather than a health hazard. Now that the danger is more fully appreciated, additional protective measures are in order. One design solution is erect meshwork screens across window openings. However this so impedes ventilation that air conditioners are required to achieve acceptable comfort levels. Other solutions involve the use either of insecticidal coils or insecticide-impregnated bed-nets (Yap et al. 1990). The coils are simply spirals made from sawdust impregnated with a *pyrethroid* insecticide and if lit during the evening will continue to release a protective vapour throughout the night. Mosquito nets in their turn are now routinely treated with similar *pyrethroid* compounds and are far more effective than the earlier untreated nets (Schreck and Self 1985; Robert and

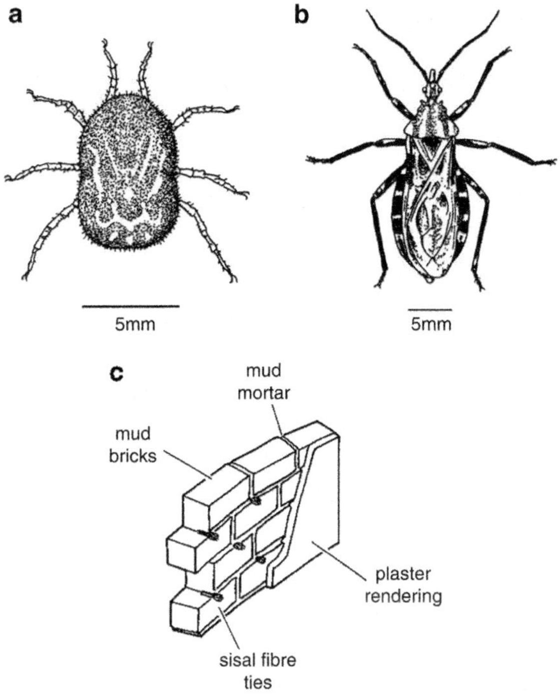

Fig. 4.15 Crevice-dwelling arthropods. (**a**) Soft tick (*Ornithodrus moubata*), (**b**) cone-nosed bug (*Triatoma infestans*), (**c**) use of a plaster skim to eliminate crevices from interior walls (**c**, after Carroll 1990)

Carnevale 1991). In regions where they have been are deployed, childhood malaria rates may fall by as much as 60%.

If not kept in good repair, the fissured walls of mud-brick houses can generate problems of a different kind. The cracks that form serve as refuges for disease-transmitting arthropods (Walton 1962; Kettle 1984), most of them species that normally live in the walls of animal burrows (Smith 1973). In houses they lurk in the crevices during the day and emerge at night to feed on sleeping room-occupants. In Africa, soft ticks are the main culprits and in the drier regions of South America their place is taken by cone-nosed bugs (Fig. 4.15a, b). The two organisms function respectively as vectors of relapsing fever and Chagas' disease. Carlos Chagas was instrumental in unraveling the transmission cycle of this latter infection when in 1908 he sought to explain the high incidence of heart disorders amongst construction workers on the Rio-Amazonia railroad. The cause proved to be a trypanosome parasite transmitted by cone-nosed bugs frequenting the workers' makeshift houses. Typically the human host becomes infected when contaminated faeces from the bug are scratched into the puncture wound. A relatively simple technical solution is available for both problems; it being quite feasible to eliminate the crevices by applying a plaster skim to interior walls (Fig. 4.15c). The best mix combines cement, lime and sand in the proportions of 1:2:9 (Carroll 1990), although a mixture of earth and cow dung and sand will serve almost as well. Spraying the walls with insecticide offers an additional safeguard.

Apparently, some of the affected householders in South America had been unaware of the significance of the bugs and were inclined to view their presence as a good omen. East African householders are apparently more knowledgeable about their soft ticks and some groups have even adopted the habit of transferring ticks to new houses; the rationale being that this will maintain the householder's immunity in the face of new and possibly more virulent local strains of relapsing fever (Good 1978).

4.6.2 Technological Shortcomings

Traditional builders are certainly capable of making some strategic errors. The author well remembers visiting villages in the tropical forest zone of Nigeria where local people originating from further north had tried to build mud-walled houses without wall top protection in climatic conditions too wet for this type of construction. The predictable result was that the first rains caused substantial wall collapses. In drier conditions there is usually a more general awareness of the threat posed by water damage, but often a lack of suitable indigenous techniques to address it. This is a situation where modern materials can make a big difference. Maintaining mud-brick structures certainly becomes noticeably more difficult once rainfall exceeds a threshold level of around 500 mm/annum (Dayton 1991). In Northern Nigeria for example, wall tops suffer damage every rainy season and have to be repaired. Adding a small quantity of a stabilizing material such as bitumen or cement to the mud mix can however virtually solve the problem. Such additives need amount to no more than 3% of the total mass (Warren 1999).

Likewise the problem of over-heating of traditional tropical forest houses in windless conditions can be at least partially remedied by installing a fan in the ceiling of the main living room. This can be powered from a small solar panel, a solution widely adopted in Indonesia. Another possibility is to position on the roof, a miniature version of the kind of rotating ventilator seen on factory roofs. This too can be induced to operate on windless days using power from a small solar panel (Ismail and Rahman 2008).

4.6.3 Shortages of Raw Materials

Population growth has had the effect of increasing pressure on supplies of traditional building materials, sometimes to the extent that they cease to be available. This can lead to inferior substitutes being used with negative consequences for the durability of the buildings. Often the widespread destruction of local natural vegetation seriously aggravates the problem. In lowland Malaysia for example, the replacement of much of the original lowland dipterocarp forest with oil-palm plantations has made it far more difficult to obtain either termite-resistant timbers or caterpillar-resistant palm leaves. Indeed nowadays, building one's house entirely from traditional materials is viewed as a sufficiently notable feat to warrant an article in the local newspaper. Attempts to build with unprotected soft-woods almost always end badly and in one survey, over 65% of village houses thus constructed were found to have suffered serious termite damage (Ahmad and Rahim 1986).

Using inferior materials has created similar problems in the tropical highlands. In Tanzania, Fleuret (1980) reports how the virtual disappearance of the tree ferns (*Cyathea usambarensis*), formerly used as house posts, has caused a switch to timber from the introduced black wattle tree (*Acacia mearnsii*). This has a useful life of only 5–10 years, compared with the tree fern's 30–40 years. A further difficulty in some African savannah regions is that large scale cattle-ranching has reduced the availability of good-quality thatching grasses (Wells 1995). On the desert fringe, vital materials are often squandered for purposes not actually requiring their special properties. For instance, many groves of *doum* palms have been cut down simply to provide fuel for steam locomotives (Wells 1995). Attempting to use alternative materials has caused difficulties here too. In Mali for example, after the useful *doum* and *borassus* palms had all been cut, builders turned to stems of the shrub *Calotropis procera*. Whilst these had the merit of repelling termites by virtue of their poisonous milky sap, they lacked mechanical strength and soon disintegrated when employed as roof supports (Wells et al. 1998).

Where householders have given up the unequal struggle and have switched to modern materials, the extra expenditure unfortunately offers no certainty of increased levels of comfort. For instance, incorporating bricks and concrete into a Malay house can make it uncomfortably hot at night. This is because the modern materials have high thermal capacities and having accumulated heat over the course of the day release it in an unwelcome fashion into the house interior at night (Hanifi 1994; Rahman 1995). The only practical solution is then to install an air conditioner.

There has been a similar tendency in Africa to replace traditional round huts with rectangular houses with brick or block walls and corrugated-iron roofs. Although such structures cost more to build, there is no guarantee that they will be more comfortable to live in. Iron-sheeted roofs have poor insulating properties, causing rooms seriously to overheat in summer. They also have the disadvantage of reverberating noisily during rainstorms.

Yet more instances of misplaced expectations can be found in arid regions where there has been much supposed upgrading of houses using preformed concrete components. Unless these dwellings are also provided with air conditioners, indoor temperatures are liable to fluctuate quite wildly and be much less tolerable than those experienced in traditional mud-walled structures (Fig. 4.12). Of course, there can be perfectly legitimate reasons for people wishing to modernize their houses, including easier maintenance, better fire-resistance and increased privacy, not to mention keeping up with the neighbours (Lim 1987). However it is unfortunate that builders are being forced down this route simply because they cannot obtain their customary constructional materials.

Occasionally novel solutions present themselves. In some regions, material from agricultural crops can be brought into service. For instance where date palms (*Phoenix dactylifera*) are grown, their trunks can be used as roof beams and their leaf mid-ribs as cross-members. Excellent, mud-roofed buildings of this type are commonplace for example in Oman. Another imaginative approach is to allow people living in the vicinity of National Parks to harvest thatching grass from within their boundaries, a process that also benefits tourists by making the wildlife easier to see. There have been notable initiatives of this kind both in Zimbabwe (MacKinnon et al. 1986) and in the *terai* region of Nepal (Sunquist 1984).

Community forests can also play their part. Although their main function has usually been to generate firewood, there is no reason why they should not also be harnessed to produce building materials. In Southern Africa for example, schemes are afoot to plant *mopane* trees (*Colophospermum mopane*) in such forests. These will yield not only serviceable building poles and firewood, but with any luck supplies of that much-prized edible caterpillar, the "*mopane* worm" (Musvoto et al. 2006).

4.7 A Future for Traditional Designs?

In practice there is considerable scope for combining traditional and modern design features. A good example is provided by the demonstration house complex created at the National Training Centre for Rural Women at Melfort in Zimbabwe. Here the decision was taken to retain the traditional thatched roof and mud-brick walls but to add to the latter an inner plaster skim to control ticks and an outer one to reduce rain damage (Fig. 4.16a, b). Another important innovation was to equip the kitchen hut with a chimney, thereby addressing the problem of indoor smoke-pollution. Sleeping quarters have been made rectangular rather than round; the better to accommodate conventional beds, but the temptation to replace thatch with corrugated iron has been resisted. Although easier to maintain, an iron roof needs an underlying insulating ceiling if it is to provide adequate temperature-buffering. Such a roof is also expensive in village terms, costing upwards of US$120 to install. A project in Botswana similarly shuns the iron roof and uses a 5% cement stabilizer to protect mud-brick walls. In this case the problem of durable-timber shortages has been addressed by reducing the overall number of timber components and by insect-proofing those that are used (Ngowi 1997).

In a parallel project in Malaysia, commercial house builders have attempted to create an updated version of the *kampung* house (Fig. 4.16c, d). This time the corrugated iron roof has been retained, but an underlying insulating layer added to reduce sound and heat passage. The shortage of durable timbers has been addressed by using tanalised components (impregnated with a protective mixture of copper, chromium and arsenic compounds) and the problem of overheating on windless

Fig. 4.16 House designs combining traditional and modern features (**a**) Demonstration house complex at the National Training Centre for Rural Women at Melfort in Zimbabwe, showing *left to right*, granary, sleeping hut and kitchen hut, (**b**) kitchen hut interior showing plastered walls and stove connected to an outdoor chimney, (**c**) a modified Malaysian *kampung* house, (**d**) interior view showing tanalised rafters and plaster-board ceiling insulation

days by installing a ceiling fan in the main room. Kitchen and bathroom facilities have been placed at ground level for ease of construction and plumbing, although this prevents the traditional free-flow of air beneath the house. Indeed, overall this design arguably involves more compromises than would have been necessary had traditional building materials been more readily available.

In spite of the difficulties a strong case can be made, not only for persisting with traditional house designs in their original settings, but also for incorporating some of their features into Western buildings (p.210). Strong incentives for doing this are offered by current drives to economize on fossil-fuels and to maximize the use of locally-sourced constructional materials.

References

Ahmad SS, Rahim S (1986) Response of the drywood termite *Cryptotermes cynocephalus* (Isoptera: Kalotermitidae) to extracts from five Malaysian hardwoods. Malay For 49:470–475
Aradeon SB (1987) The architecture of a Hausa master mason. Nigerian Field 52:3–18
Bahadori MN (1978) Passive cooling systems in Iranian architecture. Sci Am 238(2):144–152
Bahadori MN, Haghighat F (1985) Passive cooling in hot, arid regions in developing countries by employing domed roofs and reducing the temperature of internal surfaces. Build Environ 20:103–113
Berlin B (1992) Ethnobotanical classification. Principles of categorization of plants and animals in traditional societies. Princeton University Press, Princeton
Berlin B, Breedlove DE, Raven PH (1966) Folk-taxonomies and biological classification. Science 154:273–275
Blier SP (1987) The anatomy of architecture: ontology and metaphor in Batammaliba architectural expression. Cambridge Univerity Press, Cambridge
Bourgeois J-L (1980) Welcoming the wind. Nat Hist 89:70–75
Cain A, Afshar F, Norton J, Daraie MR (1976) Traditional cooling systems in the third world. Ecologist 6(2):60–64
Carroll RF (1990) Disease reduction by improved house construction. Overseas building note 194. Building Research Establishment, Garston
Conklin N (1991) Naga motifs. Sawaddi, Bangkok 37(3):28–31
Dayton L (1991) Saving mud monuments. New Scientist 130, 25 May, 38–42
Dentan RK (1979) The Semai, a nonviolent people of Malaya. Holt, Rinehart and Winston, New York
Denyer S (1978) African traditional architecture. An historical and geographical perspective. Heinemann, London
Eitel EJ (1984) Feng-Shui. The science of sacred landscape in old China, (reprint of 1873 edition). Synergetic Press, London
Ellis FP (1953) Tolerable and desirable levels of warmth in warm climates, with special reference to men in the Royal Navy. Ann R Coll Surg Engl 13:369–391
Fathy H (1986) Natural energy and vernacular architecture. Principles and examples with reference to hot arid climates. University of Chicago Press, Chicago
Fleuret A (1980) Nonfood uses of plants in Usambara. Econ Bot 34:320–333
Geiger R (1965) The climate near the ground. Harvard University Press, Cambridge, MA
Gibbs P (1987) Building a Malay house. Oxford University Press, Singapore
Good CM (1978) Man, milieu, and the disease factor, tick-borne relapsing fever in East Africa. In: Hartwig GW, Patterson KD (eds) Disease in African history. Duke University Press, Durham, North Carolina, pp 46–87
Gunawardene NR, Daniels DAE, Gunatilleke IAUN, Gunatilleke CVS, Karunakaran PV, Nayak KG, Prasad S, Puyravaud P, Ramesh BR, Subramanian KA, Vasanthy G (2007) A brief overview of the Western Ghats-SriLanka biodiversity hotspot. Curr Sci 93(11):1567–1572
Hanifi Z (1994) Housing design in relation to environmental comfort – a comparison of the traditional Malay house and modern housing. Build Res Inf 22(1):21–33
Harris WV (1971) Termites, their recognition and control. Longman, London
Hickin NE (1975) The insect factor in wood decay. An account of wood-boring insects with particular reference to timber indoors. Associated Business Programmes, London
IRRI (1988) Vector-borne disease control in humans through rice agroecosystem management. International Rice Research Institute, Manila
Ismail M, Rahman AMA (2008) The possibiity of using a hybrid turbine ventilator (HTV) to improve the air velocity (stack ventilation) in Malaysian building interiors. Arch Malaysia 20:60–63
Karyono TH (1996) Thermal comfort in the tropical South East Asia region. Archit Sci Rev 39:135–139

Kettle DS (1984) Medical and veterinary entomology. Croom Helm, Beckenham

Kidd J (1982) The influence of arid-zone climate on house design in the Sudan. In: Golany G (ed) Desert planning: international lessons. The Architectural Press, London, pp 113–123

Kirkpatrick TW, Simmonds NW (1958) Bamboo borers and the moon. Trop Agric Trinidad 35:299–301

Koenigsberger OH, Ingersoll TG, Mayhew A, Szokolay SV (1974) Manual of tropical housing and building. Part one: Climatic design. Longman, London

Letts JB (1999) Smoke blackened thatch. A unique source of late medieval plant remains from Southern England. English Heritage/University of Reading, London/Reading

Lewcock R, Brans G (1975) The boat as an architectural symbol. In: Oliver P (ed) Shelter, sign and symbol. Barrie and Jenkins, London, pp 107–116

Lim JY (1987) The Malay house: rediscovering Malaysia's indigenous shelter system. Institut Masyarakat, Pulau Penang

MacKinnon J, MacKinnon K, Child C, Thorsell J (1986) Managing protected areas in the tropics. International Union for Conservation of Nature and Natural Resources, Gland

Malama A, Sharples S (1997) The performance of traditional and contemporary housing in the cool season of Zambia. Build Environ 32:69–78

Moughtin JC (1982) Settlements and housing in an arid-zone developing country; case study in Hausaland, Northern Nigeria. In: Golany G (ed) Desert planning: international lessons. The Architectural Press, London, pp 145–159

Musvoto C, Mapaure I, Gondo T, Ndeinoma A, Mujawo T (2006) Reality and preferences in community mopane (Colophospermum mopane) woodland management in Zimbabwe and Namibia. Int J Hum Soc Sc 1(3):173–177

Ngowi AB (1997) A hybrid approach to house construction-a case study in Botswana. Build Res Inf 25:142–147

Nicol JF, Humphreys MA (2001) Adaptive thermal comfort and sustainable standards for buildings. In: McCartney K (ed) Moving thermal comfort standards into the 21st century. Oxford Brookes University, Oxford, pp 45–59

Norton J (1986) Building with earth: a handbook. Intermediate Technology Development Group, Rugby

Olgyay V (1992) Design with climate. Bioclimatic approach to architectural regionalism. Van Nostrand Reinhold, New York

Oliver P (1987) Dwellings. The house across the world. University of Texas Press, Austin

Palgrave KC (1983) Trees of Southern Africa. Struik Publishers, Cape Town

Parodi JL (1988) The use of palms and other native plants in non-conventional, low cost rural housing in the Peruvian Amazon. Adv Bot 6:119–129

Rahman AMA (1995) House design in relation to environmental comfort. Build Res Inf 23(1):49–54

Roaf S (1983) The windcatchers of the Middle East. In: Germen A (ed) Islamic architecture and urbanism. King Faisal University, Damman, pp 257–268

Robert V, Carnevale P (1991) Influence of deltamethrin treatment of bed nets on malaria transmission in the Kou valley, Burkina Faso. Bull World Health Organ 69(6):735–740

Rodger A (1974) The Sudanese heat trap. Ecologist 4(3):102–106

Roonwall ML, Chatterjee PN, Thapa RS (1966) Effect of leaching out water soluble substances in the bamboo, *Bambusa nutans* Wall., as protection against insect borers. Indian For Rec Entomol 10:249–257

Schreck CE, Self LS (1985) Bed nets that kill mosquitoes. World Health Forum 6:342–344

Sdei A (2006) Thermal comfort in the traditional Japanese house. Martin Centre for Architectural and Urban Studies, Cambridge University, Cambridge

Sharples S, Malama A (1997) A thermal comfort field survey in the cool season of Zambia. Build Environ 32(3):237–243

Smith KGV (ed) (1973) Insects and other arthropods of medical importance. British Museum (Natural History), London

Sunquist F (1984) New approach to conservation in the third world. Smithsonian 14(12):164–169
Symington CF (1974) Forester's manual of dipterocarps. Penerbit Universiti Malaya, Kuala Lumpur
Vellinga M, Oliver P, Bridge A (2007) Atlas of vernacular architecture of the world. Routledge, Abingdon
Walton GA (1962) The *Ornithodorus moubata* superspecies problem in relation to human relapsing fever epidemiology. In: Arthur DR (ed) Aspects of disease transmission by ticks. Zoological Society of London Symposium 6, London, pp 83–156
Warren J (1999) Conservation of earth structures. Buttterworth-Heinemann, Oxford
Waterson R (1990) The living house: an anthropology of architecture in South-East Asia. Oxford University Press, Singapore
Wells J (1995) Population, settlements and the environment: the provision of organic materials for shelter. Habitat Int 19:73–90
Wells J, Haddar F, Uhde M-L, Sy A (1998) Shelter provision in the context of deforestation: some evidence from Mali. Habitat Int 22:463–475
Wright G (1985) The stone villages of Britain. David & Charles, Newton Abbot
Yap HH, Tan HT, Abdul Malik Y, Rohaizat B, Loh PY, Chong NL (1990) Field efficacy of mosquito coil formulations containing *d*-allethrin and *d*-transallethrin against indoor mosquitoes especially *Culex quinquefasciatus* say. Southeast Asian J Trop Med Public Health 21:558–563

Chapter 5
Fuel Supplies

Whereas people in Western countries can usually take fuel and power supplies for granted, village communities in the developing world are in a very different position. There, wood is the principal fuel source and meeting daily needs typically requires constant attention and a great deal of physical toil. Lines of firewood carriers, usually women, trudging back from the forest with heavy loads are consequently a familiar feature of the village scene (Fig. 5.1a). Agricultural wastes come next in order of importance as fuel sources (Fig. 5.1b, c), although their degree of suitability varies (Barnard and Kristoferson 1985). Pea stems and stalks derived from cereals such as maize, millet or sorghum generally possess suitable burning properties, whereas wheat and rice stems are usually considered to burn too quickly, except for rapid frying in the Chinese style.

India is notable for making use of an unusual fuel of agricultural origin in the form of dried cakes of cow or buffalo manure. A rail journey through northern parts of the country is likely to be enlivened by the sight of large pyramids of dung cakes alongside the line. The silos are protected from the monsoonal rains with an outer covering of dung and the cakes are passed out as required through an aperture in the outer wall. Dried dung has the great merit of burning slowly and evenly and is therefore much favoured for the cooking of *chapatti* and *roti* breads.

Household illumination once also relied on the burning of biological materials (O'Dea 1958). Candles would be made from animal fats, beeswax or the fatty seeds of plants such as the aptly-named tallow tree *Sepium sebiferum*. Certain other plants have fruits that yield fats sufficiently liquid to be used directly as lamp oils. Examples include olive trees (*Olea europea*) in the Mediterranean region and shea butter trees (*Butyrospermum parkii*) in West Africa. Latterly, the ready availability of kerosene lamps and paraffin candles has prompted a decline in these traditional practices. Not so, however, for the crucial task of fuelling domestic fires. For this purpose a supply of wood (or its derivative charcoal) is still essential for most rural communities and collecting it a major chore.

J. Edington, *Indigenous Environmental Knowledge*,
DOI 10.1007/978-3-319-62491-4_5

Fig. 5.1 Aspects of village fuel provision in Nepal. (**a**) Bundles of firewood being carried from the forest (photograph by Anne Gee), (**b**) cakes of cattle dung drying on a house wall, (**c**) pea stalks being hung up to dry prior to burning

5.1 Traditional Wood-Harvesting

5.1.1 Species Selection

Village wood-harvesting is not simply a matter of cutting the trees closest to hand; certain species are reserved for other purposes. Wild trees bearing edible fruits and seeds are usually spared the axe. In Southern Africa, species thus protected include monkey-orange (*Strychnos cocculoides*), *baobab* (*Adansonia digitata*) and *marula* (*Sclerocarya birrea*) (Campbell et al. 1991). All three yield valuable vitamin-rich fruits. In the savannah lands of West Africa species with protected status include the locust bean (*Parkia biglobosa*), which has edible seeds and pods and the shea butter tree (*Butyrospermum parkii*), whose fatty seeds provide oil for cooking and for use in lamps. Local people are well aware that cutting such trees for firewood will seriously impair their fruit production. In parts of India, *tendu* (*Diospyros melanoxylon*), *mahua* (*Madhuca indica*) and *sal* (*Shorea robusta*) trees tend to be similarly regarded as sacrosanct (Basu 1987); the first because its leaves can be used for *bidi* cigarette wrappers; the second because it produces edible flowers and the third because its leaves can be fashioned into handy disposable food-serving "plates".

Certain other trees are avoided because their wood burns too quickly. In Southern Africa these species are usually rejected in favour of various bushwillows, leadwoods, acacias and sicklebushes. Ancient charcoal deposits show that prehistoric wood collectors made similar choices and modern research has revealed that a high concentration of calcium oxalate crystals in the wood is the key factor (Prior and Tuohy 1987; Prior and Cutler 1992). Such crystals are not universally present in the cells of plants and there is some speculation about what benefits they might confer on the species containing them. Amongst the various suggestions is that they serve to deter attack by herbivores and wood-boring termites, that they represent a form of calcium storage or that they serve to neutralise potentially-toxic oxalic acid (Franceschi and Horner 1980). Whatever the benefits to the plants themselves, by slowing the process of combustion and producing long-lasting glowing embers, they greatly commend themselves to village cooks.

Attributes perceived as undesirable in firewood include any tendency to produce acrid or unpleasant smoke. In Southern Africa, the *tamboti* tree (*Spirostachys africana*) is notorious for this, generating smoke that is sickly-sweet and nauseating. In India, similar objections are levelled against *bhimal* (*Grewia oppositifolia*), a tree whose foliage is a useful source of animal fodder but whose wood produces a markedly unpleasant smell when burned. The widespread reluctance to use introduced *Eucalyptus* as firewood is likewise based on the fact that its smoke can cause irritation to the nose and eyes. Conifers in their turn also tend to be poorly regarded because their high resin content causes pockets of volatile material to build up under the wood surface. When these pockets ignite they throw out showers of sparks, a most unwelcome property for occupants of thatched houses.

Growth form is also a major consideration when trees are being selected for firewood harvesting. Some species are capable of being cut in ways that stimulate the

Fig. 5.2 Traditional firewood harvesting methods. Each sequence shows (*left* to *right*): original tree form, initial cuts and continued shoot-harvesting regime

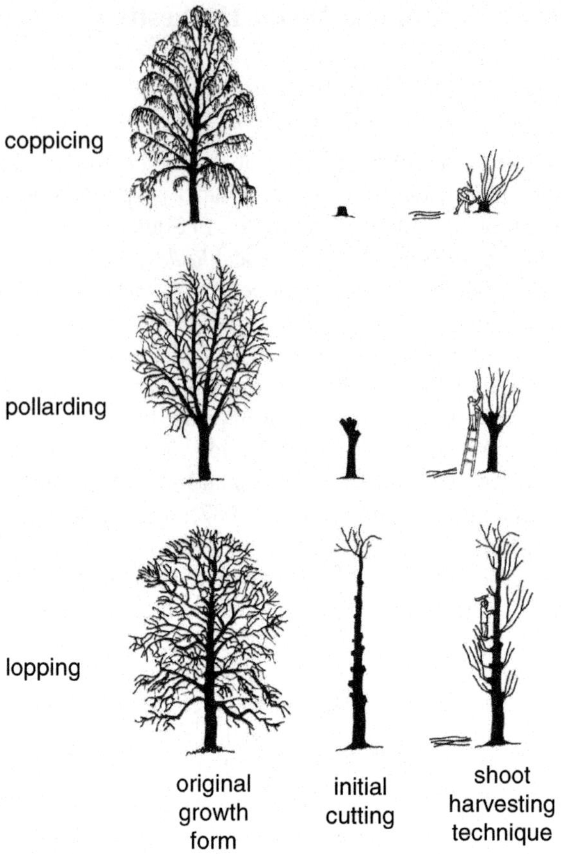

development of new shoots (Shepherd 1992). *Coppicing, pollarding* and *lopping* are the three harvesting techniques employed in this way (Fig. 5.2). Coppicing involves cutting shoots that have regrown from the trunk-base close to the ground; pollarding entails cutting the trunk above head height to stimulate shoot growth at this higher level and lopping is a similar procedure applied to side branches. Although these last two techniques have the disadvantage of requiring the firewood collector to climb the tree to do the cutting, they have the compensating merit of making the developing shoots less vulnerable to browsing animals. Wood harvesters using lopping as a technique often leave sizeable cut stubs projecting from the trunk to facilitate climbing. However caution has to be exercised with some species. *Litsea polyantha* is one such species in which the shoot bases are brittle and barely sufficient to bear a person's weight.

Properly applied, all these harvesting methods are capable of being made sustainable and, with a cutting interval of 3 years or more, individual trees will yield usable shoots at regular intervals for over a century. There are even claims that coppiced trees can survive as long as a thousand years (Rackham 1976). The use of

Fig. 5.3 Tree lopping. (**a**) Applied to *Quercus semecarpifolia* in West Bengal, (**b**) applied to *Cedrus deodara* in Himachal Pradesh

re-grown shoots has the additional merit of requiring only basic cutting equipment. The slender shoots can be cut using simple machete-like tools, and then conveniently tied into transportable bundles or fitted into carrying baskets. In villages of the African veld, typical cut stem lengths range from 600 to 1200 mm, with diameters between 50 and 250 mm (Prior and Tuohy 1987). In Nepal, where shoots are usually cut with the famous angled knife or *kukri* and transported in bamboo baskets, there is a similar restriction on size, with preferred lengths ranging between 550 and 760 mm and diameters rarely exceeding 100 mm (Joshi 1987). The other advantage of narrow, medium-length sticks is that they can be fed progressively into the mouths of traditional clay stoves (p.113).

Whilst in Europe all three techniques have been in widespread earlier use (Rackham 1976), in the developing world lopping continues to be the favoured method. With species such as evergreen oaks, the procedure inevitably alters the shape of the trees, converting them from a spreading to a columnar form (Fig. 5.3a). This transformation need not however seriously threaten the trees' survival unless shoots and twigs are removed so frequently that leaf production is impaired. With a very few exceptions, attempting to coppice or continuously remove side branches from conifers is a futile task because regeneration fails to take place (Fig. 5.3b). The monkey puzzle (*Araucaria araucaria*) and the coast redwood (*Sequoia sempervirens*) are two of the few conifers able to regrow after such treatment.

A rare instance of a sustainable cutting technique being successfully applied to another conifer is to be found in the Auvergne region of France. There, bakers obtain wood for their ovens by periodically cutting the side branches of pine trees (*Pinus sylvestris*). A special technique has to be employed to make this practice sustainable and results in forests – known as *pins de boulange* – in which the trees develop a curious but characteristic candelabra-like shape (Fig. 5.2).

5.1.2 Conservation Practices

In view of the widespread pessimism expressed about the viability of traditional village wood-supply systems (Eckholm 1975, 1976), it is appropriate to look for evidence of any inbuilt conservation mechanisms. To an extent, coppicing, pollarding and lopping all qualify, as long as they are applied to appropriate species and with sufficient intervals being allowed for regrowth. It is also the case that entire areas may be excluded from cutting operations. In India and Nepal for example, it is not unusual to come across patches of forest that have remained virtually intact in otherwise deforested landscapes. These remnants are generally associated with religious shrines (Gadgil and Vartak 1976). Two such forests are shown in Fig. 5.4. The first is in Nepal, close to the city of Bhaktapur and surrounds a temple dedicated to the Hindu god, *Ganesh*. The second is in the village of Seepur near the town of Shimla in Northern India and is associated with the god *Shiva*. Incidentally this shrine is within walking distance of Wildflower Hall, onetime residence of Lord Kitchener, and now a fine hotel. Before modern state conservation measures came into force, the presence of the shrine induced local people to protect the local *deodar* (*Cedrus deodara*) forest from damage.

Sacred forests of this type may vary in size from less than one hectare to over thirty and although at such sites the cutting of mature trees is strictly forbidden, the collection of fallen dead wood and the removal of understorey stems are usually permitted (Ingles 1995). Provided that steps are taken to replace any trees that have become senile and to encourage natural regeneration, such forests can serve as valuable sources of green cuttings and germinating seedlings for planting elsewhere. Moreover, because seeds from the protected trees are also dispersed to other areas by natural agencies including wind and wild animals, the sanctuaries are in effect acting as biological reservoirs. In Nepal, the policing of such sites often comes under the control of a "forest watcher", a person who is employed for this purpose and is typically paid with contributions of grain from local households (Ingles 1995).

India provides the most notable historical example of villagers acting communally to protect their trees (Sankhala and Jackson 1985). In 1790, when the Maharaja of Jodhpur was engaged in the construction of a new palace, he caused his workmen to scour the local countryside for the wood required to fire kilns for mortar production. This brought them into conflict with the *Bishnoi*, a local tribe noted for its fiercely protective attitude towards trees in general and most especially towards the

a

Fig. 5.4 Forest areas protected from destructive wood cutting by virtue of their association with local deities. (**a**) Mixed forest surrounding shrine dedicated to *Ganesh* near Bhaktapur in Nepal, (**b**) *deodar* forest surrounding a temple dedicated to *Shiva* near the village of Seepur in Himachal Pradesh

Fig. 5.5 Tree hugging (Drawing by Mary I. Anderson)

khejri tree, a species that still meets many everyday needs. Seeking to protect the trees, village women linked hands around their trunks in an attempt to prevent them from being felled. This was apparently to no avail and tree felling went ahead with a number of women reportedly being killed in the process. "Tree hugging" (Fig. 5.5) later achieved wider fame in India when it was adopted by the *chipko* forest-protection movement during the 1970s and 1980s (Jain 1984; Shiva and Bandyopadhyay 1986). The movement started in 1971 with village demonstrations in the Himalayan district of Gopeshwar. These were directed against a Forest Department proposal to sell felling rights for a local ash forest to a tennis racket manufacturer. This forest had served for many years as a source of wood for village agricultural implements and, had the new proposals gone ahead, this traditional use would no longer have been possible. Similar challenges were subsequently mounted in other states including Himachal Pradesh, Rajasthan and Karnataka, generally resulting in climb-downs by the respective Forest Departments. Although the instigators of the first protests were men, women later became the main participants and "tree hugging" their most potent weapon.

In parts of Africa certain forest tracts are likewise strongly protected on religious grounds. In the Yoruba-speaking regions of Nigeria, such taboos are applied to forests that screen shrines and traditional burial grounds, and also to forest areas surrounding landscape features regarded as sacred (including rivers, waterfalls, mountains and notable rock outcrops) (Kayode 2003). Amongst the *Aouan* people of the Ivory Coast, forests on erosion-prone slopes are similarly regarded as unavailable for firewood cutting, in this case because they are considered to come under the protection of the goddess *Assie* (Shepherd 1992). Whatever the underlying motivation, few ecologists would question the desirability of the outcomes. All the practices have the merit of protecting forest areas that might otherwise have been destroyed. What is harder to find, however, are instances of village firewood-harvesting being purposely regulated to maintain extraction rates within the regenerative capacity of the forest.

5.2 Disruptive Influences

5.2.1 Population Growth

As with so many village operations, human population growth has had a strongly disruptive effect on village wood-supply arrangements (Agarwal 1986). Dwindling supplies have made the task of satisfying daily firewood demands increasingly difficult. Because they are the ones typically charged with the task of collecting wood, it is usually the women and girls of the village who face the challenges. The problems are well illustrated in a study from the Indian state of Gujarat (Nagbrahmam and Sambrani 1983). At sites where village forests had remained intact, it was noted that firewood collection was needed only every fourth day, with the total distance walked amounting to less than one kilometre. However, where forests had been damaged through overpopulation and overuse, the collecting trips had to be stepped up to one every other day and the distance walked increased to 4–5 km. In the most extreme situations where forests had been virtually destroyed, the women were reduced to digging up the roots of previously-felled trees and collecting roadside weeds to use as fuel, a task that occupied them upwards of 4 h every day.

5.2.2 Annexations of Forest Land

Another problem is that tracts of forest land traditionally used by village firewood collectors frequently become annexed for other purposes. For instance, in colonial Tanganyika, not only were large former forest areas transferred to European settlers for the production of cash-crops such as coffee, tea and sisal, most of the remnants were designated as "forest reserves", with access to them being denied to local people (Barraclough and Ghimire 1996). In India similarly, the Imperial Forest Department, through the Indian Forest Acts of 1865 and 1878, assumed control of all "valuable" forest areas capable of yielding timber or secondary forest products such as oils and resins. In consequence, villagers frequently found themselves fenced out of forest areas to which they had previously enjoyed free access. Indeed it was explicitly stated as a matter of policy that the mere "accident of a village being situated close to a forest should not be allowed to interfere with the exclusive right of the Forest Department to extract forest products in the national interest" (Guha 1983). In only a minority of cases were villagers allowed to enter these "forest reserves" to collect firewood.

5.2.3 Urban Wood Use

To Westerners conveniently connected to electricity and gas distribution networks, it may come as a surprise to learn that many cities in the developing world still make extensive use of wood fuels (Fig. 5.6). For instance even in Kampala, the capital of Uganda, a city with many modern trappings, until very recently obtained 90% of its fuel from wood. The wood is imported from the surrounding countryside either in its raw form or after being processed into charcoal. Indeed, excessive woodcutting has produced a halo of deforested land around many African cities, often extending as far as 70 km in every direction (Agarwal 1986). In India, a similar depletion of forest cover has been reported in the vicinity of 27 of the 30 major cities (Bowonder et al. 1987).

Villages adjacent to cities are frequently blighted by these urban fuel wood demands, with a domino effect operating whereby villagers, whose own forests have been degraded, transfer their attention to those of their neighbours in the next circle outwards. Such a process can convert existing village wood surpluses into deficits. This phenomenon was well illustrated in an investigation of the wood supply situation in a small village close to the city of Kathmandu in Nepal (Joshi 1987). Left to its own devices this village of 61 people would have enjoyed an annual firewood production surplus of 175 metric tonnes, based on the growth of accessible small wood within the local forest. However, inroads made by invading firewood collectors, which the villagers were generally powerless to resist, resulted in this surplus being converted into an annual deficit of 99 metric tonnes, a situation with negative implications for the long term survival of the forest.

This is not to say that villagers never themselves participate in firewood trading. In the absence of other sources of income, they may feel obliged to sell firewood cut from their own forests. In India this occupation is known as "head loading". It is undertaken principally by women and can involve a long (8–10 km) trek into the forest to harvest wood, followed by a truck or train journey into the city in search of buyers (Basu 1987). The money earned is meagre (equivalent to US$ 0.6 for a 20 kg load) in relation to the labour involved and the consequences for the long-term survival of the forest frequently disastrous. Incidentally, it is not only for use on domestic fires that wood is imported into cities. It also serves to fuel a wide range of industrial and commercial operations, including metal-working, clothes-laundering, bread-baking, brick-firing and lime-burning. Indian independence brought no

Fig. 5.6 The urban use of wood fuel. (**a**) Wood market in the north Indian town of Darjeeling, (**b**) roadside sales of charcoal directed at urban buyers in Ethiopia

immediate solution to these various difficulties, with the Forest Department often attempting to sell to private companies, the rights to extract timber from forest areas still regarded by tribal communities as belonging to them.

5.3 Alternative Energy Sources

One suggested remedy for all these problems is to transfer attention to energy sources not dependent on wood. Amongst the possible options are solar and hydro-electric devices and the use of biological digesters to generate combustible methane gas from organic wastes. It is certainly possible to harness solar power for cooking by using a parabolic reflector to focus the sun's rays on a cooking vessel (Fig. 5.7a). The device illustrated was being used in Nepal both for cooking and to boil water for washing purposes; both operations which would otherwise have made serious

Fig. 5.7 Possible substitutes for wood-burning energy systems. (**a**) Parabolic solar reflector being used in the mountains of Nepal (Photograph by David Evans), (**b**) floating-bell biogas digester fuelled by manure from three family cows in Nepal, (**c**) underground chambered biogas digester in China showing from *left* to *right*, charging port, cap with gas outlet and port for sludge removal (Photograph by Brian Brinkworth), (**d**) kitchen range powered by methane gas from a digester of this type (Photograph by Shai-wai Sim)

inroads into limited firewood supplies. Unfortunately, a solar device of this kind can only function when the sun is shining and in the village in question, cooking by this method became impossible once the sky had clouded over, which it frequently did in the afternoons. Similar difficulties impede the use of solar reflectors in China where the preferred time for the main meal of the day is in the evening and therefore usually after dark. None of which precludes the use of solar energy for other functions such as lighting and the powering of communication systems.

Certain drawbacks are also encountered with other alternative energy devices. Although many hill villages already use simple water wheels to grind corn, small hydroelectric installations fall far short of being able to generate sufficient power to operate an electric cooker in every household. Biogas digesters also have certain inherent limitations. They are certainly capable of converting human and animal wastes into methane gas that can be burned in kitchen stoves (Fig. 5.7b–d) and of yielding a sludge that can be used as a fertilizer. There is, however, the underlying complication that the bacteria driving the underlying chemical reactions are temperature-sensitive and cease to work effectively when reaction temperatures fall below 35 °C or thereabouts. Consequently digesters will not function during winter unless a substantial proportion of the gas they produce – as much as a third of the total – is burned to keep them warm.

Another potential obstacle is that for each household to operate its own digester, at least three cows or their equivalent are needed to produce the necessary amount of manure (Moulik 1985). Relatively few individual families can afford to acquire or maintain this number of animals. Nor have initiatives for communal digesters been widely successful, mainly because of difficulties in achieving the necessary level of social organisation. Whilst such co-operation may have been forthcoming in Chinese communes, it has proved elusive elsewhere (Van Buren 1979; Gupta 1983). Overall then, it must be regretfully concluded that, in the crucial matter of providing energy for cooking, many of the innovative new technologies fall considerably short of expectations.

Another set of options hinges on switching to energy sources such as kerosene, bottled gas or centrally-generated electricity. These energy sources have the great merit of being cleaner and more convenient to use. They have, however, from the standpoint of rural households, the disadvantage of being costly, certainly when set alongside the possibility of harvesting firewood at no cost from local forests. Extending electricity grids into remote rural areas is an extremely costly operation, with estimates for extending the main transmission line for 1 km and creating 100 house connections ranging from US$ 15,000–58,000.

On balance, the significance of departures from wood based fuel systems is arguably not so much when they take place in the villages themselves, but when they occur in nearby towns and where the changeover reduces the pressure on village forests. In Ethiopia a campaign to address this kind of situation was mounted in the 1980s and involved persuading the citizens of Addis Ababa to switch from wood to kerosene. This is claimed to have produced a 95% response and to have reduced firewood cutting in local forests by 417,000 tonnes a year (Dang 1993). Similarly in South Korea, town dwellers have been weaned from the use of wood fuels, in this case by banning all firewood sales and also by subsidizing coal purchases (Foley and Barnard 1984). In India where the strategy adopted has been to subsidize gas and kerosene supplies, this is calculated to have reduced urban firewood consumption from 16.5 million tonnes in 1978 to 9.5 million tonnes in 1984 (Agarwal et al. 2000).

A quite different option canvassed recently has been to set up hybrid mini electrification grids in individual villages. These combine the use of renewable sources such as solar, wind and hydroelectric with back-up provided by a diesel generator (Alliance for Rural Electrification 2011). It is anticipated that the main functions for these systems will be to provide power for lighting, water-pumping and for communication devices such as radios, mobile phones and televisions. The problem of providing sufficient power for electric cookers still remains however. Therefore in this latter context a strong case can be made for rehabilitating traditional firewood systems.

5.4 Better Ways of Burning Wood

There are various ways of improving traditional methods of domestic wood- burning. Some existing cooking devices amount to little more than a triangle of stones or a metal frame on which the cooking pot is perched (Fig. 5.8a–c). These are not effective ways of transferring heat from fire to cooking vessel because of the large heat losses involved. The whole process can be made substantially more efficient by enclosing the fire within a baked clay stove or *chula* (Fig. 5.8d). Such stoves are constructed in much the same way as mud-bricks, from a mixture of sand and clay particles (typically in the ratio 3:1), together with an added binding material such as straw or animal manure. The ideal design incorporates both a chimney and a fire-gate damper (Fig. 5.9). As well as dispersing smoke from the house interior, the chimney facilitates airflow through the device by creating a pressure differential. The fire-gate damper in its turn has the merit of allowing air flow to be reduced once the wood is fully alight, thereby reducing fuel consumption. These simple modifications can produce fuel savings of between 20% and 50% (Baldwin et al. 1985). The lower values apply where certain wood burning operations such as pottery-firing and beer-brewing are carried out independently of the stove and are not therefore susceptible to the same economies.

Local resistance to such innovations comes from various quarters. In part it derives from a local perception that retaining smoke inside the house usefully conserves indoor heat especially in winter. There is also a prevalent local view that it helps to suppress domestic insect pests such as fleas, lice and mosquitoes. In consequence, donated chimney sections are often redeployed as drainage channels or simply sold. Resistance to closable fire-gates is usually prompted by the extra labour required to cut firewood into the shorter lengths suitable for feeding into the stove mouths.

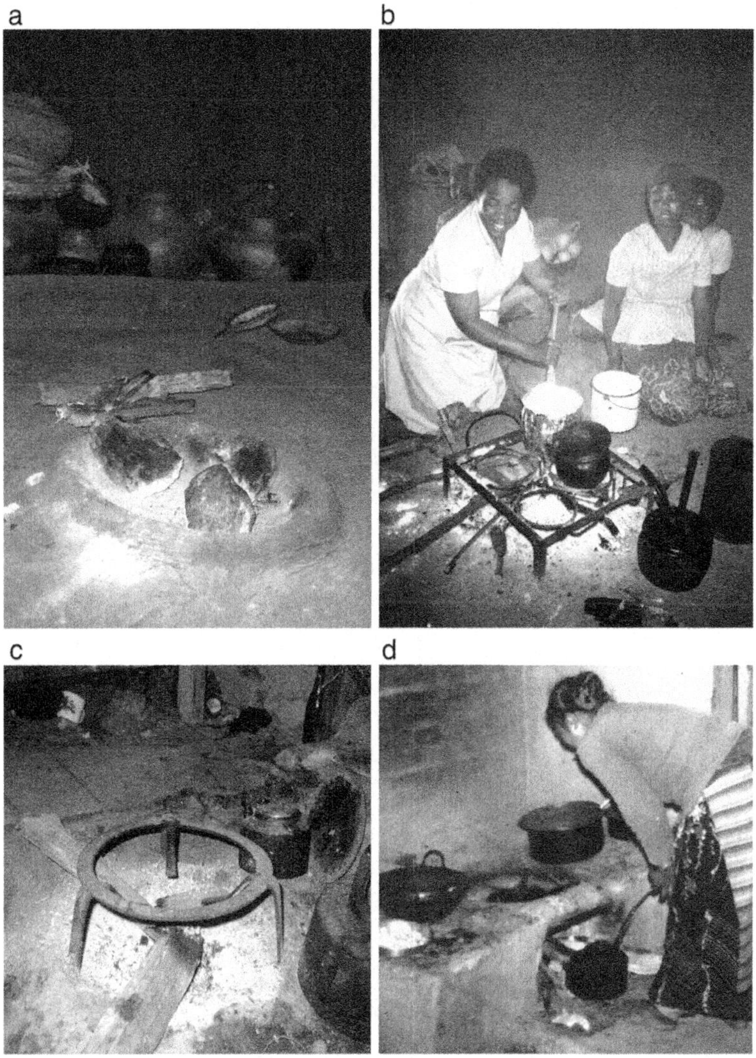

Fig. 5.8 Some traditional wood burning arrangements. (**a**) Three stone hearth (Zimbabwe), (**b**) rectangular metal pot stand, the operation being performed is the preparation of maize meal porridge or *sadza;* note the thick smoke filling the hut interior (Zimbabwe), (**c**) metal tripod pot stand (Nepal), (**d**) enclosed baked-clay stove or *chula* (Nepal)

5.4.1 Health Issues

These various reservations are not entirely without foundation, but on balance there is a case for upgrading wood-burning facilities if only on health grounds (Reid et al. 1986). With open fires there is an ever-present risk of children at the crawling stage being scalded or burned. Stoves with enclosed fires and holes into which pots can be

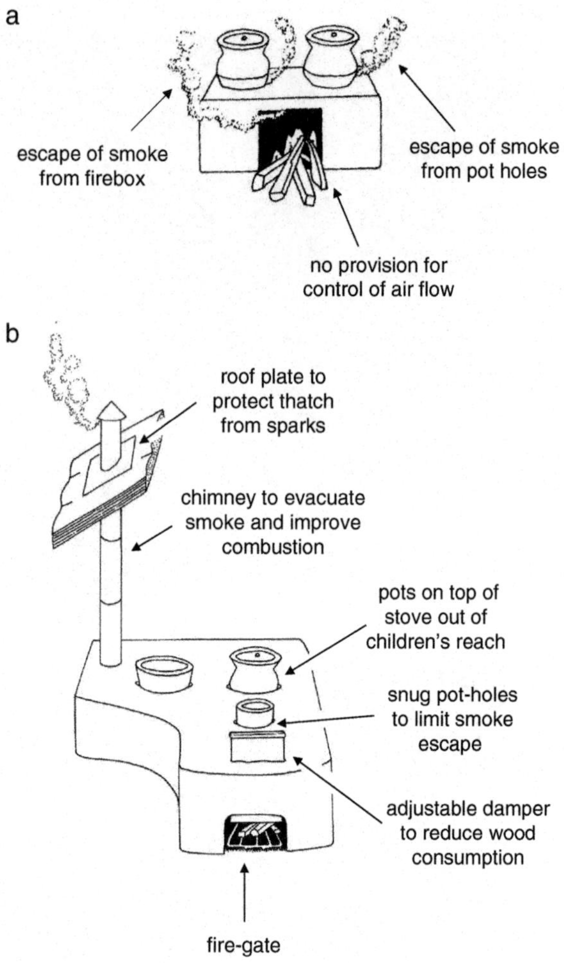

Fig. 5.9 (**a**) Traditional clay stove and (**b**) modifications designed to make it safer and more efficient

snugly fitted, effectively solve this problem as well as reducing smoke leakage from around the pot perimeters (Fig. 5.9).

If stoves lack chimneys, serious hazards can also be generated from indoor smoke pollution. The smoke has been shown to be a major cause of bronchitis and other respiratory problems (Pandey 1984a, b; Ezzati and Kammen 2001); to result in persistent eye disorders (Cleary and Blackburn 1968; De Koning et al. 1985) and even to have a negative effect on birth rates (Mishra et al. 2004). Moreover, there is an alarming correlation between smoke pollution and the incidence of nasal and pharyngeal carcinomas. These tumours seem to have a dual cause and be triggered by the combination of viral infections and environmental toxins. In the highlands of East Africa, the most likely toxic agents are the various aromatic hydrocarbons (especially *benzopyrene* and *benzoanthracene*) contained in smoke from cooking fires (Clifford 1972). In Kenya the problem becomes more serious at elevations above 2000 m, presumably because at these altitudes indoor fires are kept going for most of the day to provide room heating (Clifford and Beecher 1964).

According to one estimate, indoor smoke pollution accounts annually for 1.8 million premature deaths in rural areas of the developing world (Bruce et al. 2000). A welcome recent initiative has been the plan mounted in 2010 by the United Nations Foundation (an international charity) to supply 100 million households with safe stoves by the year 2020. It is recognized that to gain acceptance the new designs will have to be both affordable and attuned to local cooking habits. The significance of this latter consideration was well illustrated by an earlier project to design a stove for use in the refugee camps of Darfur (Anon 2008). The stoves had to be radically modified after it was discovered that the vigorous pot stirring required to produce dough for the local staple food *assida*, caused them to overturn. Likewise in India, some new stove designs were rejected because they never got hot enough to cook traditional Indian breads effectively. On the question of cost, the instigators of these various schemes are insistent that the stoves should not simply be given away. Putting a price on them not only results in them being more highly valued by users, it also allows local manufacturers to profit from their production.

Incidentally, the practice of using chimneys to minimize indoor smoke pollution has an interesting global history. It seems to have started in Northern Italy during the thirteenth century and then to have spread, first to the rest of Europe and considerably later to North America, colonial South Africa and Australia (Vellinga et al. 2007). A snapshot of this transition is to be found in the authentic period houses re-erected in the National History Museum of Wales near Cardiff. The sixteenth century house has only a simple open hearth and when a fire is lit, smoke fills the rooms and disperses only very slowly through window slats and gaps in the roof thatch. By contrast, the seventeenth century house boasts a well-constructed hearth and chimney stack, and consequently the living quarters of this house remain virtually smoke-free (Nash 1995).

5.5 Rehabilitating Forests

It is one thing to make stoves more efficient, quite another to ensure that enough wood is available to keep them going. Recent experience suggests that these problems might prove less intractable than first appeared. In situations where forests were appropriated for commercial timber extraction, logging operations were often so destructive that much of the land was left in a semi-derelict state. In India it is calculated that as much as 10% of all useable land falls into this category. Another 15% is estimated to consist of community-owned land in which forests have been poorly managed, mainly because of pressures caused by population growth. In total this adds up to a very large area indeed and, even if only a fraction of it were to be rehabilitated, this would go a long way towards addressing local firewood needs.

Such a task does, however, require close co-operation between village communities and state Forest Departments. In the past, the choice of tree species has been a serious bone of contention. Fast-growing species of *Eucalyptus* imported from Australia have been much favoured by professional foresters, but are very poorly regarded by villagers. The latter considered that the wood burned too quickly and produced unpleasant acrid smoke. They also pointed out that the oiliness of *Eucalyptus* leaves caused them to be rejected as fodder by domestic animals. It was considerations such as these that caused villagers in the Southern Indian state of Karnataka during the 1980s to march on Forest Department nurseries and to uproot *Eucalyptus* seedlings, replacing them symbolically with seedlings of traditional local species (Joyce 1988). The preferred trees all serve multiple functions. They include *honge* (*Pongamia pinnata*) which yields firewood, fodder and lamp oil; *neem* (*Azadirachta indica*) which provides firewood, toothsticks and seed oil with insecticidal properties, and *tamarind* (*Tamarindus indica*) which yields timber for agricultural implements and also nutritious fruits (Shiva et al. 1981, 1982). Arguably the appropriate place for *Eucalyptus* is not in the village forest, but in the Forest Department's own commercial plantations and those of larger landowners interested in supplying pulp to local paper and rayon factories.

The state Forest Department eventually took these points on board and started to grow in their nurseries the tree species favoured by the villagers and to make tree shoots available for planting in *community forests* (Fig. 5.10). More strategically they have entered into a variety of collaborative forest rehabilitation schemes (Table 5.1). Basu (1987) describes a typical community forest project from the Indian State of Bihar. In this particular venture, the Forest Department provided the saplings and the fencing for establishing a new forest on 24 ha of village-owned wasteland, whilst the village on their part undertook to plant and protect the trees. In such schemes the villagers select the tree species and enjoy exclusive use of the firewood and any other materials produced. In other variants, for instance where the Forest Department owns the land, both partners have a say in these matters and the eventual financial benefits are shared.

In a review of collaborative ventures in India it was estimated that over 13 thousand Community Forest Management schemes have been put into operation, and over 36 thousand Joint Forest Management schemes (Ravindranath et al. 2000). The main difference between the two is that in the community projects the villagers have a greater degree of autonomy. These schemes have been generally successful in curbing destructive over-grazing and uncontrolled wood-harvesting and most of the new forest plots are reported to be growing well (Ravindranath et al. 2000). An interesting variant in parts of Himachal Pradesh is for the Forest Department to extract firewood from local deodar (*Cedrus deodara*) forests in a carefully controlled fashion and to deliver the wood to village households at a subsidised rate. In this way it is hoped to make use of this important ecosystem without destroying it.

The neighbouring mountain kingdom of Bhutan offers a notable example of a successful community forest system (Temphel and Beukeboom 2007). This is designed to provide both firewood and building timbers. Each household is allowed 2.5 ha of forest and when all these options are eventually taken up, something like 10% of the nation's forest area will have been dedicated to these purposes. Local Forest Management groups receive professional advice about planting and

Fig. 5.10 The new face of community forestry in the Indian state of Karnataka. (a) Tree seeds being sown in a Forest Department nursery, (b) bundles of tree shoots being readied for distribution and planting, (c) an established community forest, in addition to trees to provide firewood and building materials, sisal (*Agave sisalana*) has been planted as a fibre source

Table 5.1 Village and forest department inputs in initiatives aimed at increasing tree cover in rural India

Input or benefit	Project type: Commercial forest on forest department land	Jointly-managed forest on forest department land	Community forest on village land	Tree planting on private farms
Species choice	F	FV	V	V
Seedling provision	F	F	F	fV
Planting operations	F	FV	V	V
Weeding and protection	Fv	FV	V	V
Receipt of benefits	Fv	FV	V	V

V, v = major and minor village inputs, F, f = major and minor Forest Department inputs

harvesting from the Department of Forests and at the last count, 36 community forests had been established covering 2914 ha and serving 1664 households. Under this regime, tree cover has become more complete, the incidence of destructive harvesting has been reduced and water catchments have been better protected from erosion. Most users have been able to meet their firewood and building-timber needs from local forests, but in cases where there are shortfalls the Department of Forests has stepped in to provide firewood supplies at a nominal rate from their own reserve areas.

5.5.1 Tree Planting on Farmland

Another encouraging trend in parts of Asia has been the readiness of farmers to plant trees on their own land. In some areas of Nepal, for example, this kind of farm planting has resulted in a threefold increase of tree cover (Gilmour 1988, 1997; Carter and Gilmour 1989). The sites used include stream banks, eroded gullies and the walls of crop terraces (Fig. 5.11) and the density of new trees can reach 84–192 trees/hectare on crop terraces and 440–1371 trees/hectare along stream banks. Some of the new tree stock derives from trees that have colonised the land naturally; the remainder comes from tree seedlings that have been collected from local forests and replanted. The upper limit for tree planting on crop terraces is reached when shading threatens to reduce yields from field crops. Although a number of the species planted are intended as firewood sources, others have additional or alternative uses (Table 5.2). Although this kind of extra tree-planting now tends to go under the fashionable modern name of "agroforestry", the initial impetus for many of the schemes came from the farmers themselves.

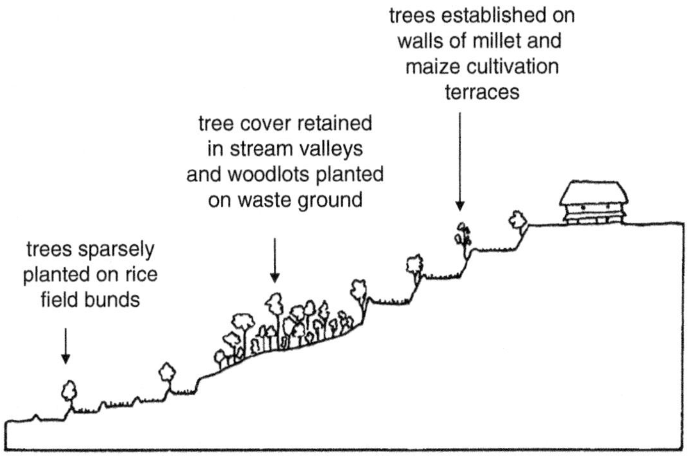

Fig. 5.11 Options for increasing tree cover on farm land in Nepal (Based on Carter and Gilmour 1989)

Table 5.2 Use of farmland trees in hill districts of Nepal

Species	Fuel	Fodder	Fruit	Construction
Prunus cerasoides (wild cherry)	++	+++	++++	+
Betula alnoides (alder birch)	++	+++		+
Alnus nepalensis (Nepali alder)	+			++
Ficus nemoralis (fig)		+		
Ficus semicordata (fig)		+	++	
Litsea monopetala (*kutmero*)		+		
Psidium guajava (guava)			+	
Pinus roxburghii (chir pine)	++			+
Bauhinia variegata (*koiralo*)		+	++	
Shima wallichii (*chilaune*)	++	+++		
Bridelia retusa (*khaja*)		+	+++	

++++ = most use, + = least use (from Carter and Gilmour 1989)

Sometimes for mainly sociological reasons, tree planting on farms does little to alleviate firewood shortages. In the Kenyan Highlands, for example, procuring firewood remains the women's task, but tree-planting is an activity reserved by the men and done mainly for income generation (Bradley et al. 1985, Bradley 1991). Because women, for reasons of social convention, are generally forbidden to plant or cut trees on farmland, increasing tree cover on such areas does little to ease their firewood supply problems.

5.6 Overall Sustainability of Firewood Production Systems

Trees and shrubs grow a small amount every year but ultimately a forest can remain in existence only if the rate of harvesting does not exceed this rate of incremental growth. Assessing firewood consumption in villages is a relatively easy task. It simply involves measuring or weighing loads or bundles as they come into the village. The only difficulty likely to be encountered is where collectors decline to co-operate because they fear being taken to task for contravening some forest regulation, such as cutting green wood in a national park. Measuring annual production rate is a somewhat more complicated task, but still achievable. It can be done by aging the trees and shrubs in village collecting areas (using increment corers or counts of growth rings on cut stems); working out the overall standing crop (by multiplying tree basal-areas by height for each species) and then finally dividing the second value by the first (Newbould 1967). In this way it is possible to arrive at an overall figure for the production of the forest expressed either as m^3/annum or, if wood specific gravity is taken into account, as tonnes/annum.

A complicating factor is that village wood-harvesting tools generally take the form of machetes or slashing knives, rather than saws. Thus makes the wood from larger trees virtually inaccessible. Studies from Nepal in which this factor was taken into account concluded that one village had a shortfall in the production of available thinner stems (less than 10 cm diameter), but an excess of thicker largely-inaccessible ones (McBride and McBride 1987; Joshi 1987). A second village faced a similar situation, but one that was aggravated by the fact that large quantities of small wood were exported from the village to supply neighbouring villages and urban areas (p.109).

Village communities are unlikely to make calculations of this kind on their own behalf (or even to embrace the concepts underlying them), but can be enabled to do so with outside help. In the case of the two Nepalese villages, a switch to power saws would have given access to larger trees, but without proper control could ultimately have destroyed the forest cover. More generally, the options for addressing a situation where consumption is outstripping production are variously: to increase the general forest area (for example by planting on degraded land or field margins); to improve wood-stove efficiency or to switch in whole or in part to other fuel sources. In situations where populations have seriously outstripped forest resources, it may be regretfully concluded that this last option is the only practical one.

5.6.1 Social Aspects of Wood Fuel Use

Optimism about the prospects of rehabilitating village wood supply systems must be tempered with a caveat about the amount of physical labour involved. This burden currently falls disproportionately on village women and girls. Not only do they carry most of the wood, they often have to climb the trees to cut it. A study by Pande

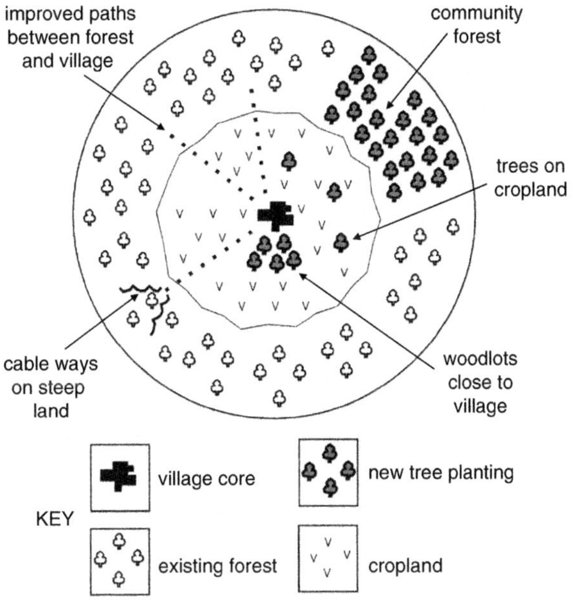

Fig. 5.12 Land management strategies designed to increase the sustainability of village firewood harvesting, whilst at the same time reducing labour expenditure

(1996) in Northern India showed that firewood collection represented a major chore for village women occupying them on average 2.03 h a day. Other components in their daily work load included 4.84 h of farm work, 1.96 h of fodder collecting, 1.42 h of water carrying and 1.21 h of flour grinding. Together with other domestic duties this added up to a women's work total of 15.04 h a day (105.28 h a week), well over twice the male total.

Carrying firewood bundles weighing upwards of 30 kg is not only a very fatiguing activity that can lead in the long term to skeletal malformations (Levy 1968; Kennedy 1989) it may also involve safety hazards. Such is the momentum of a swinging load attached to a head strap that an accidental slip can result in a spinal fracture (Scher 1978). Typical village layouts often further aggravate the problem by obliging wood collectors to traverse the inner cultivated zone on their way to and from outlying forest areas.

Entrenched social attitudes are not easy to dislodge and it is probably unrealistic to expect village men folk to shoulder their share of firewood harvesting in the near future. However, the increased representation of women on village *panchayats* – in some instances resulting in the women themselves becoming village heads – should make it easier to institute measures that make firewood carrying a less onerous task, whoever is doing the work. Such measures include establishing groves of firewood trees close to village centres thus shortening the distances loads have to be carried and also improving pathways to and from the forest (Fig. 5.12). In Northern India, the Central Himalayan Environment Association has been active in promoting both kinds of initiatives. A more radical option in hilly terrain would be to erect cableways to extract the wood (p.187).

In the context of climate change, biofuels of all kinds have the intrinsic merit of producing no more carbon dioxide on combustion than was previously assimilated during growth. In this sense and if properly managed such plant formations are carbon neutral. This contrasts with fossil fuels, which although they also came into existence as the result of biological processes, the attendant extraction of carbon dioxide from the atmosphere happened many millions of years ago and is of no help to us now. The advantages of wood fuels are no less relevant to Western situations, although, as will be seen (p.213) complications can arise that are not entirely dissimilar from those experienced by traditional communities.

References

Agarwal B (1986) Cold hearths and barren slopes. The woodfuel crisis in the third world. Zed Books, London

Agarwal A, Narain S, Sen S (eds) (2000) State of India's environment 5. The Citizens Fifth Report. Centre for Science and Environment, New Delhi

Alliance for Rural Electrification (2011) Hybrid mini-grid for rural electrification: lessons learned. Alliance for Rural Electrification, Brussels and USAID, Washington, DC

Anonymous (2008) Fresher cookers. De Economist 6:3–4

Baldwin S, Geller H, Dutt G, Ravindranath NH (1985) Improved woodburning cookstoves: signs of success. Ambio 14(4–5):280–287

Barnard GW, Kristoferson L (1985) Agricultural residues as fuel in the third world. Earthscan, London

Barraclough SL, Ghimire KB (1996) Deforestation in Tanzania. Beyond simplistic generalizations. Ecologist 26:104–109

Basu NG (1987) Forests and tribals. Manisha Granthalaya, Calcutta

Bowonder B, Prasad SSR, Unni NVM (1987) Deforestation around urban centres in India. Environ Conserv 14:23–28

Bradley PN (1991) Woodfuel, women and woodlots. In: The foundations of a Woodfuel Development Strategy for East Africa, vol 1. Macmillan. London

Bradley PN, Chavangi N, Van Gelder A (1985) Development research and energy planning in Kenya. Ambio 14(4–5):228–236

Bruce N, Perez-Padilla R, Albalak R (2000) Indoor smoke pollution in developing countries: a major environmental and public health challenge. Bull World Health Organ 78(9):1078–1092

Campbell BM, Clarke JM, Gumbo DJ (1991) Traditional agroforestry practices in Zimbabwe. Agrofor Syst 14:99–111

Carter AS, Gilmour DA (1989) Increase in tree cover on private land in Central Nepal. Mt Res Dev 9:381–391

Cleary GJ, Blackburn RB (1968) Air pollution in native huts in the highlands of New Guinea. Arch Environ Health 17:785–794

Clifford P (1972) Carcinogens in the nose and throat: nasopharyngeal carcinoma in Kenya. Proc R Soc Med 65:682–686

Clifford P, Beecher JL (1964) Nasopharygeal cancer in Kenya, clinical and environmental aspects. Br J Cancer 18:25–43

Dang H (1993) Fuel substitution in Sub-Saharan Africa. Environ Manag 17:283–288

De Koning HW, Smith KR, Last JM (1985) Biomass fuel combustion and health. Bull World Health Organ 63:11–26

Eckholm EP (1975) The deterioration of mountain environments. Science 189:764–770

Eckholm EP (1976) Losing ground, environmental stress and world food prospects. Worldwatch Institute, W.W. Norton, New York

Ezzati M, Kammen DM (2001) Indoor air pollution from biomass combustion as a risk factor for acute respiratory infections in Kenya: an exposure response study. Lancet 358:619–624

Foley G, Barnard G (1984) Farm and community forestry. Earthscan Publications, International Institute for Environment and Development, London

Franceschi VR, Horner HT (1980) Calcium oxalate crystals in plants. Bot Rev 46(4):361–427

Gadgil M, Vartak VD (1976) The sacred groves of Western Ghats in India. Econ Bot 30:152–160

Gilmour DA (1988) Not seeing the trees for the forest: a re-appraisal of the deforestation crisis in two hill districts of Nepal. Mt Res Dev 8:343–350

Gilmour DA (1997) Rearranging trees in the landscape in the Middle Hills of Nepal. In: Arnold JEM, Dewees PA (eds) Farms, trees and farmers. Earthscan Publications, London, pp 21–42

Guha R (1983) Forestry in British and post-British India. *Economic and Political Weekly,* 29 October, 1882–96

Gupta R (1983) Are community biogas plants a feasible proposition? World Health Forum 4:358–361

Ingles AA (1995) Religious beliefs and rituals in Nepal. Their influence on forest conservation. In: Halladay P, Gilmour DA (eds) Conserving biodiversity outside protected areas. The role of traditional agro-ecosystems. IUCN, Gland/Cambridge, pp 205–224

Jain S (1984) Standing up for trees, women's role in the Chipko movement. Unasylva 146:12–20

Joshi AR (1987) A study of the environmental relationships of certain village communities in the central development region of Nepal. Doctoral Thesis, University of Wales, Cardiff

Joyce C (1988) The tree that caused a riot. New Sci 117:54–59

Kayode J (2003) Conservation and Yoruba forest taboos. Nigerian Field 69:53–61

Kennedy KAR (1989) Skeletal markers of occupational stress. In: Işcan MY, Kennedy KAR (eds) Reconstruction of life from the skeleton. Alan R. Liss, New York, pp 129–160

Levy LF (1968) Porter's neck. Br Med J 1968(2):16–19

McBride JR, McBride SM (1987) Tree growth and firewood production near Syabru, Nepal. University of California, Berkeley

Mishra V, Dai X, Smith KR, Mika L (2004) Maternal exposure to biomass smoke and reduced birth weight in Zimbabwe. Ann Epidemiol 14:740–747

Moulik TK (1985) The biogas program in India and China. Ambio 14(4–5):288–292

Nagbrahmam D, Sambrani S (1983, January 8) Women's drudgery in firewood collection. *Economic and Political Weekly*

Nash GD (1995) Timber-framed buildings in Wales. National Museum of Wales, Cardiff

Newbould P (1967) Methods of evaluating the primary production of forests, International biological programme handbook 2. Blackwell Scientific Publications, Oxford

O'Dea WT (1958) The social history of lighting. Routledge and Kegan Paul, London

Pande PN (1996) Drudgery of the hill women. Indus Publishing Company, New Delhi

Pandey MR (1984a) Prevalence of chronic bronchitis in rural commmunities of the hill region of Nepal. Thorax 39:331–336

Pandey MR (1984b) Domestic smoke pollution and chronic bronchitis in a rural community of the hill region of Nepal. Thorax 39:337–339

Prior J, Cutler D (1992) Trees to fuel Africa's fires. New Sci 135:35–39

Prior J, Tuohy J (1987) Fuel for Africa's fires. New Sci 115:48–51

Rackham O (1976) Trees and woodland in the British landscape. J.M. Dent, London

Ravindranath NH, Murali KS, Malhotra KC (eds) (2000) Joint Forest management and community forestry in India. An ecological and institutional assessment. Oxford & IBH Publishing Company, New Delhi

Reid HF, Smith KR, Sherchand B (1986) Indoor smoke exposures from traditional and improved cookstoves comparisons among rural Nepali people. Mt Res Dev 6(4):293–304

Sankhala KS, Jackson P (1985) People, trees and antelope in the Indian Desert. In: McNeely JA, Pitt D (eds) Culture and conservation: the human dimension in environmental planning. Croom Helm, London, pp 205–210

Scher AT (1978) Injuries to the cervical spine sustained while carrying loads on the head. Paraplegia 16:94–101
Shepherd G (1992) Managing Africa's tropical dry forests: a review of indigenous methods. Overseas Development Institute, London
Shiva V, Bandyopadhyay J (1986) The evolution, structure and impact of the Chipko movement. Mt Res Dev 6:133–142
Shiva V, Sharatchandra HC, Bandyopadhyay J (1981) Social, economic and ecological impact of social forestry in Kolar. Indian Institute of Management, Bangalore
Shiva V, Sharatchandra HC, Bandyopadhyay J (1982) Social forestry – no solution within the market. Ecologist 12(4):158–168
Temphel KJ, Beukeboom HJJ (2007) Community forests: supporting Bhutan's national and MDG goals while protecting forests. In: Oberndorf R, Mahanty S, Burslem K, Suzuki R (eds) A cut for the poor. Food and Agriculture Organization Bangkok, Regional Community Forestry Training Centre Bangkok, Netherlands Development Organisation Hanoi, Hanoi, pp 114–125
Van Buren A (1979) A Chinese biogas manual. Intermediate Technology Publications, London
Vellinga M, Oliver P, Bridge A (2007) Atlas of vernacular architecture of the world. Routledge, Abingdon

Chapter 6
Herbal Medicine

6.1 Mixed Expectations

Travellers to the jungles often bring back stories of their ailments being successfully treated by a local healer or *shaman.* Indeed there is quite a widespread notion that the knowledge possessed by such healers has yet to be fully utilized in the West. Undoubtedly there have been some notable previous success stories. One of these is *quinine*, a natural drug extracted from the bark of the South American *Cinchona* tree (Fig. 6.1a) and universally recognized as a useful treatment for malaria. It was first discovered by Peruvian Indians and brought to Europe by Jesuit missionaries in the 1630s (Vogel 1970). More recently, now that the use of synthetic malarial prophylactics has been disrupted by parasite drug-resistance, another plant-derived remedy has come to the fore. This is *artemisinin*, a compound obtained from a weedy plant known as sweet wormwood (*Artemisia annua*) (Fig. 6.1b) and used to combat malaria in China for over 400 years (Klayman 1985). Likewise the drug *digitoxin*, widely used in modern medicine to treat heart irregularities and obtained from the leaves of the foxglove (*Digitalis purpurea*), started life as a village remedy. It was employed in eighteenth century England to treat *dropsy*, a condition of water retention caused by faulty heart action (Aronson 1985).

In fact it has been estimated that as many as 25% of modern medicines owe their origins to traditional herbal remedies, either by direct adoption or through the imitation of their chemical configurations (Farnsworth 1994; Cox 1994). Examples in this latter category include *aspirin,* which is modelled on a compound from the willow tree (*Salix alba*) and used to treat pain relief, and *amiodarone,* which is modelled on a compound from Bishop's weed (*Ammi visnaga*) and used to treat the heart condition angina pectoris. Many countries now have active screening programmes designed to identify further plant materials with therapeutic potential (Prance et al. 1994).

J. Edington, *Indigenous Environmental Knowledge*,
DOI 10.1007/978-3-319-62491-4_6

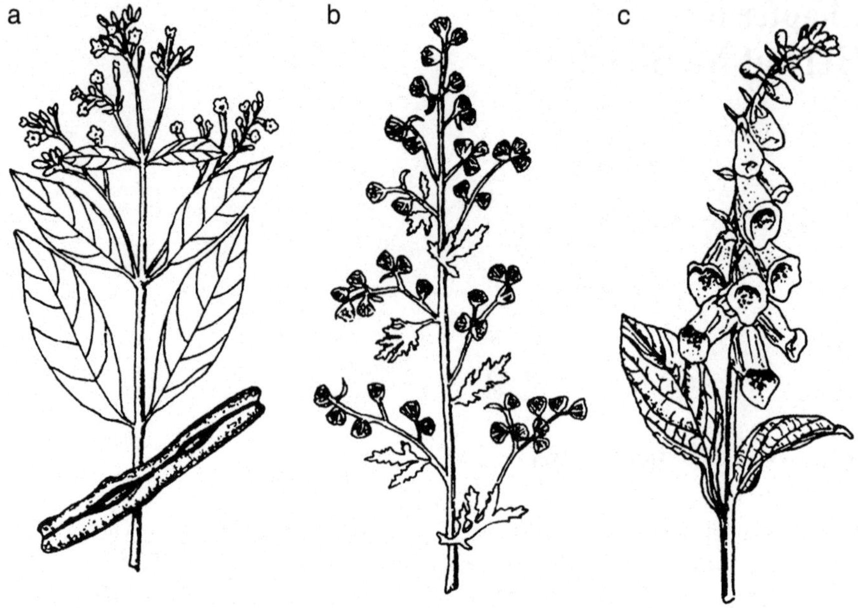

Fig. 6.1 Village remedies achieving world-wide renown. (**a**) Cinchona (*Cinchona ledgeriana*), (**b**) sweet wormwood (*Artemisia annua*), (**c**) foxglove (*Digitalis purpurea*)

6.2 Patterns of Village Use

Narrowing the focus to medicinal plant-use in individual communities, we set out to explore this issue in certain villages in the Middle Hills of Nepal (Joshi 1987; Joshi and Edington 1990; Joshi and Joshi 2007). Almost invariably in village settings the local healer is the first point of contact for the sick and according to one estimate such practitioners outnumber their Western-trained counterparts by a factor of thirty (Tsey 1997). Herbs feature prominently in their treatments, with the necessary raw materials being typically drawn from the local environment.

In one village, the local medicine man or *jhankri* was found to be using 71 herbal remedies and to derive the necessary raw materials from some 48 wild plants. The first column in Table 6.1 classifies the remedies according to the type of problem ostensibly being treated. Predictably in this harsh mountain environment with limited water supply and sanitation facilities, much attention is given to intestinal disorders and rheumatic pains, closely followed by skin and respiratory problems. Although none of the plant species employed has featured in a full clinical trial, most of them (or at least their near-relatives) have been subjected to some form of chemical analysis and to have had their chemical constituents screened for pharmacological activity.

On this basis it was possible to reach the tentative conclusion that 46% of the treatments were likely to function effectively in their assigned roles (Table 6.1

Table 6.1 Herbal remedies used in a Nepalese hill village, classified according to illness treated and probable effectiveness

	Number of treatments in each category				
Illness treated	Total used	Probably effective	Lacking validation	No data available	Questionable remedies
Intestinal disorders	17	9	3	3	2
Rheumatic pains	10	4	4	1	1
Skin conditions	8	4	2	2	0
Respiratory disorders	8	4	2	1	1
Headache	5	1	4	0	0
Kidney disorders	4	3	1	0	0
Fevers	2	0	1	1	0
Minor burns/ wounds	3	2	0	1	0
Asthma	2	1	1	0	0
Bleeding	2	1	1	0	0
Nervous disorders	2	0	1	1	0
Leprosy	1	1	0	0	0
Poor circulation	1	1	0	0	0
Mouth ulcers	1	1	0	0	0
Internal injuries	1	1	0	0	0
Liver disease	1	0	1	0	0
Eye infections	1	0	0	1	0
Warts	1	0	1	0	0
Need of tonic	1	0	1	0	0
Totals	71	33	23	11	4
Percentages	100	46	32	16	6

Based on Chopra et al. (1956, 1969), Leung (1980), Joshi (1987), Joshi and Edington (1990) and Reynolds (1996)

column 2). Successful remedies of this kind included the use of oil from fragrant wintergreen (*Gaultheria fragrantissima*) (Fig. 6.2a) as a soothing rub to relieve rheumatic pain. This plant is known to contain *methyl salicylate*, a compound closely related to the active ingredient of *aspirin* (Wade 1982). Likewise there is firm scientific support for the local practice of smoking the leaves of angel's trumpet (*Datura stramonium*) (Fig. 6.2b) to relieve the symptoms of asthma. This species contains the alkaloids *atropine* and *hyoscyamine*, which block the action of the parasympathetic nerves supplying the muscles of the bronchus. Smoking the leaves can thus be expected to reduce the distressing spasms associated with asthma attacks (Jentzch 1953). Especially interesting was the local use of water pennywort (*Centella asiatica*) (Fig. 6.2c) to treat the skin lesions caused by leprosy. This is a small herb growing in damp places and, quite independently, work by biochemists in Europe has shown it to contain a glycoside (*asiaticoside*) capable of killing the

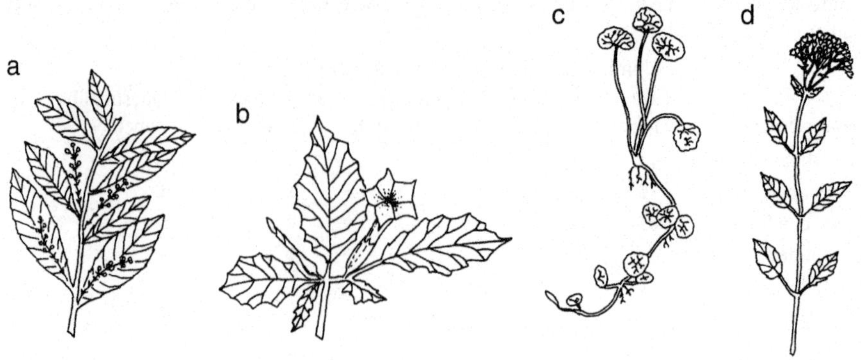

Fig. 6.2 Effective herbal remedies employed in a Nepalese village. (**a**) Fragrant wintergreen (*Gaultheria fragrantissima*), (**b**) angel's trumpet (*Datura stramonium*), (**c**) water pennywort (*Centella asiatica*), (**d**) *banmara* (*Eupatorium adenophorum*)

leprosy bacterium (*Mycobacterium leprae*). This it does by dissolving the waxy coating of the bacterium's cell walls (Boiteau and Ratsimamanga 1956).

Other appropriate remedies in the village repertoires included the use of leaf and stem extracts from the weed *Eupatorium adenophorum* (Fig. 6.2d) to stop bleeding, which fits in with the known presence of the haemostatic compound *ayapanin* in this genus (Bose and Roy 1936), and the use of the leaves of dandelion (*Taraxacum offininale*) as a diuretic, which has equally good experimental support (Rácz-Kotilla et al. 1974). Two materials with known antiseptic properties, resin from the bark of *chir* pine (*Pinus roxburgii*) and leaf extracts from the shrub *Lyonia ovalifolia* (Chopra et al. 1956), were likewise appropriately employed to treat skin infections. It is a measure of the value attached to these various preparations that they feature in similar roles in other regions of Nepal (Joshi and Joshi 2000; Joshi 2004).

At the other end of the spectrum, for a handful of remedies (about 6% of the total in this particular list), queries arise about the advisability of continued use. For example the widespread employment of the bark, leaves and fruits of the chinaberry tree (*Melia azedarach*) (Fig. 6.3a) to treat worm infections is called into question by the presence of a toxic resin in most parts of the plant (Chopra et al. 1956; Lewis and Elvin-Lewis 1977). Indeed, village respondents have themselves reported that swallowing the fruits can cause abdominal pain, vomiting and respiratory distress (Joshi and Joshi 2000). There are also doubts about the wisdom of treating various disorders with extracts from the rhizome of sweet flag (*Acorus calamus*) (Fig. 6.3b). Preparations from this plant have been shown to produce intestinal tumours in animals (Taylor et al. 1967) and it is for this reason that one of its constituents, the compound *β-asarone* has been banned in Western countries even as a food-flavouring agent (Hall 1973; Shellard 1966).

This left a puzzling but substantial number of remedies (amounting to 32% of the total) with no obvious biochemical or pharmacological justification for their use (at least on the basis of current information), but equally no reason to regard them as

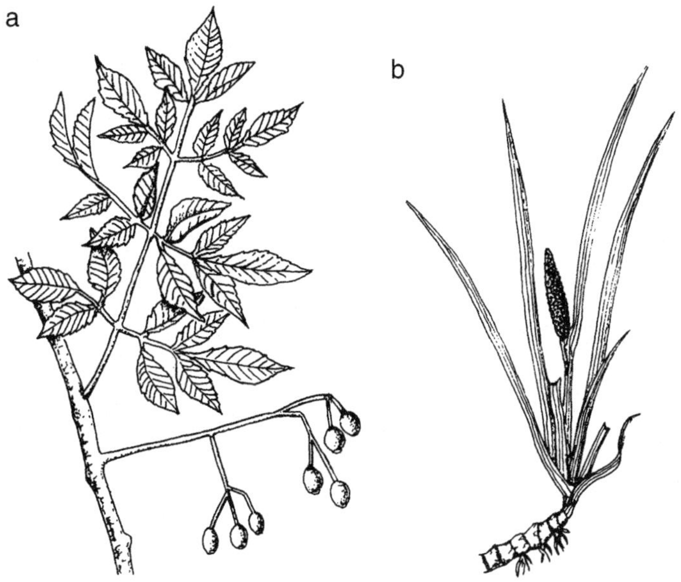

Fig. 6.3 Two village herbal remedies whose safety has been called into question. (**a**) Chinaberry (*Melia azedarach*), (**b**) sweet flag (*Acorus calamus*)

harmful (Table 6.1, column 3). Here, conceivably, we are entering the realm of *placebos*. These can be defined as inert preparations lacking any biochemical potency but which, if proffered as part of a consultation with a trusted medical practitioner, may still be capable of bringing benefits. Western studies suggest that the proportion of patients capable of experiencing such benefits may be as high as 35% (Beecher 1955). The exact nature of the phenomenon is not fully understood, but the prevailing theory is that the consultation in itself reduces stress and stimulates the release of opiates in the brain (Brown 1998; Benedetti 2008). In effect, the consultation process can be regarded as mobilizing the patient's own internal pharmacy (Meissner et al. 2011).

Following this line of reasoning, at least some of the remedies listed in column 3 of Table 6.1 could have been acting as placebos. It may be significant that village consultations frequently include dramatic elements that might be calculated to increase the strength of placebo effects. For instance, when medicine men in Nepal are dealing with patients they often don special clothing including a head dress of peacock feathers and a coat ornamented with bells. Moreover their deliberations are usually accompanied by an assistant beating rhythmically on a drum (Fig. 6.4).

Viewed solely on the basis of this particular case history, the picture that emerges is quite a positive one. An estimated 46% of the *jhankrie's* remedies seemed capable of producing the outcomes expected of them and on the basis of Beecher's placebo success rate of 35%, at least a proportion of the treatments with no known biochemical potency could still have been producing benefits. Widening the discussion

Fig. 6.4 The "consultation" (Drawing by Mary I. Anderson)

to other practitioners and other regions does however reveal some significant complications.

6.3 Residual Difficulties

6.3.1 The Problem of Unsound Remedies

Various instances have come to light of village herbalists using remedies that not only fail to have the curative effects expected of them, but also to pose a threat to the patient's well being. The widespread use of herbal purgatives to treat children suffering from diarrhoea is a case in point. Such a treatment is usually coupled with an intentional reduction of fluid intake. Presumably the rationale for this is to limit the volume of diarrhoeal discharges. In fact, both practices are undesirable. Purgatives are always difficult to use safely because of the risk of toxic effects arising from excessive dose levels and, far from withholding fluids from diarrhoea sufferers, it is vital in these circumstances to increase their fluid intakes.

Another instance of a dangerous practice is the West African "cow's urine" regime administered to children suffering from fever-induced convulsions. The children are persuaded to swallow a concoction prepared by soaking various leaves in a solution of salt and cow's urine. Typically, the leaves used are derived from tobacco (*Nicotiana tabacum*), basil (*Ocimum viride*) and garlic (*Allium sativum*) and the results are frequently fatal. During 1965–1966, a study by the University College Hospital at Ibadan in Nigeria attributed 130 child deaths to this cause and similar case-rates were still being recorded 10 years later. It is not entirely clear which component is the harmful one, but *nicotine* from the tobacco leaves is the most likely culprit (Oyebola and Elegbe 1975).

Elsewhere in Africa, fatalities frequently result from women using plant preparations in attempts to procure abortions. According to a dossier on poisonings compiled by staff at the National Herbarium in Zimbabwe, at least five plant preparations are implicated (Gelfand et al. 1985). In some cases, the victim selected the plant material herself and the problem may thus have been one of misidentification. In other cases and more worryingly, the plant selection was supervised by a recognised traditional healer (or *n'anga*).

African traditional herbalism may also be at the root of a problem that has emerged on some of the Caribbean islands. A practice probably introduced by African slaves is to treat sick children with "bush teas" made from locally-gathered herbs. When the mixtures include the yellow wild pea *Crotalaria fulva,* the consequences can be disastrous. Compounds known as *pyrrolizidine alkaloids* contained in the tissues of this plant can cause progressive liver damage and have fatal consequences in as many as 30% of cases (Bras et al. 1957).

Herbal preparations made from various species of ragwort (*Senecio*, Asteraceae) are another cause of liver damage, this time leading to tumour development (Schoental 1972). Various tribal groups in Southern Africa routinely use ragwort extracts to treat a wide range of disorders including migraine, swollen gums, colic, chest pains and syphilis (Watt and Breyer-Bandwijk 1962). Yet *Senecio* extracts contain *alkaloids* similar to those in the yellow wild pea and have been shown to produce liver tumours in experimental animals (Cook et al. 1950; Schoental 1968; Harris and Chen 1970). This finding will come as no surprise to horse owners in Britain, who know that allowing their animals to graze on ragwort can make them ill and sometimes cause their deaths (Cooper and Johnson 1984). Employing medicines derived from these plants would therefore seem highly ill-advised in any circumstances.

Traditional herbal remedies are also implicated in the development of another kind of tumour in Africa, namely Burkitt's lymphoma. This tumour affects mainly children and develops particularly in the jaw region. It is prevalent in areas where medicinal use is made of extracts from the milk bush (*Euphorbia tirucalli*). The many conditions ostensibly being treated include hemorrhoids, rheumatism, toothache and ulcers, and the damaging side effects arise in a somewhat roundabout way. The principal cause of the tumour is a virus (known as the Epstein-Barr virus). This causes genetic damage that leads to the proliferation of blood cells known as *B-lymphocytes*. The body can usually correct this problem, but apparently not in the presence of a compound, *4-deoxyphorbol ester,* contained in milk bush extracts (Osato et al. 1987; Aya et al. 1991).

The overall picture that emerges is therefore quite alarming and prompts one to wonder how it is that dangerous remedies in herbalists' repertoires have come to survive alongside effective ones; if healers can recognize useful remedies why can't they spot the dangerous ones too? The likely explanation is that medicine is a field where it can be exceedingly difficult to identify the true links between causes and effects. In agriculture, if crops have been planted in the wrong places this soon becomes apparent when they fail to thrive; likewise if village houses are constructed from timbers vulnerable to termite attack, their collapse soon offers a clear signal that a mistake has been made. However in health matters such certainties are much harder to come by. In some forms of cancer, for example, there can be a time lag of 20 years between the initial triggering factor and the development of the disease. Even with liver damage, a delay of 10 years is not unusual. Whilst hospitals with modern technical facilities and trained epidemiologists are equipped to unravel such complexities – and even they find it difficult – village practitioners are not in a position to do so. There is also the problem that a particular symptom, a severe headache for example, can have any one of a number of underlying causes, yet distinguishing between the various possibilities without the help of appropriate diagnostic equipment poses substantial difficulties.

6.3.2 Confusing Generalizations

The problems are further aggravated by what can be termed "confusing generalizations". Sociologists tell us that when human societies lack empirical knowledge, they are prone to fill the gaps with generalizations that have the appearance of serving a unifying purpose, but in practice are often seriously misleading (Glasser 1988). This is especially the case when knowledge is transmitted orally and where a certain "mental economy" is exercised to render information-transfer more manageable (Balée 1999).

One such simplifying notion is the idea that illnesses and remedies can be divided into "hot" and "cold" categories and that treatment should involve balancing one influence against the other. This concept has already been seen to bring confusion to some aspects of food selection (p.60). In the context of herbal medicine, it requires that "hot" diseases be countered with "cold" herbs, and "cold" diseases with "hot" ones. Examples drawn from traditional Chinese medicine include an insistence that a supposedly "cold" disorder such as diarrhoea be treated with a "hot" plant such as aconite (*Aconitum carmichaeli*), and that a "hot" disorder such as an inflamed throat be treated with a "cold" plant such as honeysuckle (*Lonicera japonica*). In fact no firm scientific basis exists for such notions. "Hot" plants do not, as a category, differ from "cold" ones in obviously measurable ways. Neither is the system geographically consistent, with the same medicinal herbs being placed in opposite categories by different cultures; for instance the hot herbs of Thailand are the cold ones of Bangladesh and vice versa.

Equally questionable is the so-called "doctrine of signatures". This is based on the idea that the appearance or life-style of a plant provides the clue to its appropriate medical use. This was a concept prevalent in mediaeval Europe, it being believed that the Creator had provided such indications as a useful service. Thus the folded kernel surface of the walnut (*Juglans regia*) was believed to advertise its value as a treatment for brain disorders, an organ with a similarly convoluted structure. By the same token, plants such as saxifrages (*Saxifraga* species) which grow amongst rocks were recommended for use in eliminating kidney stones, and plants with red flowers for treating disorders of the blood (Arber 1953). Such ideas are of course fanciful. In reality, natural selection causes plants to evolve features that are advantageous to themselves. For example the shapes and colours of flowers have usually evolved to facilitate pollination by insects, and certainly not to provide a convenient *aide-memoire* for herbalists.

Nonetheless, the idea that plants commonly provide helpful indications about their appropriate therapeutic use prevails in much of the developing world. In parts of Africa for example it is still believed that plants with milky saps are useful in stimulating milk flow in nursing mothers, that plants with red flowers are useful for treating anaemia, and that yellow-flowered plants can be used for treating jaundice. In the same fashion an extract from the forest shrub *Palisota hirsuta,* which typically has swollen stem nodes, is considered to be useful for treating swollen knees. Likewise a preparation derived from the oil bean tree (*Pentaclethra macrophylla*), a plant that frequently grows crooked, is judged suitable for treating stiff backs (Oliver-Bever 1986).

Properties are attributed to animal materials on a similar basis. Thus because of their shape snail shells are used in powdered form to treat earache and the knobbly skin from the back of toads to treat warts. Even modern inventories of Chinese medicines include prescriptions of this sort (Beijing Medical College 1984). Hence the cast skins of the noisy cicada are recommended for voice loss and the powdered scales of the flexible-bodied pangolin for stiff joints. By a similar token, preparations made from the tightly-interlocked plates of turtles are viewed as a suitable treatment for babies whose skull bones are proving slow to fuse.

Apart from their ineffectiveness for their intended purpose (except conceivably as placebos) these unsubstantiated notions can pose a threat to various endangered wildlife species. For instance there is a widespread notion in parts of Asia that powdered extracts from the limb bones of tigers will relieve rheumatic conditions and "general weakness". The dubious rationale is that the strength of that animal will be transferred to the patient and that the bone preparations, perceived as having a "warm" quality, would be suitable for countering "cold" conditions such as rheumatism. These spurious beliefs and the commercial opportunities they generate have unfortunately provided poachers throughout Asia with an incentive to slaughter tigers, a species already seriously endangered by habitat loss (Mills and Jackson 1994). Similar threats are generated towards various species of rhinoceros on the basis of an unfounded "law of signatures" belief that preparations made from powdered rhino horns are effective as aphrodisiacs (Leader-Williams 1992);. This has had the effect of raising the market price of the horn to US$60,000 a kilo (higher than the price of gold) and has provided poachers with a strong incentive to slaughter rhinos illegally. Sharks too have started to feature in these discussions. Large numbers are slaughtered in the mistaken belief that preparations such as shark's fin soup can prevent cancer. This unfounded notion (Ostrander et al. 2004) has had the effect of depleting shark populations and destabilizing coastal communities including coral reefs in which sharks function as "keystone" species (Ferretti et al. 2010). Clearly, strenuous action is needed to dispel myths that needlessly endanger important wildlife species.

6.3.3 The Attribution of Illness to Supernatural Causes

When asked to comment generally on the causes of illness, Nepalese *jhankries* unhesitatingly draw a distinction between conditions they consider to be of natural origin and those they attribute to supernatural causes. The supernatural influences were perceived to emanate variously from ancestral spirits or farmland and forest gods (Joshi and Edington 1990). Likewise in Zimbabwe, when Professor Michael Gelfand was allowed to eavesdrop on 289 consultations carried out by traditional *n'angas*, he noted that they attributed 47% of the illnesses to natural causes and 53% to supernatural ones. The supernatural category was further subdivided into disorders considered to be the result of witchcraft (29%) and those thought to be caused by the displeasure of ancestral spirits (17%) or other spirits of the dead (7%) (Gelfand et al. 1985). In Ghana there is a remarkably similar breakdown into natural and supernatural attributions, with 47% of diseases being perceived to fall into the first category and 39% into the second. The remaining 14% were attributed to one or other category according to individual circumstances (Fosu 1981).

Supernatural attributions appear in a particularly interesting form amongst the indigenous peoples (or *Orang Asli*) of Malaysia (Werner 1986). In these communities, every illness is considered to have a spirit counterpart and when the local healer or *poyang* is called upon to treat a sick person he carves a miniature wooden model of the spirit in question. This is known as a *sepili* and, when applied to the skin of the patient, is believed to draw the harmful spirit into itself. The final step is to cast the model into a river so that, in the process of being washed downstream, the spirit contained in it is prevented from causing further harm.

Such notions are consistent with a world-view in which humans, non-humans, deities and spirits are all linked together in a complex web of relationships. The idea has some redeeming aspects. For instance the notion that one's actions are continually being scrutinized by the gods and ancestral spirits can offer a strong deterrent against anti-social behaviour. There may even be some bonuses in medical matters. It is possible, for example, that at least a proportion of the mentally-ill can be helped by being presented with a narrative that involves malevolent spirits being driven from the body, a notion according well with local belief systems (Lambo 1956; Maclean 1971).

There are however some seriously negative aspects. Diagnoses may be made on the basis of divination rituals such as throwing down bones or carved wooden plaques, sometimes without the patient even being present (Chiwuzie et al. 1987) and postulations of witchcraft are almost invariably damaging. These can affect both those who perceive themselves to be targeted (who may suffer psychological and sometimes physical collapse) and those who are accused of being the perpetrators. As was once widely the case in Europe, it is not uncommon in villages of the developing world for mildly eccentric but innocuous individuals to be accused of witchcraft, and to have their welfare and sometimes their lives placed in jeopardy. In parts of South Africa, for example, it has been necessary to set up special villages to serve as refuges for men, women and children driven from their homes following accusations of witchcraft. One such refuge in Venda Province, known as *Motonawabaloi* or "place of the witches", sheltered on the last count over a hundred such refugees (Minnaar et al. 1992). Similarly, in Northern Ghana, the town of *Gambaga* contains a refuge harbouring women who have fled their homes following accusations that they have employed witchcraft to cause diseases such as meningitis (de Trey-White 2006). Some of the refugees have been in the centre as long as 13 years. In its turn, Kinshasa, the capital of the Congo, has recently become a refuge for hundreds of street children evicted from their homes following witchcraft accusations. In most instances the children had been implausibly blamed for causing HIV/AIDS amongst family members. The intractability of the situation was unwittingly demonstrated by a pronouncement from a Congolese official to the effect that *"the biggest problem is knowing how to sort the real witches from the others"* (Anonymous 2002).

6.4 Remedial Strategies

6.4.1 Restricting Harmful Treatments

Where it is necessary to dissuade local healers from persisting with dangerous remedies, the task is made easier if effective alternatives can be suggested. A good example is provided by the action of the Ghanaian Ministry of Health in organizing workshops to dissuade traditional healers from using *nicotine* preparations for the treatment of feverish childhood convulsions. They are taught instead to employ a simple but effective cold-water sponging technique (Warren et al. 1982).

Since its original use to treat cholera (Nalin et al. 1968), *oral rehydration therapy* has likewise shown itself to be infinitely preferable to purging herbal techniques as a diarrhoeal treatment and far less dangerous. It involves regularly dosing the patient with boiled water containing a few pinches of salt and a larger amount of sugar. In an imaginative initiative from UNICEF (United Nations Children's Fund) in Nepal, local *jhankries* have been supplied with attractive multi-coloured cards the size and shape of playing cards (Fig. 6.5a). One side carries a portrait of the warrior goddess

Fig. 6.5 UNICEF initiatives designed to promote the use of oral rehydration therapy in Nepal. (**a**) Instructional cards (front and reverse sides), (**b**) merit badge for participating scouts and girl guides, (**c**) the oral hydration message featured on a dress fabric (All courtesy of UNICEF Nepal)

Durga – much revered for her battles against the forces of evil – the other provides instructions for preparing the salt and sugar mixture needed to rehydrate patients. The success of the playing-card strategy stems partly from the high quality and attractiveness of the cards themselves and partly from the *kudos* gained by local *jhankries* being in a position to distribute them. Other ingenious schemes promoted by UNICEF are to produce merit badges for local scouts and guides who are active in promulgating the rehydration therapy message and also to produce a fabric for dresses and shirts with the rehydration formula emblazoned across it (Fig. 6.5b, c).

There have been similar initiatives in Ghana and Swaziland aimed at providing instructional workshops for local healers (Warren et al. 1982; Green and Makhubu

1984; Hoff and Maseko 1986). In Mexico, the issue has been approached in a different but equally ingenious way by the non-governmental agency *Project Piaxtla.* Children are encouraged to mount puppet-show dramas recounting how initially dubious parents and grandparents eventually become persuaded about the merits of rehydration therapy for children. The demonstration is made all the more telling by arranging for the gourd representing the baby to be linked with water tubes that allow it to "vomit" and to produce muddy water to simulate diarrhoeal discharges (Werner and Bower 1982).

6.4.2 *Increasing Coverage*

A scrutiny of national herbal inventories reveals that not all common conditions are covered (Table 6.2). In such cases, modern drugs are needed to fill the gaps. One problem is that drug companies have historically given rather little attention to manufacturing remedies for the less well-known tropical diseases; another is that even the available products are usually prohibitively expensive by local standards. Fortunately there are signs of progress on both fronts, with partnerships being set up

Table 6.2 Extent of coverage by herbal remedies of common diseases and disorders in two tropical countries

	Availability of herbal remedy	
Disease or disorder	Nigeria	Belize
1. Roundworm infection	o	o
2. Louse infection	+	+
3. Scabies	+	o
4. Ringworm fungus	+	+
5. Wounds and skin infections	+	+
6. Skin irritation	+	+
7. Malaria	o	o
8. Asthma	+	+
9. Trachoma (eye infection)	o	o
10. Intestinal spasms	+	+
11. Acid indigestion	o	o
12. Diarrhoea	o	o
13. Hemorrhoids	o	o
14. Allergic and shock reactions	o	o
15. Pain or fever	+	o
16. Swallowed poison	+	o
17. Hemorrhage during childbirth	o	o
18. Anaemia and nutritional deficiency during pregnancy	o	o

Based on Sofowara (1982), WHO (1985), Oliver-Bever (1986) and Arvigo and Balick (1993)
+ = herbal remedy available, o = herbal remedy unavailable

between drug companies, governments and private foundations. In this way development costs are shared and the drugs made cheaper.

Unfortunately the gap-filling process can sometimes take an unhelpful turn, with modern pharmaceuticals being drawn back into misleading classification systems. Thus, a study from Sierra Leone reported how a "doctrine of signatures" concept caused red-coated pills to be used for treating blood disorders. This was quite contrary to their intended application, which was either as a diuretic or a vitamin supplement (Bledsoe and Goubaud 1985). Similarly in Ghana, black and red capsules of *ampicillin* were found to be locally favoured for menstrual problems, which was quite contrary to their proper use as an antibiotic (Wolf-Gould et al. 1991). Such misuse of antibiotics is doubly unfortunate, not simply because they are ineffective for the conditions ostensibly being treated, but also because the practice aggravates problems of drug-resistance. The confusion is accentuated by the fact that modern drugs often find their way from hospital pharmacies to market stalls and, in the process of being repackaged into smaller lots for sale, frequently become divorced from their original instructional leaflets (van der Geest 1987; Foster 1991).

Persistent "hot" and "cold" concepts likewise retain their capacity to confuse. In parts of rural Malaysia, the malaria prophylactic *chloroquine* is often withheld from children suffering from measles on the grounds that it is a "hot" drug, which should not be combined with a "hot" disease such as measles (McKay 1980). In another unhelpful ramification, *chloroquine* use tends to be discontinued during the fruiting season of the popular *durian* (*Durio zibethinus*), because this is a "hot" fruit. A way out of this particular dilemma might be to persuade local people to reclassify *chloroquine* as a "cold" drug. The difficulty with any such approach is that it might be seen as offering support to the "hot" and "cold" system in general.

Even some governments have failed in the past to act decisively to replace dubious traditional remedies with effective modern ones. This was formerly the case with control actions against HIV/AIDS taken in many Southern African countries, where infection rates can be as high as 18%. The use of herbal preparations from the African yellow star plant (*Hypoxis hemerocallidea*) or the so-called cancer bush (*Sutherlandia fructescens*) receives no clear support from biochemical studies, indeed there is evidence that these materials interfere with the action of proven antiretroviral drugs (Mills et al. 2005a, 2005b).

6.4.3 Improving Co-ordination

In spite of the optimistic prognostications of the World Health Organization (Akerele 1984, 1987), substantial obstacles hamper attempts to integrate traditional and modern medicine (Velimirovic and Velimirovic 1980). A critical step in the process appears to lie in refining the traditional healer's initial judgment about which conditions he can treat himself and which he needs to refer to a Western-trained doctor. Even if the disorder has been correctly identified at the outset and is one that the healer feels competent to deal with, the process can still falter if an inappropriate herbal remedy is used, or the right herb in the wrong concentration. Communication difficulties can also occur between traditional and modern practitioners. For instance, Western-trained physicians are understandably disconcerted when they receive gravely ill patients for whom no records of previous treatments are available. Some countries have made considerable progress in addressing these issues. Swaziland, for example, has developed an effective scheme for using nurses as go-betweens (Hoff and Maseko 1986; Upvall 1992) and Ghana has improved the performance of its traditional healers by providing them with periodic instructional workshops (Warren et al. 1982).

6.4.4 Refining Effective Remedies

Even traditional remedies that have been fully validated as safe and effective may still require fine-tuning in such matters as dose levels and modes of preparation (Croom 1983). In the Philippines, this kind of exercise has involved collaboration between the Ministry of Health and scientists from the University at Los Baños. It has resulted in a list of effective traditional remedies accompanied with detailed instructions for use. Thus the seeds of the Rangoon creeper (*Quisqualis indica*) are identified as a suitable treatment for intestinal worms (the effective component is an alkaloid) and an appropriate dose is defined. This specifies that 8–10 of the dried seeds should to be chewed after the main evening meal and followed by a glass of water. Similarly, the leaves of the candle bush (*Cassia alata*) are identified as a suitable treatment for ringworm (a condition caused by a fungal infection) and the method of preparation and application is spelled out. Apparently one cupful of chopped leaves should be boiled in two cups of water for 15 min and the decoction applied to the affected area three times a day.

When remedies are demonstrably effective, it may also be necessary to take steps to ensure that the plant populations supplying the raw materials are protected from overuse. Threats can come from various quarters. Many medicinal plants derive from forest areas, 40–50% of them according to some estimates, and are thus highly vulnerable to the effects of deforestation (Joshi 2004; Joshi and Joshi 2007). Intense grazing can also deplete pastures of useful medicinal plants, although it has been observed that certain species with effective defensive mechanisms such as spines or chemical deterrents are left alone by grazing animals with the consequence that their numbers increase under grazing pressure rather than declining. In Nepal, *Mahonia* and *Berberis,* sources of *berberine* an important eye-treatment, come into this category and in South Africa the same principle applies to various lily species (Cunningham 1991).

Particular problems arise where there is an active export trade for particular herbs (Akerele et al. 1991). Exercising restraint in wild-plant harvesting doesn't seem to come naturally to most collectors and external guidance or control is usually required (Cunningham 2001). An option being pursued in some regions is for farmers to grow medicinal herbs on their land and to manage them as crops. There are good examples of this approach being followed in both Sri Lanka and India (Lokubandra 1991; Kala 1997).

Whilst our attention so far has been focused on the strengths and weaknesses of traditional medicine in its original settings, there are also some interesting modern ramifications. In recent years traditional remedies or their derivatives have become surprisingly popular in some sectors of Western society. In the United States for example the proportion of the population using such therapies rose from 33.8% to 42.1% between 1990 and 1997 (Eisenberg et al. 1993, 1998). Similarly in Britain the percentage of users increased from 14 to 28.3%, between 1985 and 1999 and in France, where homoeopathy is the favoured alternative therapy, its use more than doubled from 16 to 36% between 1982 and 1992 (Fisher and Ward 1994). We will discuss later (p.220) possible reasons for this trend and raise the possibility that some of the revived therapies might meet needs not currently covered by conventional medicine.

References

Akerele O (1984) WHO's traditional medicine programme: progress and perspectives. WHO Chron 34(2):76–78

Akerele O (1987) The best of both worlds: bringing traditional medicine up to date. Soc Sci Med 24:177–181

Akerele O, Heywood V, Synge H (1991) Conservation of medicinal plants. Cambridge University Press, Cambridge

Anonymous (2002) Sad little sorcerers. *The Economist,* December 7, 2002, 68

Arber A (1953) Herbals, their origin and evolution, a chapter in the history of botany 1470–1670. Cambridge University Press, Cambridge

Aronson JK (1985) An account of the Foxglove and its medical uses 1785–1985. Oxford University Press, London

Arvigo R, Balick M (1993) Rainforest remedies. One hundred healing herbs of Belize. Lotus Press, Twin Lakes

Aya T, Kinoshita T, Imai S, Koizumi S, Mizuno F, Osato T, Satoh C, Oikawa T, Kuzumaki N, Ohigashi H, Koshimizu K (1991) Chromosome translocation and c-MYC activation by Epstein-Barr virus and *Euphorbia tirucalli* in B lymphocytes. Lancet 337:1190

Balée WL (1999) Footprints of the forest. The historical ecology of plant utilization by an Amazonian people. Columbia University Press, New York

Beecher HK (1955) The powerful placebo. J Am Med Assoc 159:1602–1606

Beijing Medical College (1984) Dictionary of traditional Chinese medicine. The Commercial Press, Hong Kong

Benedetti F (2008) Placebo effects. Understanding the mechanisms in health and disease. Oxford University Press, Oxford

Bledsoe CH, Goubaud MF (1985) The reinterpretation of western pharmaceuticals amongst the Mende of Sierra Leone. Soc Sci Med 3:275–282

Boiteau P, Ratsimamanga AR (1956) Asiaticoside isolé de *Centella asiatica* et ses emplois thérapeutiques. Therapie 11:125–150

Bose PK, Roy AC (1936) The constitution of ayapanin. J Indian Chem Soc 13:586–587

Bras G, Berry DM, Gyorgyi P (1957) Plants as an aetiological factor in veno-occlusive disease of the liver. Lancet:960–962

Brown RJ (1998) The placebo effect. Sci Am 278(1):90–95

Chiwuzie J, Ukoli F, Okojie O, Isah E, Eriator I (1987) Traditional practitioners are here to stay. World Health Forum 8:240–244

Chopra RN, Nayar SL, Chopra IC (1956) Glossary of Indian medicinal plants. Council of Scientific & Industrial Research, New Delhi

Chopra RN, Nayar SL, Chopra IC (1969) Supplement to glossary of Indian medicinal plants. Council of Scientific and Industrial Research, New Delhi

Cook JW, Duffy E, Schoental R (1950) Primary liver tumours following feeding with alkaloids of *Senecio jacobaea* Lin. Br J Cancer 4:405–410

Cooper MR, Johnson AW (1984) Poisonous plants in Britain and their effects on animals and man. HMSO, London

Cox PA (1994) The ethnobotanical approach to drug discovery; strengths and limitations. In: Prance G, Chadwick DJ, Marsh J (eds) Ethnobotany and the search for new drugs. Wiley, Chichester, pp 25–36

Croom EM (1983) Documenting and evaluating herbal remedies. Econ Bot 37:13–27

Cunningham AB (1991) Development of a conservation policy on commercially exploited plants: a case study from South Africa. In: Akerele O, Heywood V, Synge H (eds) Conservation of medicinal plants. Cambridge University Press, Cambridge, pp 337–358

Cunningham AB (2001) Applied ethnobotany. People, wild plant use & conservation. Earthscan Publications, London

De Trey-White S (2006) The witches of Gambaga. Geogr Mag 78(8):48–53

Eisenberg DM, Kessler RC, Foster C, Norlock FE, Calkins DR, Delbanco TL (1993) Unconventional medicine in the United States. Prevalence, costs, and patterns of use. N Engl J Med 328:246–252

Eisenberg DM, Davis RB, Ettner SL, Appel S, Wilkey S, Van Rompay M, Kessler RC (1998) Trends in alternative medicine use in the United States, 1990–1997. Results of a follow-up national survey. J Am Med Assoc 280:1569–1575

Farnsworth NR (1994) Ethnopharmacology and drug development. In: Prance G, Chadwick DJ, Marsh J (eds) Ethnobotany and the search for new drugs. Wiley, Chichester, pp 42–51

Ferretti F, Worm B, Britten GL, Heithaus MR, Lotze HK (2010) Patterns and ecosystem consequences of shark declines in the ocean. Ecol Lett 13:1055–1071

Fisher P, Ward A (1994) Complementary medicine in Europe. Br Med J 309:107–111

Foster S (1991) Supply and use of essential drugs in sub-Saharan Africa: some issues and possible solutions. Soc Sci Med 32:1201–1218

Fosu GB (1981) Disease classification in rural Ghana: framework and implications for health behaviour. Soc Sci Med 32:1201–1218

Gelfand M, Mavi S, Drummond RB, Ndemera B (1985) The traditional medical practitioner in Zimbabwe. His principles of practice and Pharmacopoeia. Mambo Press, Gweru

Glasser M (1988) Accountability of anthropologists, indigenous healers, and their governments: a plea for reasonable medicine. Soc Sci Med 27:1461–1465

Green EC, Makhubu L (1984) Traditional healers in Swaziland: toward improved cooperation between the traditional and modern health sectors. Soc Sci Med 18:1071–1079

Hall RL (1973) Toxicants occurring naturally in spices and flavors. In: Toxicants occurring naturally in foods. Natural Academy Press, Washington DC, pp 448–463

Harris PN, Chen KK (1970) Development of hepatic tumors in rats following ingestion of *Senecio longilobus*. Cancer Res 30:2881–2886

Hoff W, Maseko DN (1986) Nurses and traditional healers join hands. World Health Forum 7:412–416

Jentzch K (1953) Production of alkaloids in Solanaceae. Sci Pharm 21:285–289

Joshi AR (1987) A study of the environmental relationships of certain village communities in the central development region of Nepal. Doctoral Thesis, University of Wales, Cardiff

Joshi K (2004) Documentation of medicinal plants and their indigenous uses in Likhu Sub-Watershed, Nepal. J Non-Timber Prod 11(2):86–94

Joshi AR, Edington JM (1990) The use of medicinal plants by two village communities in the central development region of Nepal. Econ Bot 44:71–83

Joshi AR, Joshi K (2000) Indigenous knowledge and uses of medicinal plants by local communities of the Kali Gandaki Watershed Area, Nepal. Ethnopharmacol 73(1–2):175–183

Joshi AR, Joshi K (2007) Ethnomedicinal plants used against skin diseases in some villages of Kali Gandaki, Bagmati and Tadi Likhu Watersheds of Nepal. Ethnobot Leaflets 11:235–246

Kala CP (1997) Ethnobotanical survey and propagation of rare medicinal herbs for small farmers in the buffer zone of valley of flowers National Park, Garhwal Himalaya. Wildlife Institute of India, Dehradun

Klayman DL (1985) *Qinghaosu* (Artemisinin): an antimalarial drug from China. Science 228:1049–1055

Lambo TA (1956) Neuropsychiatric observations in the western region of Nigeria. Br Med J ii:1388–1396

Leader-Williams N (1992) The world trade in rhino horn: a review. TRAFFIC International, Cambridge

Leung AY (1980) Encyclopedia of common natural ingredients used in food, drugs and cosmetics. John Wiley, New York

Lewis WH, Elvin-Lewis MPF (1977) Medical botany, plants affecting Man's health. John Wiley, New York

Lokubandra WJM (1991) Policies and organization for medicinal plant conservation in Sri Lanka. In: Akerele O, Heywood V, Synge H (eds) Conservation of medicinal plants. Cambridge University Press, Cambridge, pp 241–252

Maclean U (1971) Magical medicine, a Nigerian case study. Allen Lane, The Penguin Press, Harmondsworth

McKay DA (1980) Foods, illness and folk medicine: insights from Ulu Trengganu, West Malaysia. In: Robson JRK (ed) Food, ecology and culture. Readings in the anthropology of dietary practices. Gordon and Breach Science Publishers, New York, pp 61–66

Meissner K, Kohls N, Colloca L (2011) Introduction to placebo effects in medicine: mechanisms and clinical implications. Philos Trans R Soc Lond B 366:1784–1789

Mills JA, Jackson P (1994) Killed for a cure. A review of the worldwide trade in tiger bone. TRAFFIC International, Cambridge

Mills EJ, Foster BC, van Heeswijk RP, Wilson KEP, Kosuge K, Kanfer LB (2005a) Impact of herbal medicine on antiretroviral metabolism. AIDS 19:95–97

Mills EJ, Cooper C, Seely D, Kanfer I (2005b) African herbal medicines in the treatment of HIV: Hypoxis and Sutherlandia. Nutr J 4:19–34

Minnaar A, Offringa D, Payze C (1992) To live in fear: witchburning and medicine murder in Venda. Human Sciences Research Council, Pretoria

Nalin DR, Cash RA, Islam R, Molla M, Phillips RA (1968) Oral maintenance therapy for cholera in adults. Lancet 1968(2):370–373

Oliver-Bever B (1986) Medicinal plants in tropical West Africa. Cambridge University Press, Cambridge

Osato T, Mizuno F, Imai S, Mizuno F, Aya T, Koisumi S, Kinoshita T, Tokuda H, Ito Y, Hirai N, Hirota M, Ohigashi H, Koshimizu K, Kofi-Tsekpo WM, Were JBO, Muganbi M (1987) African Burkitt's lymphoma and an Epstein-Barr virus-enhancing plant *Euphorbia tirucalli*. Lancet 1987(i):1257–1258

Ostrander GK, Cheng KC, Wolf MJ, Wolfe MJ (2004) Shark cartilage, cancer and the growing threat of pseudoscience. Cancer Res 64:8485–8491

Oyebola DDO, Elegbe RA (1975) Cow's urine poisoning in Nigeria. Trop Geogr Med 27:194–202

Prance G, Chadwick D, Marsh J (eds) (1994) Ethnobotany and the search for new drugs. Wiley, Chichester

Rácz-Kotilla E, Rácz G, Solomon A (1974) The action of *Taraxacum officinale* extracts on the body weight and diuresis of laboratory animals. Planta Med 26:212–217

Reynolds JEF (ed) (1996) Martindale. The extra pharmacopoeia, 31st edn. London, Royal Pharmaceutical Society

Schoental R (1968) Toxicology and carcinogenic action of pyrrolizidine alkaloids. Cancer Res 28:2237–2246

Schoental R (1972) Prevention or cure? Trop Geogr Med 24:194–198

Shellard EJ (1966) Medicinal plants and the pharmacist. Pharm J 237:495

Sofowara A (1982) Medicinal plants and traditional medicine in Africa. John Wiley, Chichester

Taylor JM, Jones WI, Hagan EC, Gross MA, Davis DA, Cook EL (1967) Toxicity of oil of calamus (jammu variety). Toxicol Appl Pharmacol 10:405

Tsey K (1997) Traditional medicine in contemporary Ghana: a public policy analysis. Soc Sci Med 45:1065–1074

Upvall MJ (1992) Nursing perceptions of collaboration with indigenous healers in Swaziland. Int J Nurs Stud 29(1):27–36

van der Geest S (1987) Self-care and the informal sale of drugs in South Cameroon. Soc Sci Med 25:293–305

Velimirovic H, Velimirovic B (1980) Do traditional plant medicines have a future in third world countries? Curare 3:173–191

Vogel VJ (1970) American Indian medicine. University of Oklahoma Press, Norman

Wade A (ed) (1982) Martindale, the extra pharmacopoeia, 28th edn. London, The Pharmaceutical Press

Warren DM, Bova GS, Tregoning MA, Kliewer M (1982) Ghanian national policy towards indigenous healers. The case of primary health training for indigenous healers (PRHETIH) program. Soc Sci Med 16:1873–1881

Watt JM, Breyer-Bandwijk MG (1962) The medicinal and poisonous plants of Southern and Eastern Africa. Livingstone, Edinburgh

Werner R (1986) Bomah-Poyang. Traditional medicine and ceremonial art of the aborigines of Malaysia. University of Malaya, Kuala Lumpur

Werner DB, Bower BL (1982) Helping health workers learn. The Hesperian Foundation, Palo Alto

WHO (1985) The use of essential drugs. Technical Report 722, World Health Organization, Geneva

Wolf-Gould CS, Taylor N, Horwitz SM, Barry M (1991) Misinformation about medications in rural Ghana. Soc Sci Med 33:83–89

Chapter 7
Water Supply and Waste Disposal

7.1 Water Use

In the developing world, every person uses on average 10–50 l of water a day (White 1977). About 2 l of this amount needs to be drunk directly to fulfill a variety of physiological functions (McCance et al. 1971). These include evacuating wastes, regulating body temperature through sweating and replacing water losses from the lungs incurred during respiration. Humans also need water for other purposes, many of them with health implications. For instance water is required to cook food – certain foods are virtually inedible uncooked – and also to wash one's person and one's clothes. Failure to do this washing increases the likelihood of flea and lice infestations, thereby increasing susceptibility to the diseases such as typhus and plague that these insects can carry. Likewise having sufficient water to wash the face reduces the risk of infection from the bacterium *Chlamydia trachomatis* responsible for the eye disease *Trachoma*. In village situations water may also be required for such other miscellaneous purposes as fabric dyeing, beer brewing, brick making and pottery manufacture.

Domestic animals require water for much the same physiological functions as humans, but with consumption rates closely linked to body size. Thus a sheep or a goat needs 4–5 l a day and a cow around 30–40 l (Williamson and Payne 1978). Camels are in a class of their own and can survive without water for up to 14 days although, when they do find some, will drink as much as 186 l at a time. Having such consumption figures to hand does not however automatically allow predictions to be made about how much water the farmer needs to provide. This is because animals frequently meet part of their requirement by drinking unaided from natural water sources such as rivers and ponds. Crop plants also require water for functional reasons, including the somewhat different one of maintaining the turgor pressure and rigidity of stems and leaves. A square metre area of crop plants consumes on average 3 l of water a day. Some crop plants are able to exist solely on direct rainfall; others require special watering provisions to be made for them by the farmer.

J. Edington, *Indigenous Environmental Knowledge*,
DOI 10.1007/978-3-319-62491-4_7

Locating sufficient water for these various functions can be a complicated task. Wild plant species have sometimes been used to indicate the presence of water near the surface. Reeds and rushes often function in this role, as so too do certain tree species. In Southern Africa such indicator trees include the sycamore fig (*Ficus sycomorus*) and the water berry (*Syzygiun cordatum*). Likewise in the drier regions of New Mexico and Arizona, both native North Americans and early settlers were aware that a grove of plains cottonwood trees (*Populus sargentii*) signified a water source close to the surface.

Tapping deeper aquifers presents more difficulties. Sinking wells is the principal method, but in the absence of geological information and the ability to make test borings, there is no guarantee of success. The only consolation is that once a well strikes water, the disposition of aquifers is such that other wells sunk nearby and to a similar depth are also likely to prove successful. Even in the West, harnessing geological knowledge for these purposes was quite a late development, with its use first formally recorded in 1785 when Erasmus Darwin sank a well with the expressed intention of striking the junction between permeable and impermeable rocks, a location he correctly predicted would yield water (Bromehead 1942). More generally appreciated is the fact that springs tend to arise along geological junctions on hillsides, a situation that has given rise to so-called "spring-line" villages (Stamp 1969).

In other village settings the situation tends to be more complicated. Surprisingly, settlements are not universally sited directly alongside water sources. This is because other factors intervene. These include historic considerations of defensibility and very commonly a preference for living on fairly flat areas that lend themselves to easy house construction and crop cultivation. Occasionally the decisive factor is a totally unexpected one. There are for example African villages whose positioning coincides with rocky outcrops of dolerite or granite. The attraction of such sites is that they lend themselves to the creation of communal grain-grinding pits (Quin 1959).

Where water sources and points of use are separated, it is often possible for local communities to construct channels to allow a gravity feed of water between the two. If this is not feasible, the default mode applies and supplies have to be carried manually in containers. This is a chore that frequently falls on the shoulders of village women and is one that can occupy them in excess of an hour daily. In such circumstances it is not surprising that the amount of water used is closely linked to its availability. Hanlon (1983) nicely illustrates this point in his study of some villages in Mozambique. In communities having access to standpipe supplies, consumption ran at 11 l/person/day whereas in villages where water-collection required a 5 h round trip to a distant spring, consumption predictably fell to a mere 4 litres.

Many village water systems are cleverly designed to serve double or triple functions, serving humans and animals, humans and plants and often all three together. Figure 7.1 shows a system in Rajasthan where oxen are being used to raise a well bucket to serve household needs, but where spillage from the upper platform is being directed into an animal drinking trough.

Fig. 7.1 Well in Rajasthan designed to serve both community and livestock water needs

7.1.1 The "Tanks" of Sri Lanka as Water Storage Systems

In areas where there is a marked seasonality of rainfall, as is the case in much of Sri Lanka, rice farmers have responded by constructing elaborate systems of water-storage "tanks". These trap water from rivers and streams swollen by wet-season rains and then subsequently release supplies in a controlled manner throughout the year (Fig. 7.2). Flow is regulated through channels, sluice gates and diversion weirs, with the weirs often being carved asymmetrically in wood to guarantee each farmer his allotted share of the water (Leach 1968). These complex arrangements predate modern technology by many centuries and archaeological evidence shows that "tank" building in Sri Lanka reached its zenith during the fourth and fifth centuries, when an estimated 10,000 such structures were in operation (Codrington 1947). The water in the tank is generally not clean enough to drink directly, not least because buffaloes are allowed to wallow in it. However, the fact that water seeping under the dam becomes clarified by passage through subsoil deposits makes it feasible to obtain clean water by sinking wells in this downstream area (Fig. 7.2).

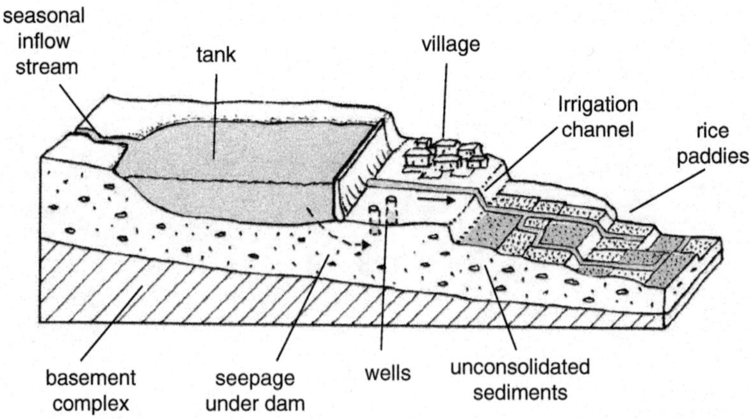

Fig. 7.2 A Sri Lankan "tank" (Developed from Ulluwishewa 1991)

7.1.2 The Multiple-Function Qanãts of Arabia

Another example of an ingenious water system serving multiple purposes is provided by the *qanãt* systems of the Middle East (Biswas 1970; Wulff 1968; Garbrecht 1983). In this region, fertile soils are frequently restricted to valley floors and coastal plains and are often frustratingly distant from natural water sources. Such sources typically take the form of rainwater accumulations within subterranean gravel deposits underlying distant mountain ranges. The solution, first recorded from prehistoric Persia, has been to build sloping underground tunnels or *qanãts*. These serve to tap water from the gravels and to convey it to lowland settlements and farms (Fig. 7.3). The tunnels are excavated through a series of shafts. These are capped and the excavated soil is heaped around the shaft-tops to prevent entry of potentially-contaminated surface water (Fig. 7.4a). The water issuing from the tunnels is cool and silt-free and is used initially for drinking, then to provide washing facilities and finally to irrigate crops (Fig. 7.4b, c). In the crop areas, removable stoppers in the channel walls are used to control the allocation of water to individual plots. The fact that the water reaching the farms is mildly sullied with soap and detergents is no disadvantage because these serve as an extra source of nutrients. All of which speaks for a level of local ingenuity on a par with that applied to such enterprises as farming and house construction. Unfortunately the picture changes when attention is focused, not on the volumetric manipulation of water, but on local judgments about its suitability for human consumption.

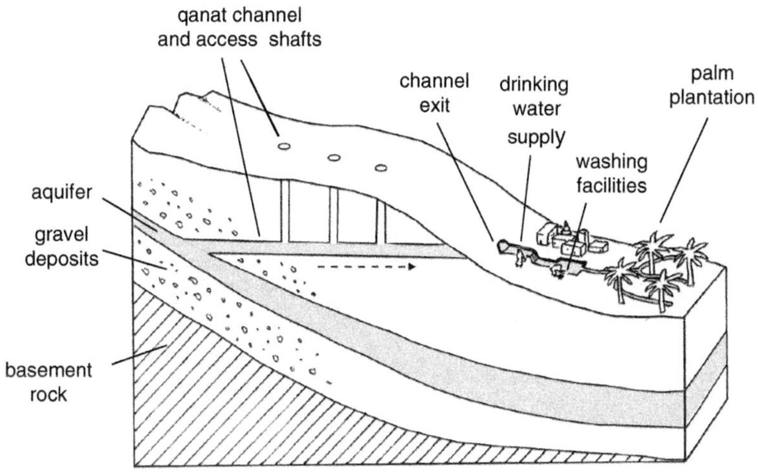

Fig. 7.3 A Middle Eastern *qanāt* (Developed from Wulff 1968)

7.2 Water Quality Issues

7.2.1 Physical and Chemical Properties

Local communities do not entirely ignore issues of water quality. For instance water containing a high concentration of dissolved iron is readily detected by taste and widely rejected except occasionally for therapeutic purposes. There is also a general aversion to drinking water containing large quantities of silt and various local techniques are employed to remove such material. These include filtering the water through a piece of cloth (Fig. 7.5a) or arranging for it to pass through the porous walls of an earthenware pot (Fig. 7.5b). This latter procedure leaves the silt behind and allows the clean (and cooled) water to seep through the pot walls into a receiving vessel beneath. In India, various plant materials have been used to settle sediment from cloudy water; two of the flocculating agents employed include crushed seeds from the drumstick tree (*Moringa oleifera*) and crushed fruits from the aptly named clearing nut tree (*Strychnos potatorium*).

7.2.2 Bacteriological Purity

Water is also particularly susceptible to pollution by pathogens derived from human wastes. Modern research has produced a comprehensive list. They span a range of biological groups including, on a descending scale of size, protozoa, bacteria and viruses (Fig. 7.6). The diseases they cause are all diarrhoeal in character and the resulting disorders can be sufficiently severe to be life-threatening (Table 7.1).

Fig. 7.4 The components of a *qanāt* system in Oman. (**a**) Shaft tops surrounded by excavated earth, (**b**) basin providing drinking and washing water, (**c**) channel directing water into a date palm plantation

Anthropological investigations reveal that local notions about the causes of diarrhoea are often incomplete or mistaken. A study from Peru revealed that diarrhoeal episodes are attributed locally to the consumption of "cold" water or "cold" foods, with 80–90% of mothers subscribing to this opinion (Escobar et al. 1983). In

Fig. 7.5 Traditional methods of removing silt from drinking water. (**a**) Filtering well water through a cloth (Ethiopia), (**b**) passing water through a porous earthenware pot (the Philippines)

warmer parts of the world the opposite view tends to prevail, with "hot" rather than "cold" influences being thought responsible. For instance, a survey from the Sind region of Pakistan revealed that "hot" foods were most frequently cited by mothers as the cause of diarrhoea in their children (Mull and Mull 1988). These notions are similar to the "humoral" theories of ancient Greece and may indeed share a common origin with them. They rarely refer to temperature as such, or even to the commonplace notion of spicy foods being hot, and certainly have little bearing on the potability of water.

Another idea prevalent in traditional societies is that childhood diarrhoea is linked with teething, the underlying rationale being that the emerging teeth displace tissue material that the body then needs to get rid of. Surveys have shown that 75% of respondents in Nepal and 44% in Nigeria held this view (Stapleton 1989; Adetunji 1991). This is a more understandable misconception. By the time children are teething, they are usually starting to crawl about on the house floor and, where this is made of beaten earth and possibly contaminated with the excreta of other children, the possibility of picking up infections is high. An additional hazard is that the weaning foods being introduced to children at this stage have often been prepared using contaminated water.

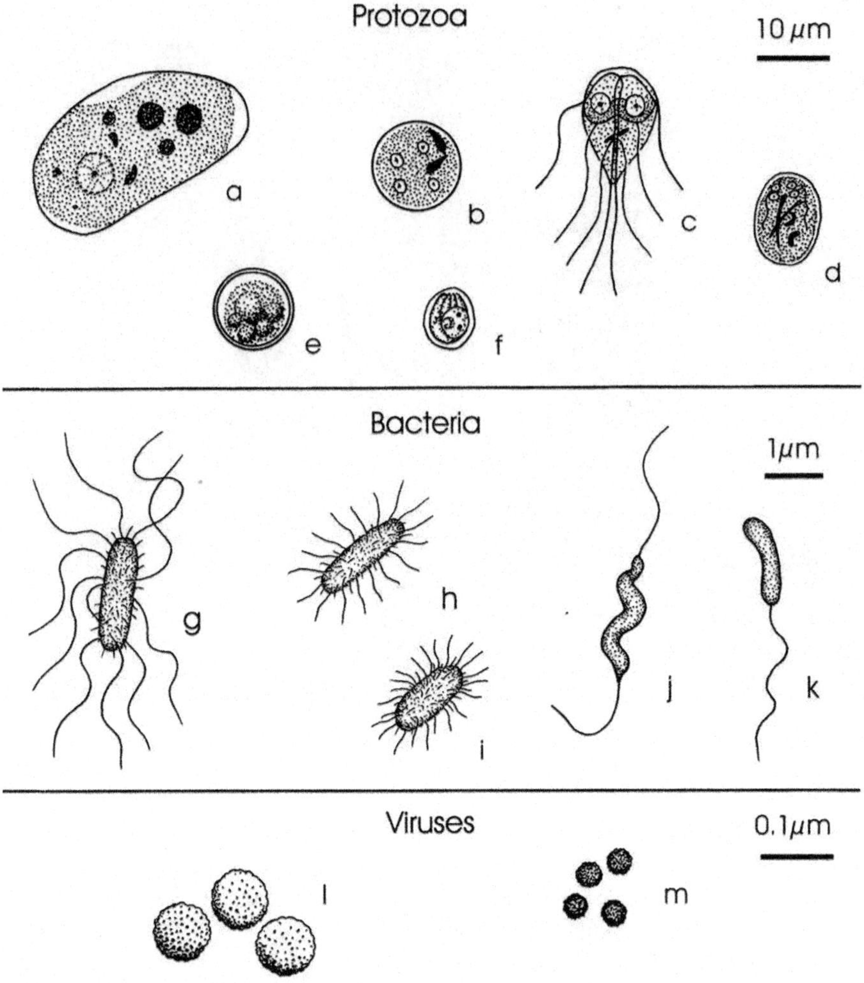

Fig. 7.6 Important water-borne pathogens. (**a**, **b**) *Entamoeba histolytica,* feeding stage and cyst; (**c**, **d**) *Giardia lamblia,* feeding phase and cyst; (**e**) *Cyclospora cayetanensis,* cyst; (**f**) *Cryptosporidium parvum,* cyst; (**g**) *Salmonella typhi*; (**h**) *Escherichia coli*; (**i**) *Shigella flexneri*; (**j**) *Campylobacter jejuni*; (**k**) *Vibrio cholerae*; (**l**) Rotavirus; (**m**) Noro virus (the scale of the three sets of drawings increases incrementally)

Witchcraft may also be regarded as a cause of enteric disorders. Its effects are claimed to be brought about, either through the operation of the *evil eye* whereby people behaving as witches cast envious glances at handsome children, or because ancestral spirits have chosen to punish people for social misdemeanors of various kinds (Patel et al. 1988; Weiss 1988). In one survey from the Indian state of Orissa, when 1000 mothers with children suffering from diarrhoea were questioned about its likely cause, 65% cited "casting of the evil eye" as the crucial factor. Teething

Table 7.1 Organisms responsible for diarrhoeal diseases and capable of being transmitted in faecally-contaminated drinking water

Disease	Causative organism
PROTOZOA	
Amoebic dysentery	*Entamoeba histolytica*
Giardiasis	*Giardia lamblia*
Cyclosporiasis	*Cyclospora cayetanensis*
Cryptosporidiosis	*Cryptosporidium parvum*
BACTERIA	
Typhoid fever	*Salmonella typhi*
Gastroenteritis	other *Salmonella* species
Gastroenteritis	enterotoxic *Escherichia coli*
Bacterial dysentery	*Shigella* species
Infantile gastroenteritis	*Campylobacter jejuni*
Cholera	*Vibrio cholerae*
VIRUSES	
Viral enteritis	Rotavirus
Viral enteritis	Norovirus

Based on McFeters (1990), WHO (2002)

and "hot" foods also featured in the explanations offered, but further down the list (Mohapatra 1989).

In short, on the question of the circumstances likely to precipitate a bout of diarrhoea, a large gulf exists between prevailing village notions and their science-based equivalents, with the role played by contaminated drinking water being especially poorly appreciated. Virtually the only point of convergence between the two systems is that some traditional societies accurately regard flies and "dirt" as potential threats. Some anthropologists argue in favour of embracing traditional concepts rather than seeking to replace them (Nichter 1985). A much-quoted example is the practice of boiling water. When highland peoples in Peru do this to counteract water's undesirable "cold" properties, they are fortuitously achieving something that any health professional would approve of, namely the killing of pathogens (Wellin 1955).

Whilst this would normally be applauded, doing it on the basis of "hot" and "cold" theories has its attendant pitfalls. For instance, one would not wish to perpetuate the other local practice of denying fruit to children with diarrhoea on the grounds that these are "cold" foods likely to aggravate a "cold" disease (Escobar et al. 1983). One of the dangerous consequences of diarrhoea is that it reduces potassium levels in the body and this in turn can interfere with heart action, digestion and muscle tone. Oranges and bananas are both rich sources of potassium and therefore any belief system that forbids their consumption by diarrhoea-sufferers is manifestly unhelpful.

Before congratulating ourselves too readily on the superiority of Western awareness in these matters, we would do well to recollect that it was only quite recently that Europeans managed to shake off their own brand of medical mysticism. Until the middle of the nineteenth century it was firmly believed that diseases such as cholera and typhoid fever (and for that matter malaria) were generated by clouds of noxious gaseous materials or *miasmata* (singular *miasma*) arising from decaying organic matter. Attempts to clear this offending material fortuitously brought some improvements in sanitary practices albeit for the wrong reasons. However, full clarification had to await investigations carried out in London in 1854 by the epidemiologist John Snow (Hempel 2007). He was able to show that cholera epidemics were typically triggered by the consumption of water contaminated with faecal pathogens. These observations, together with similar ones on typhoid fever made by William Budd in Bristol, paved the way for a more general "germ theory" of disease. Although blatantly obvious all this might seem to us now, its basic tenets were not generally accepted even in Western circles until the end of the century. The central proposition is that all infectious diseases are caused by specific organisms and from no other cause. This principle underlies the design of all modern water-supply and waste-disposal systems and in Western countries the successful operation of these facilities can usually be taken for granted.

7.2.3 Implications for Village Water Supplies

However in rural areas of the developing world, blind spots about the importance of water purity are still much in evidence. The modern yardstick for assessing the quality of drinking water is the *faecal coliform* test. The organisms featured are for the most part harmless inhabitants of the human intestinal tract and in this test feature as *indicators*. Although recently one kind of faecal coliform – *Escherichia coli 0157* – has been revealed as a serious pathogen, this does not detract from the general utility of the test; although it does put a premium on careful laboratory technique (Locking and Cowden 2009).

The higher the coliform count for a water sample, the greater its level of faecal contamination and therefore the greater the likelihood that it will contain dangerous pathogens. The World Health Organization recommends that for water from a piped distribution system to be considered fit to drink, no faecal coliforms whatsoever should be detectable in a 100-millilitre sample (WHO 1993). Although the same overall criterion applies to supplies obtained from local sources such as wells and springs, a occasional count of up to 10 coliforms/100 ml is generally regarded as permissible, although no cause for complacency (WHO 1985). The fact that counts of hundreds and even thousands of faecal coliforms can be routinely recorded from village wells, springs and surface watercourses used as drinking water sources therefore signals a serious problem (Agarwal and Narain 1997). A common reason for surface sources being of poor quality is that they are allowed to become polluted with run-off from land contaminated with animal and human faeces. The other point to make is that the pressure on water resources is such that even persons with some appreciation of water-purity issues may still be unable to obtain safe supplies. It is hardly surprising in these circumstances that the incidence of gastro-enteritis in rural communities often exceeds 40% (Taylor 1951; MacFarlane 1981).

7.3 Other Water and Waste Related Hazards

7.3.1 Schistosomiasis

Schistosomiasis, also popularly known as Bilharzia, is a disease affecting something like 200 million people world-wide, mainly in Africa, but also in certain parts of Asia and South America. The critical point of infection occurs when the free-swimming larvae of the *Schistosoma* parasite bore through the skin of persons using infected waters for bathing, clothes-washing or water collection (Figs. 7.7 and 7.8). The adult worms (male and female together) live in the blood vessels of the liver,

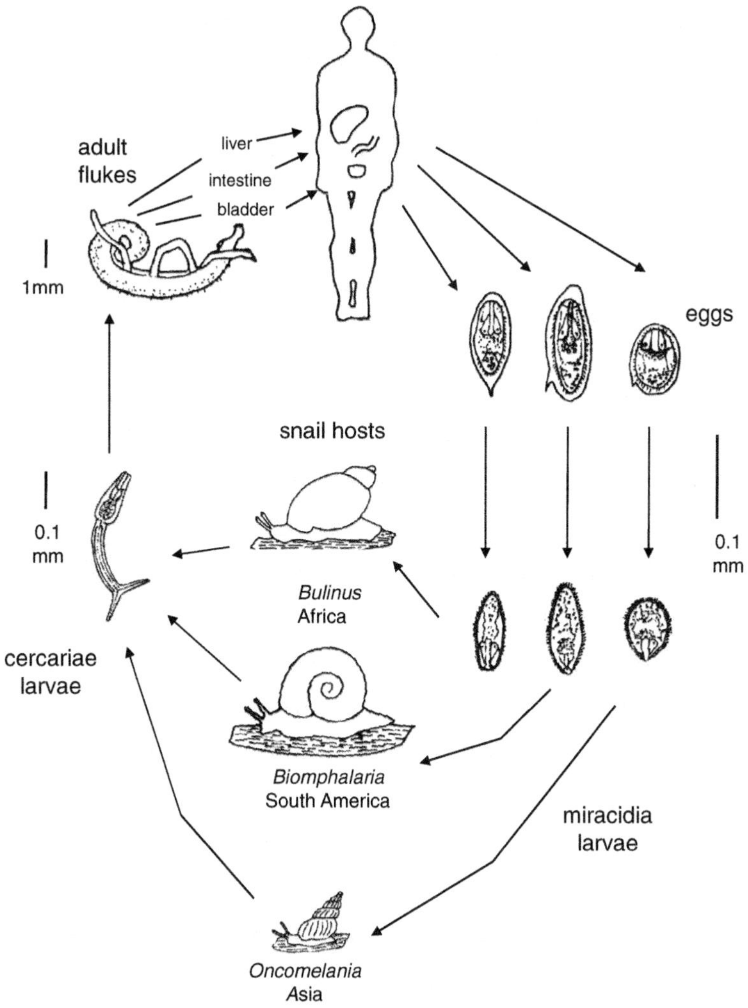

Fig. 7.7 The life cycle of the various *Schistosoma* species showing typical snail hosts

Fig. 7.8 (**a**) An irrigation scheme posing a Schistosome hazard because it serves as a human water source, receives faecal material and harbours infected snail hosts, (**b**) snails grazing algae along the water line of an irrigation channel

intestines or bladder and the transmission sequence is initiated when human faeces or urine containing parasite eggs are allowed to pass into water. The eggs first hatch into *miracidia* larvae that enter the tissues of freshwater snails where they undergo an additional multiplicative phase. This process generates large numbers of free-swimming infective *cercariae* larvae, with each snail being capable of producing as many as 2000 of these daily. The larvae are programmed to penetrate the skin of persons in contact with water and they use a combination of mechanical and biochemical means to achieve this. After 24 h the parasites will have entered the blood stream and be in transit to the various parts of the circulatory system where they mature into adults. Schistosome eggs are spiked, with the position of the spike varying according to species. Possibly this feature evolved to allow eggs to pass through blood vessel walls into the lumen of the bladder and alimentary canal and thence to the outside environment. It is also responsible for some of the more serious symptoms of the disease. Eggs that have failed to become discharged to the exterior circulate in the blood stream and are prone to become lodged in small capillary vessels. The body treats such eggs as foreign bodies and proceeds to encapsulate them. Unfortunately the nodules so formed can cause serious problems by interfering physically with the functioning of vital organs such as the bladder and the brain (Gray et al. 2011).

This is a complex lifecycle and, perhaps unsurprisingly, local knowledge about it tends to be incomplete. A survey from Zimbabwe revealed that 70% of respondents were either unable to offer an explanation of how the disease was contracted or believed it to be caused by drinking dirty water (Taylor et al. 1987). Only 24% came anywhere near the truth in thinking that it was associated with a particular pond or river. A later post-educational study yielded only marginally better results; 11.8% of the people asked, now appreciated the connection with snails, but none could give any specific details (Ndamba et al. 1989).

Even where progress has been made in communicating the idea of snail involvement, misleading traditional ideas still tend to persist. A fairly recent survey from Nigeria showed that whilst a quarter of the respondents correctly made the connection with water, the explanations favoured by the remainder included food and blood poisoning and the action of evil spirits (Sarkinfada et al. 2009). In another Nigerian study, only 18% of the respondents knew about the link with water and only 6% were aware of the significance of snails. The remainder continued to believe variously that the problem was caused by drinking dirty water, being bitten by mosquitoes or consuming unripe fruit (Houmsou et al. 2010).

Again in criticizing this local lack of local awareness, we would do well to remind ourselves that Westerners have often proved equally ill-informed about this particular hazard. Tourists visiting Africa frequently contract Schistosomiasis by unwittingly swimming in dangerous waters (Fuller et al. 1979; Zuidema 1981). It is also the case that European dam builders in Africa frequently ignored the problem, with the result that their constructions have often triggered significant increases in disease incidence (Stanley and Alpers 1975; Edington and Edington 1977). This unfortunate outcome could have been avoided had the main engineering works been accompanied by schemes for improved water-supply and sanitation.

7.3.2 *The Use of Human Wastes as Fishpond Fertilizers*

Damaging parasitic life-cycle progressions are likewise initiated when raw human wastes are used to fertilize fishponds. Often this is a matter of village people collaborating with urban residents to collect faecal material or "night-soil" that is then

Fig. 7.9 Aspects of Asian fishpond operations (a) Latrine discharging directly into a fishpond, (b) netting fish from a pond, (c) a good catch of grass carp (Photographs by courtesy of Shai-wai Sim)

deposited in the ponds. The problems can equally arise when villagers arrange for their own latrines to be perched above the ponds so as to discharge directly into them (Fig. 7.9a).

In many ways Asian multi-species fishponds are marvels of biological ingenuity (Fig. 7.10). The various carp species used to stock them are carefully selected to complement one another's feeding habits (Ruddle et al. 1983; Zweig 1985) and enrichment with human faecal material has the effect of increasing overall productivity. In these "mix and match" fisheries, the plant nutrients released by bacterial decomposition of organic matter serve to stimulate the growth of waterweeds and microscopic algae, which in turn provide food for grass carp (*Ctenopharyngodon idella*) and silver carp (*Hypophthalmichthys molitrix*) respectively. The algae also support populations of small planktonic animals such as water "fleas" that are eaten by bighead carp (*Aristichthys nobilis*). Finally, organic detritus sinking to the pond bottom supports various populations of invertebrate animals, which in turn serve as prey for species such as the common carp (*Cyprinus carpio*), the mud carp (*Cirrhinus moitorella*) and the black carp (*Mylopharyngodon piceus*), the last-mentioned being a specialized mollusc feeder. The grubbing of these species in the pond bottom has the additional benefit of returning nutrients back into the water. Viewed overall this is an artificial ecosystem cleverly devised to maximize fish production and in these terms it is conspicuously successful (Fig. 7.9b, c).

Unfortunately, parasitic organisms also readily become established in these pond food chains. Eggs of various parasitic worms are unwittingly introduced into the system via human faeces (sometimes from pig-manure too) and on hatching, start cycles of development that involve freshwater snails as intermediate hosts. In the later phase of these cycles, infective cysts develop (according to species) on the scales of fish, on the bodies of freshwater crabs or on the surfaces of aquatic fruits (Fig. 7.11). It is through the ingestion of these cysts on foodstuffs that human infections arise.

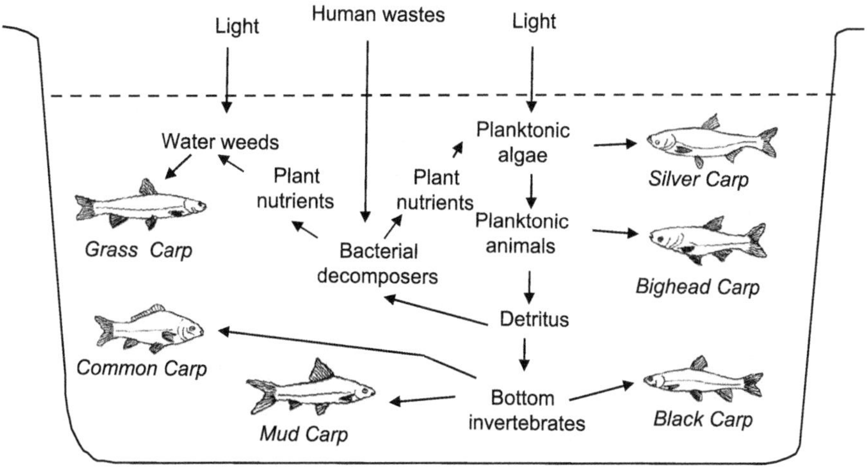

Fig. 7.10 Typical mix of carp species in a Chinese fishpond, indicating the passage of energy and materials through the food chain (Developed from Bardach et al. (1972) and Zweig (Zweig 1985))

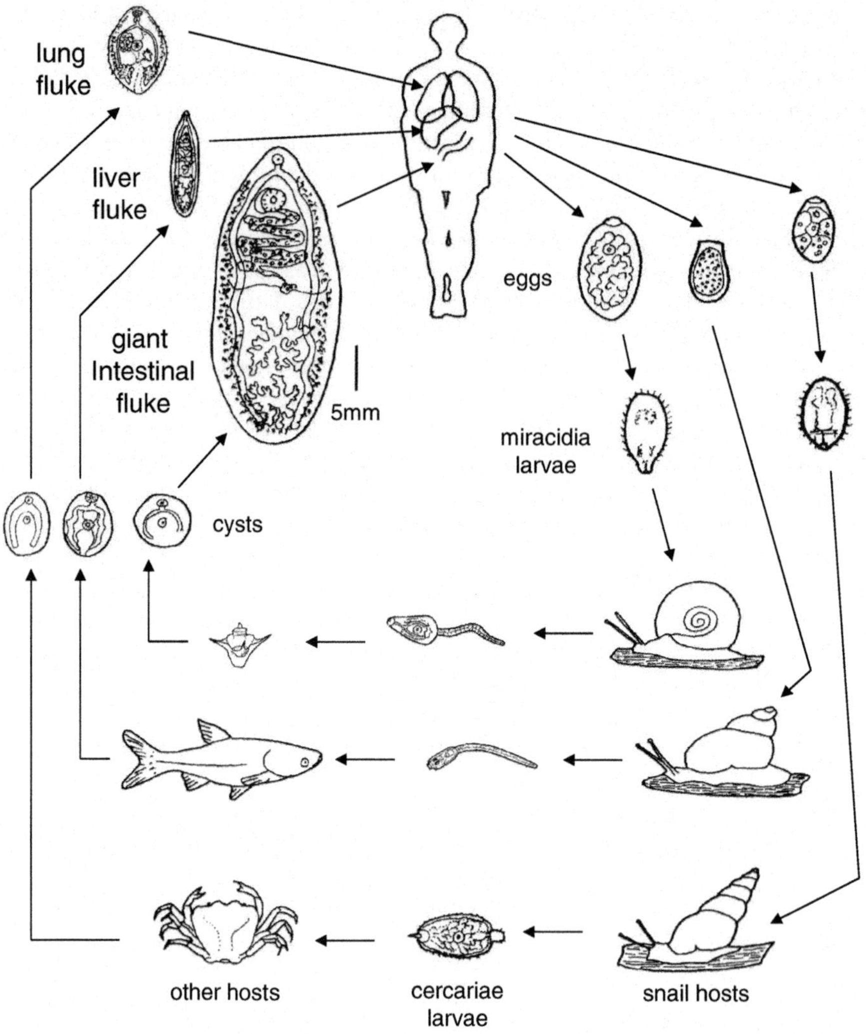

Fig. 7.11 Parasitic worm infections associated with the fertilization of fishponds with untreated human wastes

The Chinese liver fluke (*Clonorchis sinensis*) is contracted by eating parasitized and inadequately cooked fish; the lung fluke (*Paragonimus westermani*) by eating uncooked crabmeat and the giant intestinal fluke (*Fasciolopsis buski*) by consuming raw fruits of aquatic plants such as the water caltrop (*Trapa natans*). A favoured traditional way of eating fish is to dip raw pieces into hot rice soup, a process that unfortunately leaves the parasite cysts intact. The stickiness of the cysts is in all cases an aggravating factor. Besides facilitating their immediate ingestion, it causes them to become attached to knives and chopping boards and thereby to be transferred to foods previously free from contamination.

Japanese parasitologists working between 1914 and 1921 accurately described the life cycles of all these species (Kean et al. 1978) and in Japan, the introduction of strict controls on the use of untreated faecal matter has virtually eliminated the problems (WHO 1995). Similar progress has been evident in the Republic of Korea (Hong et al. 2006), but much less so in China despite some strenuous government-inspired initiatives (World Bank 1995). Major stumbling blocks include the veneration accorded to time-hallowed practices and a continued adherence to the mystical notion that consuming raw animal flesh increases one's vitality. It is for this reason that current educational programmes are being directed mainly at young people on the grounds that their culinary habits are likely to be less entrenched than those of their elders (Lun et al. 2005).

7.3.3 Human Wastes Employed as Crop Fertilizers

Applying animal manures to cropland is a procedure that has been going on at least since Roman times and one that creates relatively few problems. Using human wastes (or night-soil) for this same purpose is a more questionable practice. Nonetheless the rural inhabitants of some Asian countries have followed this latter tradition for centuries and continue to do so (McGarry 1976; Feachem et al. 1983; Bray 1984; Ling et al. 1993). In the absence of any form of pre-treatment of the wastes the practice generates significant health hazards.

Vegetables destined to be consumed uncooked are liable to become contaminated with pathogens, not just protozoa, bacteria and viruses, but also the infective eggs of parasitic worms. Two important species in this latter category are the large roundworm *Ascaris lumbricoides* and the smaller whipworm *Trichuris trichiura* (Fig. 7.12). Washing contaminated vegetables offers no certain way of removing the parasite eggs. Some inevitably remain attached to plant surfaces, whilst others become trapped in stem crevices and within folded leaves. The threat is further increased by the fact that the eggs can remain viable in the soil for a year or more (*Ascaris* has particularly robust eggs) and are likely to become regularly transferred to crops during digging operations. Not surprisingly in these circumstances, community infection rates for these two worms can reach 60% and 40% respectively (Feachem et al. 1983).

There is one further twist to the story. Some worm parasites have life cycles that follow a somewhat different course. Their eggs hatch prematurely in the soil and liberate active larvae that move freely in the damp conditions prevailing there. These larvae can gain access to new human hosts by boring through the skin, usually that of inadequately-shod feet. The parasites are, in effect, actively seeking out their targets. Important species in this category include the hookworms *Ancylostoma duodenale* and *Necator americanus* and also *Strongyloides stercoralis,* a smaller species but one with a similar lifecycle (Fig. 7.12). Because the larvae of all these

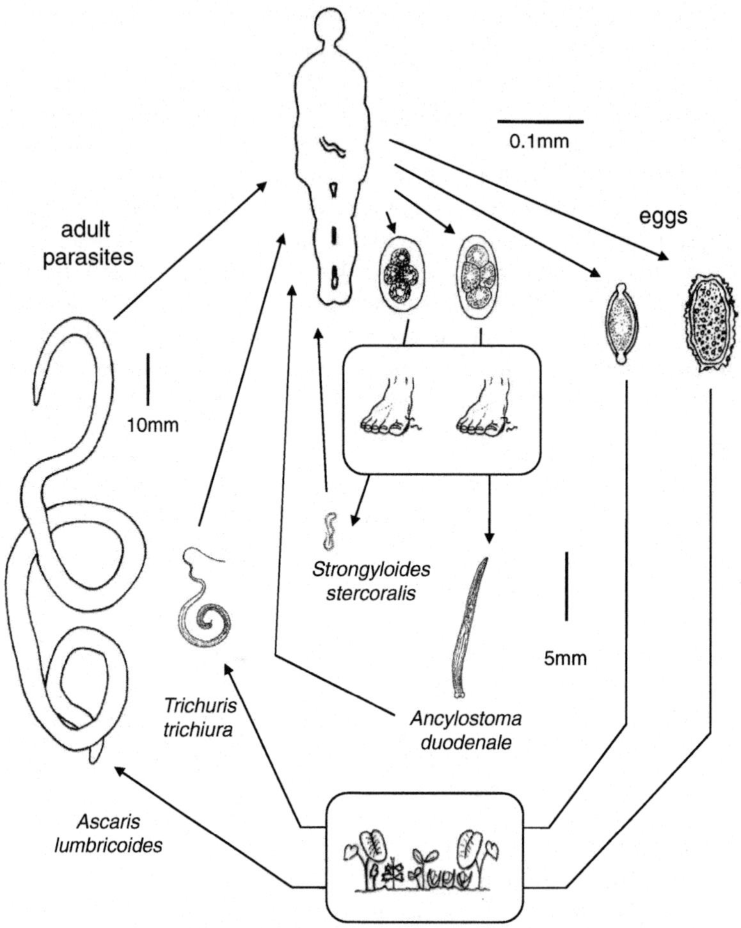

Fig. 7.12 Parasitic worms whose transmission is facilitated by the application of "nightsoil" to crop areas

species can persist in damp soil for upwards of 16 weeks, they represent a significant ongoing hazard for farm workers. The parasites' eventual destination is the human intestine and once established there they settle down to suck blood and to chew on the intestinal lining. Because they contrive for their own protection to neutralize the host's digestive enzymes they cause, not only anaemia through blood-loss, but also for malnutrition (Ash et al. 1984; Stephenson and Holland 1987). One survey from China has shown that 95% of farmers using untreated nightsoil suffer from these infections. Whilst the local claim that the nightsoil fertilization makes vegetables tastier may or may not have substance, the practice undoubtedly generates needless health hazards (World Bank 1995).

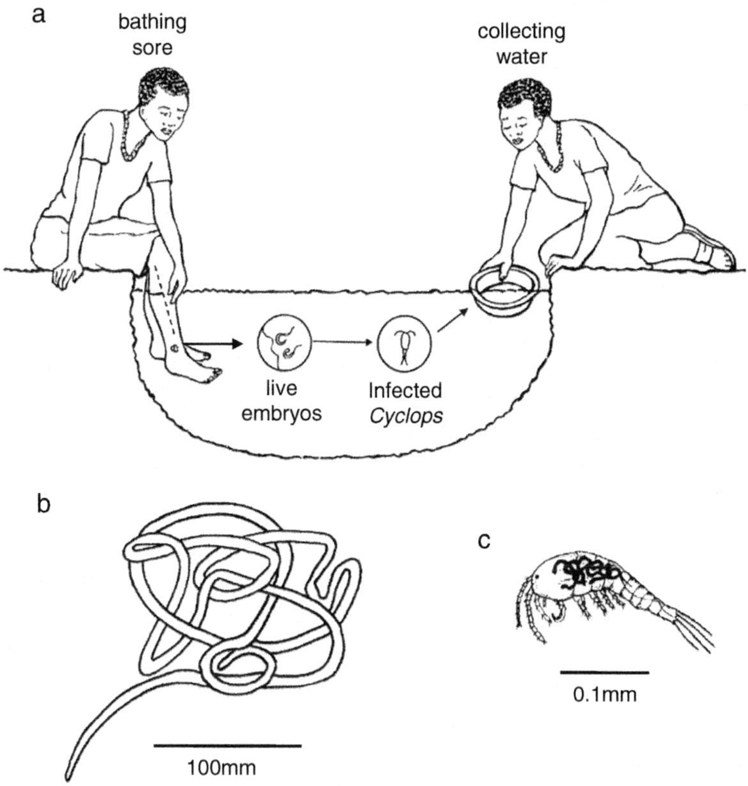

Fig. 7.13 The Guinea worm. (**a**) Infection route, (**b**) adult female worm, (**c**) *Cyclops* carrying infective embryos (Drawing by Mary I. Anderson)

7.3.4 Guinea Worm Infections

Guinea worm (*Dracunculus medinensis*) infections arise in a rather unusual way. The large adult worms (often as long as 600 mm) typically lodge in the soft tissues of the leg and because their presence causes intense irrigation it has been a common local practice to bathe the infected limb in the water of a local well or pond. This works greatly to the parasite's advantage by allowing infective embryos to be liberated into the water; a single worm being estimated to release as many as 3 million of these during its lifetime. These embryos do not infect new human hosts directly, but only after they have been ingested by small swimming crustaceans called

Cyclops. It is the act of swallowing these infected *Cyclops* in drinking water that triggers the human infection (Fig. 7.13). One of the difficulties in controlling the problem has again been the poor local appreciation of its true nature. Often the worm has been perceived as a normal part of the body such as a tendon, or failing this, as a consequence of witchcraft (Bierlich 1995). Fortunately as will be seen (p.172) this parasite is now being brought substantially under control.

7.3.5 Mosquito-Borne Diseases

Finally we come to an important category of health hazards that are water-based only in the sense that the insects carrying them complete their development in this medium. These are the diseases for which mosquitoes serve as carriers or "vectors". The *Anopheles* species transmitting malaria prefer to deposit their eggs in fairly large sunlit bodies of water. By contrast *Aedes* mosquitoes (the vectors of dengue fever, yellow fever and zika virus infections) commonly select small water accumulations of the kind found in storage pots, and in their turn *Culex* mosquitoes (the vectors of filariasis and Japanese encephalitis) are attracted to organically-enriched aquatic sites (Mattingly 1969). Unfortunately, village communities often inadvertently manage their water resources in ways that favour the populations of these disease carriers rather than limiting them. The practice of storing water in open-topped vessels is a case in point. These vessels usually take the form of earthenware pots which, because they lack effective covers, offer convenient breeding sites for *Aedes* mosquitoes. Excluding the mosquitoes is not an impossible task, but one unlikely to be undertaken unless the benefits of doing so are appreciated. Investigators in Northern Nigeria found that although pottery fragments were sometimes placed across the mouths of pots, the intention was not to exclude mosquitoes but to prevent contamination from rat and pigeon droppings, indeed because the makeshift covers often took the form of dished pottery fragments, they themselves trapped water and offered the mosquitoes yet more places to breed. On a much larger scale, the flooding of fields for rice cultivation creates precisely the open sunlight conditions favoured by *Anopheles* mosquitoes. Furthermore, where sections of the fields have been fertilized, this produces the organically-enriched conditions preferred by *Culex* mosquitoes (Wada 1988).

It can by no means be assumed that the true nature of these connections is correctly perceived at a local level. Even the fundamental link between mosquitoes and disease is not invariably made. For example in a survey from Ghana, it was revealed that 77% of respondents considered malaria to result from working long hours in the sun or of spending too much time alongside the cooking fire (Agyepong 1992). This is another manifestation of the "hot" and "cold" concept (p.135). Indeed a minority view was that it was an excess of cold rather than heat that was the problem making it inadvisable to walk barefoot in the rain or drink cold water after working in the hot sun. Only 4% of the respondents correctly believed mosquitoes to be involved. Similar "hot" and "cold" explanations for malaria have been recorded elsewhere in West Africa amongst the *Yoruba* in Nigeria and the *Mende* in Sierra Leone (Adetunji 1991).

Even where explanations involving mosquitoes have started to be accepted as the result of educational programmes, they still tend to run alongside misleading local theories, rather than displacing them altogether. In Ethiopia the unfounded theories invoke such causes as hunger, exposure to unhygienic surroundings and an excessive consumption of *eshet*, a local food prepared from fried green maize shoots (Deressa and Ali 2009). In Guatemala, whilst it is now locally accepted in many villages that malaria can be transmitted to humans by mosquito bites, 75% of the respondents believed that the mosquitoes themselves picked up the disease by biting snakes and frogs or by drinking dirty water (Ruebush et al. 1992). Another standby explanation is that over-frequent bathing constitutes a risk; a not implausible proposition because the regular exposure of bare skin during bathing would increase the chances of being bitten. Similar local misconceptions exist about the causes of yellow fever. In a survey from Tanzania in which people were questioned, 58% of the respondents could offer no explanation for the disease, or only one based on witchcraft or "cold" influences (Winch et al. 1996). This was the result obtained even when the question was posed in the leading form "can this illness be caused by mosquitoes?"

In fairness, it is again worth reminding ourselves that it was only relatively recently that Western confusion about such matters started to be dispelled. *Miasma*-based theories of malaria persisted until the end of the nineteenth century and it was not until 1898 that Ronald Ross working in India was able to demonstrate, not only that the malarial pathogens are injected in saliva when the mosquito takes a blood feed from a human subject, but also that the parasite multiplies within the body of the insect (Harrison 1978).

7.4 Remedial Strategies

7.4.1 The Application of Technology

Well-established technical solutions are available to address most water supply and waste disposal issues. Techniques available for protecting drinking water from faecal pollution include capping wells with concrete covers, enclosing springs in box-like structures and, where water cannot be protected at source, constructing sand filters to cleanse it subsequently. In its turn, schistosomiasis can be tackled by providing villages with pit latrines and piped water supplies, strategies that prevent the parasites from becoming established in local snail populations in the first place. The infection cycle can also be broken using mass drug treatments.

Various forms of pre-treatment can be applied to faecal material destined for use on farm plots or in fishponds (Mara and Cairncross 1989). The options include stabilization ponds, composting heaps, biogas digesters and double-vault pit latrines (Fig. 7.14). The last-named structures are designed to operate on a two-year cycle with one half in active use whilst the other performs its compositing and sanitizing

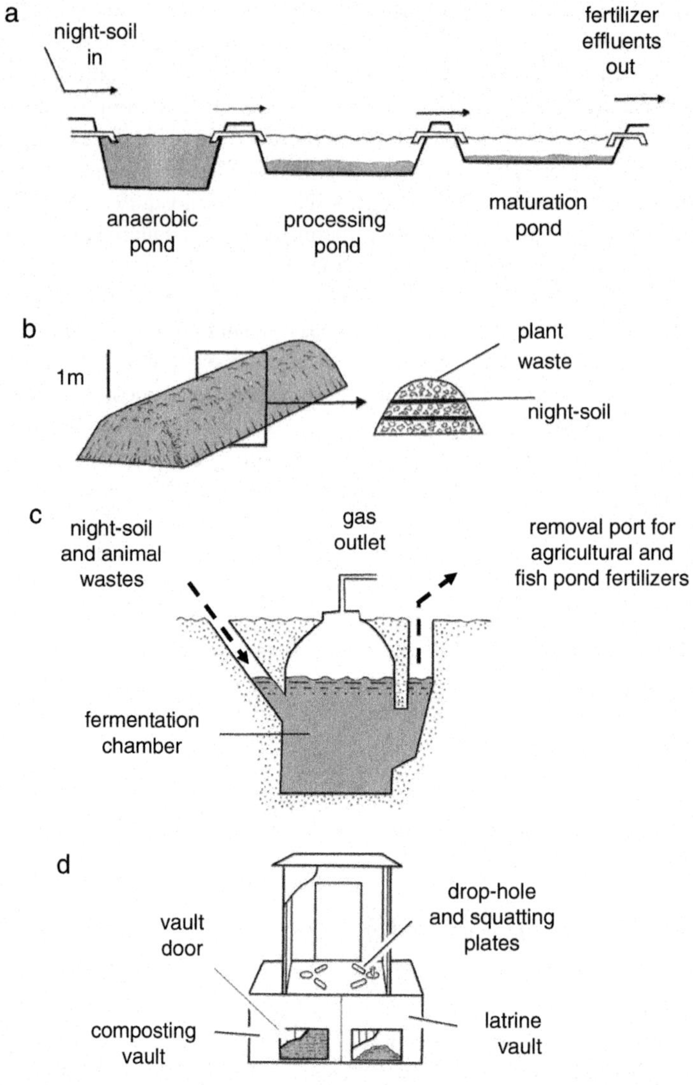

Fig. 7.14 Techniques for sanitizing human wastes prior to use. (**a**) Stabilization ponds, (**b**) composting windrow, (**c**) biogas digester, (**d**) double vault pit-latrine

function. In China, some communities have wisely adopted the practice of retaining night-soil in holding tanks prior to use, the intention being to reduce the numbers of pathogens present in the waste (Pluknett and Beemer 1981). All these techniques rely on the fact that bacterial decomposition increases temperatures and/or reduces oxygen levels in the wastes beyond levels that pathogens can tolerate (Table 7.2). Of the four methods, biogas digestion is probably the least reliable way of killing parasite eggs because retention times tend to be too short (5–30 days) and temperature

Table 7.2 Techniques for sterilizing human wastes prior to their use as crop or fishpond fertilizers

Treatment type	Temperature elevation	Oxygen depletion	Minimum conditions for pathogen destruction	Reference
Stabilizing ponds		✓	Ambient temperature for 40 days	Mara (1977)
Outdoor composting	✓		55 °C for 3 h	Feachem et al. (1983)
Biodigestion	✓	✓	45 °C for 30 days	Feachem et al. (1983)
Latrine composting	✓	✓	Ambient temperature + 2–6 °C for 60 days	McMichael (1978)

increases too modest (often failing to exceed 35 °C) (McGarry and Stainforth 1978). Recently in China there has been a concerted effort to address the problem by employing fermentation tanks to pre-treat materials destined for the digesters (Lu et al. 1983).

In the matter of excluding *Aedes* mosquitoes from water pots, something as simple as an elasticized cover (akin to a shower cap) stretched over the pot mouth will usually do the trick. Restricting the breeding of *Anopheles* mosquitoes in rice fields requires a rather grander strategy, but is still achievable. The technique is to modify the water-management regime of the rice paddies so as to create alternating periods of flooding and drying. Typically the cycle interval is of the order of 5 days, with the process being repeated some 20 times over the course of a growing season. This procedure significantly reduces breeding levels both of the *Anopheles* mosquitoes responsible for malaria and the *Culex* mosquitoes that transmit Japanese encephalitis (Mather and Trinh 1984; IRRI 1988). Perhaps surprisingly, it appears to have no adverse effects on crop production.

7.4.2 Awareness Raising

Unfortunately, it cannot be assumed that technology alone will guarantee success. If useful outcomes are to be achieved, an awareness-raising element usually needs to be part of the mix (Roundy 1985; Kolsky 1993). In Ethiopia, I examined a situation where a recently installed hand pump had ceased to function and could not readily be repaired. Desperate to get access to the water, the villagers had broken open the protective concrete cap of the well and, being apparently unaware of the dangers of contamination, had started to lower a miscellany of containers into it (Fig. 7.15). As many of these vessels were soiled with animal and possibly human faeces too, the bacteriological purity of the supply had certainly been compromised. Had the well-users appreciated the risks, they might have adopted a more hygienic approach. Even with the hand-pump out of action, they could have used a single bucket attached to a homemade windlass to withdraw water from the well. This bucket

Fig. 7.15 The unhygienic use of a damaged well in Ethiopia

could then have been used to fill other containers, but to have been itself kept clean, for example by being hung on the windlass handle.

In parts of rural Africa traditional beliefs have often seriously impeded the adoption of pit-latrines. In some regions it is believed that foul air arising from the latrine pit can enter the intestine and cause stomach pains (Bomba 1982). Another unhelpful notion was that latrines were dangerous places because they render users vulnerable to witchcraft. The underlying reasoning was that witches seeking to gain control over their victims habitually do so through such subterfuges as obtaining threads from the targeted person's clothes, locks of their hair, or in the case of pit latrines by using the same toilet facility (Mead 1953; Gelfand 1964). From this standpoint, defaecating at random in the bush was seen to have the merit of confusing the witches by introducing an element of unpredictability into the process. Residual notions of this kind need to be firmly countered, if improved facilities are to yield their intended benefits.

Some aid agencies can be congratulated for having taken this point fully on board. For example, the United Nations Children's Fund (UNICEF) is to be commended for insisting that schemes for drilling wells and installing latrines should be accompanied by health-training programmes. These are directed at receptive community members, who are then charged with passing on the messages to the rest of the village (Black 1983). In some parts of Africa, community health clubs have been formed to serve in this capacity, each weekly meeting covering a particular health

Fig. 7.16 Combating the Guinea worm hazard by passing drinking water through a fabric filter designed to remove infected *Cyclops* (Drawing by Mary I. Anderson)

theme. One programme is credited with increasing safe faecal disposal (either through latrine use or faecal burial) from 2% to 100% (Waterkeyn and Cairncross 2005).

The success of the campaign to eradicate Guinea worm from most of its former strongholds in Africa offers another striking example of what can be achieved by combining simple technology with awareness-raising. An educational programme mounted from the Carter Centre (an organization associated with the former US president), has given village communities a clearer understanding of the causes of the problem and the means available to counter it. One of these is to remove infected *Cyclops* from drinking water by filtering it through a nylon-mesh (Fig. 7.16). The other is to discourage sufferers with worms lodged in their legs from bathing the affected limbs in wells that are also used for drinking. By these means, the transmission cycle can be brought to a halt and it has proved possible to eliminate the parasite from its former strongholds in Nigeria, Ghana and Uganda, although not yet from those in South Sudan (Miri et al. 2010).

In the overall scheme of things, it is regrettable that the high levels of expertise village communities bring to such operations as growing crops, designing houses, managing firewood and manipulating volumes of water, do not extend in equal measure to managing waste and water to minimize health risks. Consequently significant blind spots persist and remedial interventions are likely to be needed for some years to come.

References

Adetunji JA (1991) Response of parents to five killer diseases among parents in a Yoruba community, Nigeria. Soc Sci Med 32:1379–1387
Agarwal A, Narain S (eds) (1997) Dying wisdom. Rise, fall and potential of India's traditional water harvesting systems. Centre for Science and Environment, New Delhi
Agyepong IA (1992) Malaria: ethnomedical perceptions and practice in an Adangbe farming community and implications for control. Soc Sci Med 35:31–37
Ash C, Crompton D, Keymer A (1984) Nature's unfair food tax. New Sci 8:17–20
Bardach JE, Ryther JH, McLarney WO (1972) Aquaculture: the farming and husbandry of freshwater and marine organisms. John Wiley, New York
Bierlich B (1995) Notions and treatment of Guinea worm in Northern Ghana. Soc Sci Med 41:501–509
Biswas AK (1970) History of hydrology. North Holland Publishing Company, Amsterdam
Black M (1983) Spreading the good news about water and sanitation. UNICEF News 116:13–24
Bomba WG (1982) Sanitation aspects of water supply and excreta disposal. In: Schiller EJ, Droste RL (eds) Water supply and sanitation in developing countries. Ann Arbor Science Publishers, Ann Arbor, pp 311–325
Bray F (1984) Joseph Needham, science and civilisation in China, vol 6, Biology and biological technology, Part II, Agriculture. Cambridge University Press, Cambridge
Bromehead CEN (1942) The early history of water-supply. Geogr J 99(141–151):183–196
Codrington HW (1947) A short history of Ceylon. Macmillan, London
Deressa W, Ali A (2009) Malaria-related perceptions and practices of women with children under the age of five years in Ethiopia. Biomed Centr Public Health 9:259
Edington JM, Edington MA (1977) Ecology and environmental planning. Chapman and Hall, London
Escobar GJ, Salazar E, Chuy M (1983) Beliefs regarding the etiology and treatment of infantile diarrhea in Lima, Peru. Soc Sci Med 17:1257–1269
Feachem RG, Bradley DJ, Garelick H, Mara DD (1983) Sanitation and disease: health aspects of excreta and wastewater management. John Wiley, Chichester
Fuller GK, Lemma A, Haile T (1979) Schistosomiasis in Omo National Park of Southwest Ethiopia. Am J Trop Med Hyg 28:526–530
Garbrecht G (1983) Ancient water works-lessons from history. Impact Sci Soc 1983(1):5–16
Gelfand M (1964) Medicine and custom in Africa. Livingstone, Edinburgh
Gray DJ, Ross AG, Li YS, McManus DP (2011) Diagnosis and management of schistosomiasis. Br Med J 342:2651
Hanlon J (1983) Water comes to the "place of the wasps". UNICEF News 116:5–7
Harrison G (1978) Mosquitoes, malaria and man: a history of the hostilities since 1880. John Murray, London
Hempel S (2007) The medical detective. John Snow, cholera and the mystery of the broad street pump. Granta Books, London
Hong ST, Chai JY, Huh S, Rim HJ, Lee SH (2006) A successful experience of soil-transmitted helminth control in the Republic of Korea. Korean J Parasitol 44(3):177–185
Houmsou RS, Kela SL, Suleiman MM, Ogidi JA (2010) Perceptions and assessment of risk factors in *Schistosoma haematobium* infection in Buruku and Katsina-Ala local government areas of Benue state-Nigeria. Int J Infect Dis 8(1):1–12
IRRI (1988) Vector-borne disease control in humans through rice agroecosystem management. International Rice Research Institute, Manila
Kean BH, Mott KE, Russell AJ (1978) Tropical medicine and parasitology. Classic investigations. Cornell University Press, Ithaca
Kolsky PJ (1993) Water, sanitation and diarrhoea: the limits of understanding. Trans R Soc Trop Med Hyg 87(S3):43–46

Leach ER (1968) Pul Eliya. A village in Ceylon. A study of land tenure and kinship. Cambridge University Press, Cambridge

Ling B, Den T, Lu Z, Min L, Wang Z, Yuan A (1993) Use of night soil in agriculture and fish farming. World Health Forum 14:67–70

Locking M, Cowden J (2009) *Escherichia coli 0187*. Br Med J 339:817–818

Lu Z, Qiang G, Dai X (1983) Management of human excreta. World Health Forum 4:279–280

Lun Z-R, Gasser RB, Lai D-H, Li A-X, Zhu X-Q, Yu X-B, Fang Y-Y (2005) Clonorchiasis: a key foodborne zoonosis in China. Lancet Infect Dis 5:31–41

Macfarlane A (1981) Death disease and curing in a Himalayan village. In: von Fürer Haimendorf C (ed) Asian highland societies in anthropological perspective. Sterling Publishers, New Delhi, pp 79–130

Mara D (1977) Wastewater treatment in hot climates. In: Feachem R, McGarry M, Mara D (eds) Water, wastes and health in hot climates. John Wiley, Chichester, pp 264–283

Mara D, Cairncross S (1989) Guidelines for the safe use of wastewater and excreta in agriculture and aquaculture. World Health Organisation, Geneva

Mather TH, Trinh TT (1984) Environmental management for vector control in rice fields. Food and Agriculture Organization of the United Nations, Rome

Mattingly PF (1969) The biology of mosquito-borne disease. George Allen and Unwin, London

McCance RA, El Neil H, El Din N, Widdowson EM, Southgate DAT, Passmore R, Shirling D, Wilkinson RT (1971) The response of normal men and women to changes in their environmental temperatures and ways of life. Philos Trans R Soc Lond B 259:533–561

McFeters GA (ed) (1990) Drinking water microbiology. Springer, New York

McGarry MG (1976) The taboo resource. The use of human excreta in Chinese agriculture. Ecologist 6(4):150–154

McGarry MG, Stainforth J (1978) Compost, fertilizer and biogas production from human and farm wastes in the People's Republic of China. International Development Research Centre, Ottawa

McMichael JK (1978) The double septic bin in Vietnam. In: Pacey A (ed) Sanitation in developing countries. John Wiley, London, pp 110–114

Mead M (ed) (1953) Cultural patterns and technical change. United Nations Educational, Scientific and Cultural Organization, Paris

Miri ES, Hopkins DR, Ruiz-Tiben E, Keana AS, Withers PC, Anagbogu IN, Sadiq LK, Kale OO, Edungbola LD, Braide EI, Ologe JO, Ityonzughul C (2010) Nigeria's triumph: dracunculiasis eradicated. Am J Trop Med Hyg 83(2):215–225

Mohapatra SS (1989) Beliefs of rural mothers about diarrhoea in Orissa, India. Dial Diarrhoea 39:7

Mull JD, Mull DS (1988) Mothers' concepts of childhood diarrhoea in rural Pakistan: what ORT program planners should know. Soc Sci Med 27:53–67

Ndamba J, Chandiwana SK, Makaza N (1989) Knowledge, attitudes and practices among rural communities in relation to *Phytolacca dodecandra* – a plant molluscicide. Soc Sci Med 28:667–669

Nichter M (1985) Drink boiled water: a cultural analysis of a health education message. Soc Sci Med 21:667–669

Patel VL, Eisemon TO, Arocha JF (1988) Causal reasoning and the treatment of diarrhoeal disease by mothers in Kenya. Soc Sci Med 27:1277–1286

Pluknett DL, Beemer HL (eds) (1981) Vegetable farming systems in China. Westview Press, Boulder

Quin PJ (1959) Food and feeding habits of the Pedi. Witwatersrand University Press, Johannesburg

Roundy RW (1985) Clean water provision in rural areas of less developed countries. Soc Sci Med 20:293–300

Ruddle K, Furtado J, Zhong GF, Deng HZ (1983) The mulberry dike-carp pond resource system of the Zhujiang (Pearl River) Delta, People's Republic of China. 1. Environmental context and system overview. Appl Geogr 3:45–62

Ruebush TK, Weller SC, Klein RE (1992) Knowledge and beliefs about malaria on the Pacific coastal plain of Guatemala. Am J Trop Med Hyg 46(4):451–459
Sarkinfada F, Oyebanji A-A, Sadiq IA, Iyasu Z (2009) Urinary schistosomiasis in the Danjarima community in Kano, Nigeria. J Infect Dev Ctries 3(6):452–457
Stamp LD (1969) Man and the land. Collins, London
Stanley NF, Alpers MP (eds) (1975) Lakes and human health. Academic, London
Stapleton MC (1989) Diarrhoeal diseases: perceptions and practices in Nepal. Soc Sci Med 28:593–604
Stephenson LS, Holland C (1987) The impact of helminth infections on human nutrition. Schistosomes and Soil-transmitted Helminths. Taylor & Francis, London
Taylor CE (1951) A medical survey of the Kali Gandak and Pokhara Valleys of Central Nepal. Geogr Rev 41:421–437
Taylor P, Chandiwana SK, Govere JM, Chombo F (1987) Knowledge attitudes and practices in relation to schistosomiasis in a rural community. Soc Sci Med 24:607–611
Ulluwishewa R (1991) Modernization versus sustainability: disintegrating village agro-ecocomplexes in the dry zone of Sri Lanka. Environ Conserv 18(2):103–109
Wada Y (1988) Strategies for control of Japanese encephalitis in rice production systems in developing countries. In: Smith WH (ed) Vector-borne disease control in humans through rice agroecosystem management. International Rice Research Institute, Manila, pp 153–160
Waterkeyn J, Cairncross S (2005) Creating demand for sanitation and hygiene through community health clubs: a cost-effective intervention in two districts in Zimbabwe. Soc Sci Med 61:1958–1970
Weiss MG (1988) Cultural models of diarrheal illness; conceptual framework and review. Soc Sci Med 27:5–16
Wellin E (1955) Water boiling in a Peruvian town. In: Paul BD (ed) Health, culture and community. Russel Sage Foundation, New York, pp 71–103
White AU (1977) Patterns of domestic water use in low-income communities. In: Feachem R, McGarry M, Mara D (eds) Water, wastes and health in hot climates. Wiley, Chichester, pp 96–112
WHO (1985) Guidelines for drinking-water quality, vol 3. Drinking-water quality control in small-community supplies. World Health Organization, Geneva
WHO (1993) Guidelines for Drinking-water quality. World Health Organization, Geneva
WHO (1995). Control of foodborne trematode infections. Technical Report 849, World Health Organization, Geneva
WHO (2002) Guidelines for drinking-water quality. Addendum, microbiological agents in drinking water. World Health Organization, Geneva
Williamson G, Payne WJA (1978) An introduction to animal husbandry in the tropics. Longman, London
Winch PJ, Makemba AM, Kamazima SR, Lurie M, Lwihula GK, Prmji Z, Minjas JN, Shiff CJ (1996) Local terminolgy for febrile illnesses in Bagamoyo District, Tanzania and its impact on the design of a community-based malaria control programme. Soc Sci Med 42:1057–1067
World Bank (1995) China: nightsoil management improvement study in Hubei. Regional water and sanitation group for East Asia and the Pacific environmental sanitation study. World Bank, Washington, DC
Wulff HE (1968) The qanats of Iran. Sci Am 218(4):94–105
Zuidema PJ (1981) The Katayama syndrome; an outbreak in Dutch tourists to the Omo National Park, Ethiopia. Tropic Geogr Magaz 33:30–35
Zweig RD (1985) Freshwater aquaculture in China: ecosystem system management for survival. Ambio 14(2):66–74

Chapter 8
Indigenous Knowledge and the Course of Development

8.1 Quality of Life Shortfalls

Notwithstanding the expertise displayed by traditional communities in such fields as farming, house design and fuel management, such communities rarely achieve the quality of life that they themselves aspire to and which we would wish for on their behalf. This conclusion is confirmed by the annual statistics collected by the World Bank and is also readily apparent from individual village studies. For example, Shepperdson's survey of a village complex in the Indian State of Uttar Pradesh revealed infant mortality levels of 206 per 1000 live births and child (age 1–5 years) mortality rates of 112 per 1000; the equivalent figures for Western countries being 6 and 7 respectively (Shepperdson 1995). He also found certain illnesses, especially those of a gastro-intestinal nature, to be extremely common and the number of working days lost annually to sickness to be five times the British average. In seeking explanations for this general state of affairs, it is important to draw a distinction between negative influences that originate mainly from outside villages and those that are generated from within them (Fig. 8.1).

8.2 External Factors

In tropical situations extreme climates can offer serious impediments to agriculture and whilst it is widely assumed that the wet tropics provide especially favourable conditions for crop-growing, this is not invariably the case. In this sense, the luxuriance of the rain forest is misleading. Whilst heat and water are readily available, forest-zone soils are often seriously deficient in nutrients (Janzen 1973; Harris 1974). This is because their antiquity has allowed ample time for depletion through the leaching effects of torrential rain (Gourou 1958; Morgan 1979). Moreover, even in drier savannah regions, the alternating processes of wetting and drying can cause

J. Edington, *Indigenous Environmental Knowledge*,
DOI 10.1007/978-3-319-62491-4_8

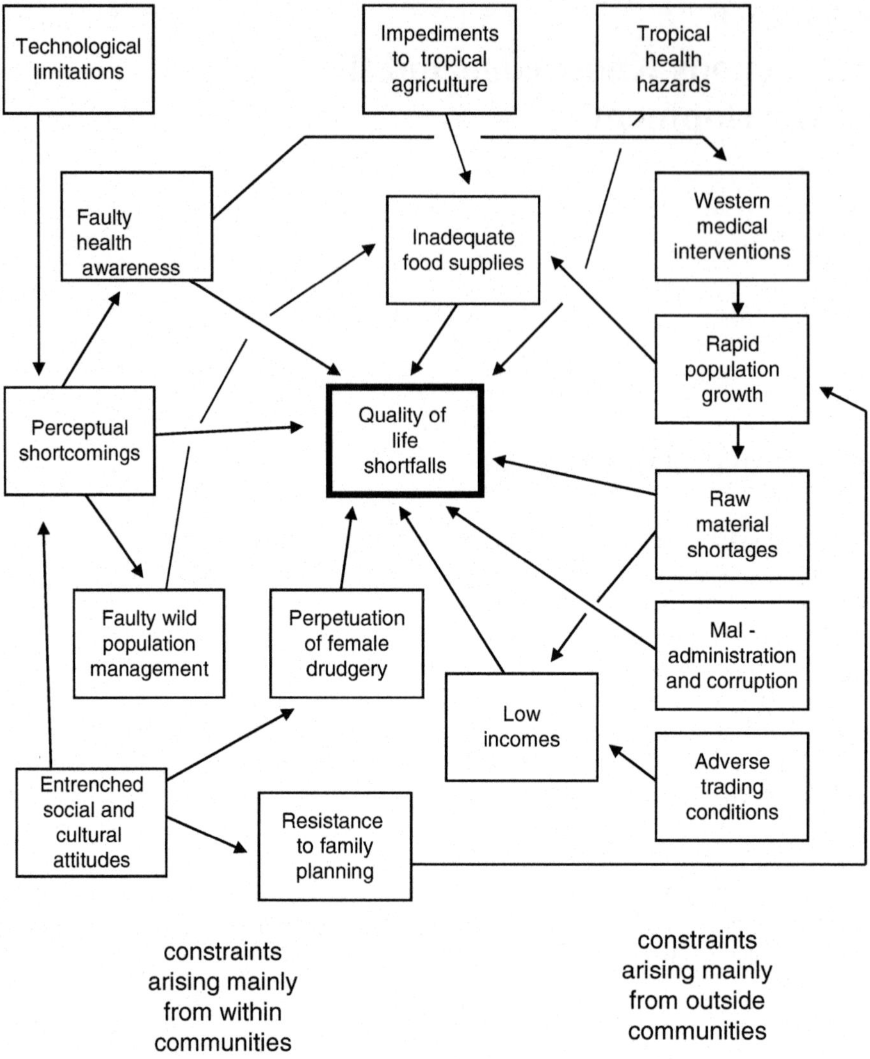

Fig. 8.1 Some of the quality-of-life impairments affecting rural communities in the developing world

soils to be converted into intractable *laterites*. In this progression valuable plant nutrients such as phosphates, potassium and calcium become replaced with agriculturally-useless oxides of iron and aluminium and the soil assumes a virtually unworkable brick-like consistency (McNeil 1964).

It is also the case that tropical climates harbour substantially more infectious diseases than do temperate ones (Table 8.1). The affiliation of these diseases with the tropics stems from a requirement for elevated temperatures on the part of the pathogens and sometimes also by their vectors and intermediate hosts. Thus, with

Table 8.1 Tropical diseases

Strictly tropical	Mainly tropical
Yellow fever	Malaria[a]
Dengue fever	New World leishmaniasis
Sleeping sickness	Bancroftian filariasis
Schistosomiasis	Chagas' disease
Brugian filariasis	
Loiasis	
River blindness	
Guinea worm infections	
Yaws	

Developed from Manson-Bahr and Apted (1982)
[a]Of the 4 (possibly 5) species of malaria only *Plasmodium vivax* can tolerate temperate conditions (Martens et al. 1997; Chin and Welsby 2004)

the mosquito-borne disease, dengue fever, not only does the causative *flavivirus* require temperatures in excess of 12–13 °C to survive inside the body of the *Aedes* mosquito, equally critical is the fact that the mosquito itself can breed only where temperatures exceed 18 °C (Martens et al. 1997). Likewise with schistosomiasis (p.159), elevated temperatures are required both by the free-living larval stages of the parasite and the freshwater snails that act as intermediate hosts (Appleton 1978).

In the days when areas of the tropics, particularly parts of Africa, were dubbed "the white man's grave" threats were perceived to arise not only from infectious diseases but also from hostile physical conditions (Huntington 1915; Stamp 1964). Since that time we have become more sanguine about these postulated physical hazards. Combating heat disorders does not in fact, as was once claimed, necessitate wearing pith helmets and flannel spine pads to protect against the dangerous rays of the sun (Allen 1979). In fact sufficient protection can usually be achieved by consuming sufficient water and salt to allow the body's temperature regulation mechanism, which is based on sweating, to function properly (Manson-Bahr and Apted 1982). Local people are equally capable of taking these precautions and often augment them by living in houses cleverly designed to resist overheating and by sensibly reducing their level of physical activity during the heat of the day. In the words of Noel Coward "only mad dogs and Englishmen go out in the midday sun".

Some external constraints arise, not from physical geography or disease agents, but from external human agencies. Typical difficulties originate from uncertainties about land tenure and there are constant complaints about the failure of governments to deliver promised infra-structural projects or to assist in the launching of trading ventures (Briceño-Garmendia et al. 2004). Whilst some of these shortcomings certainly reflect funding shortages, many are exacerbated by maladministration and corruption (Transparency International 2004, 2006). Misappropriation of funds frequently diverts finance from much needed projects, with consequent negative repercussions on the quality of peoples' lives. Furthermore, if rigged elections prevent redress through the ballot box, the scene is set for coups and uprisings which, whatever their final outcomes, are likely in the immediate term to cause further disruption and hardship for the population at large.

Sometimes external environmental and external human factors combine to aggravate one another. A good illustration is provided by David Wilkie's examination of the vicissitudes of the *Lese* people of the Congo (Wilkie 2005). Not only do they have to cope on a day to day basis with problems of tropical disease and intractable soils, they have also had the misfortune of suffering a succession of external human disturbances. These have included incursions from nineteenth century slave traders, colonial pressures to abandon their traditional agricultural practices and the destructive impacts of a succession of civil wars. The latter have caused the tribe to be regularly attacked by marauding bands, to have their crops plundered and generally to have suffered such serious disruption as to be forced to seek refuge in nearby towns.

8.3 Influences Arising from Within Communities

8.3.1 Faulty Health Awareness

For mainly technical reasons, village communities have a poor appreciation of microscopic entities and the problems they generate. These blind spots relate to pathogens in drinking water (p.155), eggs and cysts of parasitic worms in foodstuffs (p.163) and disease vectors in the general environment (p.167). The neglect of indoor smoke hazards is an instance of a physical hazard being seriously overlooked (p.114). Although the process of trial and error in the field of herbal medicine has allowed some local remedies to be appropriately matched with illnesses, the lack of biochemical screening capabilities and experimental protocols has led to difficulties and the resultant gaps in knowledge have tended to be filled on the basis of spurious classification systems (p.135) or postulations of supernatural interventions (p.138).

Apart from their immediate impacts, these shortcomings have also had the fundamental long-term effect of stimulating population growth; an ostensibly unlikely claim requiring some further explanation. By the beginning of the twentieth century, Western countries had reached the stage of applying their improved understanding of medical matters both at home and abroad. This they proceeded to do and overseas, as would be expected, such inputs were enthusiastically welcomed and, whilst not all problems were entirely solved, mortality rates for many diseases including malaria, diphtheria, measles, whooping cough, tetanus, tuberculosis and smallpox were more than halved (McCord 1978). Unfortunately family-planning provisions, which should logically have accompanied the health interventions, were not promoted with the same vigour and neither, for mainly cultural reasons, were they as readily accepted. The opportunity to reduce death rates and birth rates simultaneously was therefore missed. The net result was that populations all over

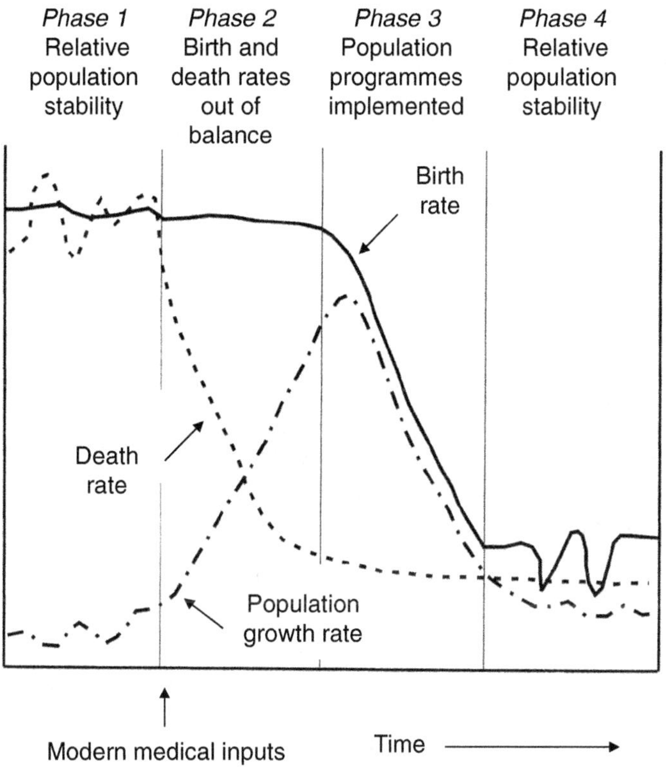

Fig. 8.2 The population consequences of a mismatch between birth and death rates. Stability can be restored by bringing the two parameters back into balance again (Developed from Lenon 1992)

the developing world became stranded part-way through a demographic transition (Clark 1967; King 1990). They had departed from an earlier state where birth and death rates were in approximate balance and had entered into one where the former greatly outstripped the latter (Fig. 8.2). Rapid population growth was the inevitable result.

Weaver and Beckerleg (1993) document the typical course of events in a case study from the Gambia. In a group of villages, a British research unit making nutritional investigations between 1949 and 1991 also undertook to provide medical care and to record its effects. Their ministrations produced spectacular reductions in childhood mortality, with death rates falling from 400 per thousand to only 20. Their efforts also, however, had the effect of stimulating rapid population growth, with numbers more than doubling from 750 to 1614 over the same period. Although family planning services had been offered, this predominantly Muslim community declined to use them; prompting concern amongst the investigators that the population increases would eventually prevent the village from being able to feed itself.

This remains a common state of affairs especially in Sub-Saharan Africa. In most village communities there is a tradition of the men of the household making the key decisions about family size, usually with the aim of maximising the number of children produced. In part, this reflects a logical attempt to compensate for the deaths of children in infancy, but it also seems to stem from a less laudable desire by the men folk to demonstrate their virility and extend their family lineages (Adongo et al. 1997). The usual consequence is that their wives, who may well have access to contraceptive methods and be inclined to use them, are dissuaded from doing so. In only a very few atypical situations can the resultant population surges be comfortably accommodated and even then, not indefinitely. This has been seen to happen in villages rendered unusually prosperous by having lucrative trading links with nearby towns (Tiffen and Mortimore 1994; Tiffen et al. 1994). Otherwise, the general effect of soaring populations has been to cause breakdowns in traditional farming practices, especially ones involving fallow periods and to put additional pressure on natural resources of every kind.

The low social status of village women has some further consequences. It frequently places them in the position of having to undertake a disproportionate share of the physically-demanding chores associated with village life (p.121). The intensity of these activities may not only result in anatomical damage, but can also have a generally erosive effect on the quality of the women's lives.

8.3.2 *Faulty Wild Species Management*

The widespread cultural respect for wild species by traditional communities unfortunately offers no guarantee that these populations will be harvested in a sustainable manner. When Indian bison hunters stampeded entire herds over cliffs (Wheat 1967; Reeves 1983), it was apparently in the belief that the gods would make good any losses by arousing spirit bison lying quiescent in the beds of lakes (Krech 1999, 2000). Much the same attitudes have been revealed in a broader dialogue with various groups of tribal hunters in Latin America. They expressed the view that the depletion of stocks should occasion no alarm because the gods could be relied on to release replacement animals from giant stockyards in the sky (Redford and Mansour 1996). In reality the only way to ensure that exploitation is sustainable, is consciously to adjust its intensity to take account of population sizes and natural recruitment rates (Sutherland 2001; Sinclair et al. 2006). The task is easiest where populations generate an identifiable surplus, say 10% per annum, then this is the amount that can be cropped without threatening the population's long term survival. In an example taken from a modern game-ranching enterprise in Southern Africa, transect surveys were made of adult and juvenile ungulates in a savannah habitat and it was concluded that 525 impala out of a population of 2100 could safely be cropped, and 32 wildebeest out of a population of 160 (Dasmann 1964). The aim of these particular operations was to supplement local meat supplies. Not all situations are quite as straightforward as this one, nor surveys as readily carried out. Some populations are already at the limits of their habitat's carrying capacity and may be producing no identifiable surplus. Then the procedure is to make an initial cull and when the population responds by extra reproduction, to devise a harvesting schedule based either on taking a fixed proportion of the population or the excess above a specified value (Sinclair et al. 2006).

These procedures are rarely adopted by indigenous communities and in consequence many natural populations become seriously damaged by over-harvesting. An instructive recent example is provided by the hunting of white-lipped and chacoan peccaries (*Tayassu pecari, Catagonus wagneri*) in the Argentine Chaco (Altrichter 2005). These pig-like animals are an important source of meat for local people, yet apparently little thought is given locally to limiting hunting to conserve stocks or to adopting such desirable practices as sparing pregnant or suckling females. Consequently the populations are rapidly declining.

Equivalent shortcomings are evident in the indigenous harvesting of some wild plants (Cunningham 2001). Local standing-crop estimates are often wildly inaccurate, a point well-illustrated by a study carried out on a Brazilian rainforest plot. The villagers' estimates of the abundance of *Caryocar villosum* trees, an important source of edible fruits and durable timber, proved to be almost seven times greater than those obtained using scientific census techniques (Shanley et al. 1997). Indeed, in traditional settings there have been relatively few recorded examples of immediate plant-harvesting opportunities being foregone in the interests of long-term sustainability. Such rare instances include the harvesting of wild tubers by Australian aborigines in which fragments are returned to the ground so that they can regenerate (Meggitt 1964) and in Asia the harvesting of palm clumps to provide roof coverings in such a way as to preserve young shoots for regrowth (Brosius 1991; Fong 1992).

Indigenous harvesters are often stimulated to become even more destructive than they might otherwise be, by a combination of external commercial inducements and increased access to modern weaponry and equipment. Numerous historical and recent examples bear testimony to this sequence of events. As well as the North American Indian participation in the excess killing of beavers in the eighteenth century (Innis 1956), mention can be made of the later Amerindian involvement in the trade in spotted-cat furs, the procurement of primates for medical research and of live birds for the pet trade (Lehman 1967; Grimwood 1968; Smith 1976). Comparable examples of damaging exploitation from Asia are provided by the aboriginal harvesting and sale of wildlife artefacts (including stuffed animals and rare mounted butterflies) in Taiwan (Wayre 1969) and of crocodile skins for the leather trade in Papua New Guinea (Bustard 1970). In all these cases, whatever attitudes of respect local people may formerly have entertained towards their wild quarry, these were evidently not proof against powerful outside inducements.

In Northern regions, serious doubts have similarly been expressed about the sustainability of *Inuit* hunting of walrus, narwhal and bowhead whales (Bockstoce 1980; Strickland 1981; Reeves and Mitchell 1981). Situations, where hunting success was originally limited by the use of flimsy skin boats and hand-thrown lances, were radically changed by the ready availability of rifles, explosive harpoons and snowmobiles. Not only did this threaten to make the hunting process more damaging to target populations, it also tended to make it more wasteful. For instance, whilst formerly all parts of a walrus carcase would be utilised, in recent years there have been instances of only the tusks being used and then frequently to produce such trivial trinkets as walrus-ivory swizzle sticks and serviette rings (Strickland 1981). Fortunately, there are recent signs of these particular abuses being brought under control (p.188).

At this point the reader might feel impelled to protest that if indigenous perceptions of health and wild-species management issues are really so badly flawed, how is it that the practices of such communities should have come to enjoy such a favourable popular reputation in Western circles? A possible explanation is that modern observers, wishing to highlight what they regard as excesses in their own societies, have been keen to find positive traditional role models in order to make unflattering comparisons (Whelan 1999).

This certainly seems to be the case in some health matters and is probably also at the root of an on-going controversy about the ability of local communities to manage protected areas unaided. Sociologists mainly concerned with equity issues make a strong case for indigenous peoples being granted a high degree of autonomy in such situations, especially if the land areas in question were previously part of their territory (Colchester 1997). Ecologists on the other hand are much less sanguine about such arrangements, and press for science-based inputs to inform planning and management actions (Spinage 1998).

8.4 Constructive Interventions

8.4.1 New Health Initiatives

In the matter new health initiatives, much benefit can be anticipated from international programmes such as the Global Fund's initiatives on malaria, tuberculosis and HIV/AIDS and the renewed attention being given to hitherto neglected tropical diseases (NTD) (WHO 2009). However, serious problems still remain at the grass-roots level. Diarrhoeal diseases fall into this category and some of the most promising initiatives combine educational programmes with extra sanitary provisions (p.171). An example of a very simple intervention, where latrines have yet to replace faecal burial, is to encourage the placing of forked sticks into the ground from which can be suspended arrays of water containers, excavating hoes and soap bars (Fewtrell et al. 2005). So far, surprisingly little effort has been made to promote the "germ" concept of disease. One of the few serious attempts has been the initiative by *Project Piaxtla* (a Mexican NGO) in which a needle and a magnifying glass are used to illustrate the fact that even at this scale of magnification, significant infective organisms could be present on the needle point yet remain invisible (Werner and Bower 1982).

Another major task is to clarify the role of traditional healers in village health care. Currently, their ubiquity is sometimes offset by their wrong-headedness, particularly when treating such conditions as childhood diarrhoea and febrile convulsions. A radical alternative is to train a body of medical auxiliaries from scratch, with their lower level of training being offset by their greater numbers. This approach was pioneered by the "barefoot doctor" system of China (Cui Weiyuan 2008), and similar programmes have since been developed in Thailand, India, Brazil and Ethiopia. It is accepted that the range of expertise possessed by these auxiliaries will necessarily be limited, but that the systems can be made effective by them being able to seek advice from fully-trained doctors and specialists. Such referrals now frequently involve the use of smart phones, not only to seek verbal advice from their supervisors, but also where appropriate to transmit photographs of their patient's symptoms (especially where skin conditions are involved).

As far as female drudgery is concerned, the introduction of relatively simple labour-saving devices can bring significant benefits. For instance, substituting a rotary grinder for a saddle quern or a pestle and mortar can greatly ease the task of grinding cereals, principally because of the extra momentum imparted by the fly-wheel (Fig. 8.3a). Equally, where women are currently in the position of using hand hoes to till and cultivate the ground, these operations can be rendered much less laborious by the introduction of ox-drawn ploughs (Fig. 8.3b). Even the chore of carrying firewood in hilly terrain could be eased by the erection of aerial ropeways (Fig. 8.3c). In India and Nepal such cableways have been developed for a variety of other purposes, including transporting people and agricultural produce across river gorges. However, as yet, little attempt has been made to employ them for moving firewood down slope, notwithstanding the fact that simple workable hoist designs are available (Parish 1998).

As always, there are social as well as technical dimensions to such issues and frequently one of the problems is to persuade male heads of households to release funds for labour-saving devices. If such entreaties fall on deaf ears, assistance has to be sought from outside sources in the form of international agencies or local non-government organisations (Sandhu and Sandler 1986). In the Cameroon for example a programme has been devised to teach women to substitute ox-drawn ploughs for their hand hoes – learning how to handle the animals being a vital part of the process – and in Zambia an appropriate technology organisation has been set up to inform village women about the various labour-saving devices available to them. However such is the power of cultural inertia that one cannot be very optimistic, at least in the short term, that male household members will spontaneously opt to shoulder their share of the burdens.

Alleviating health problems can also go a long way towards addressing population issues. If mothers can be assured that their existing children will survive, they may feel less impelled to produce more of them. The other issue to be tackled in some regions is the (mainly male-inspired) preference for large families on status grounds. Experience has now shown that this matter can be addressed, firstly by having the means for contraception freely available, and secondly by raising the social and educational status of women (Harrison 1992). When this is done, they almost invariably opt for smaller families. The case of India is instructive. States with high levels of female literacy have consistently shown low fertility rates, and *vice versa*. Thus in the 2001 census, Kerala had a female literacy of 88% and a fer-

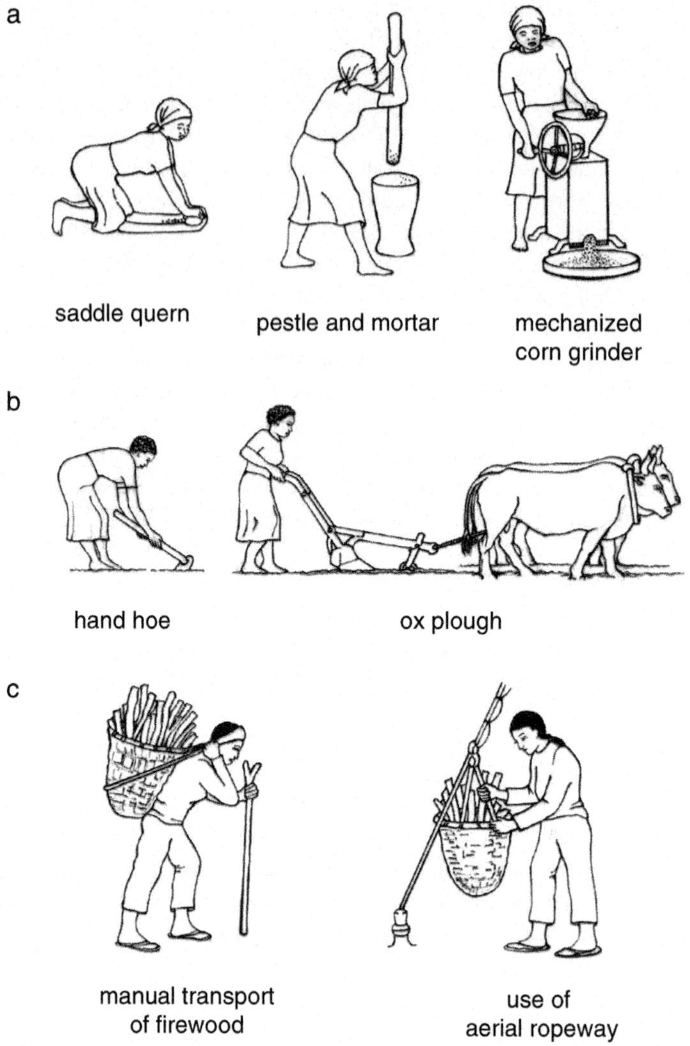

Fig. 8.3 Addressing the problems of female drudgery. (**a**) Replacement of a saddle quern or a pestle and mortar with a rotary corn grinder, (**b**) replacement of a hand hoe with an ox-drawn plough, (**c**) alleviating the burden of carrying firewood by means of an aerial ropeway (Drawings by Mary I. Anderson)

tility rate of only 1.8, whereas in a poorly-performing state such as Bihar the equivalent figures were 35% and 4.8 (Drèze and Sen 2002). Thailand offers a shining example of an effectively-applied population strategy based mainly on increased public awareness, with birth rates falling from 6–7 in 1972, to a replacement level of 2 in 1990 (Hirschman et al. 1994). Indeed most countries (with the notable exception of those in Sub-Saharan Africa) have made good progress, with average births rates worldwide falling from 4.45 in 1970, to 2.45 in 2011. The ultimate aim is for all populations eventually to make the transition to the final phase of the sequence shown in Fig. 8.2 – the one almost universally enjoyed by the Western World – in which birth rates balance death rates with both parameters running at low levels. However, such is the time lag in population systems that nothing will stop world populations rising from the current 7 billion to around 9 billion by 2050, but by then growth rates should have levelled off.

8.4.2 Increasing the Sustainability of Wild-Species Harvesting

One way of improving the sustainability of wild-species harvesting is to combine the tracking and identification skills of local people with inputs derived from scientific ecology. These latter inputs typically involve the definition of safe harvesting levels, but may also include the designation of protected zones and proposals for seasonal hunting bans. Examples of successful integration are provided by the arrangements put in place to harvest down from the nests of eider ducks (*Somateria mollissima*) in the Canadian low Arctic (Nakashima 1993; McDonald and Fleming 1993) and to harvest other duck species for meat in Northern Alaska (Huntington 1992). Even the vexed question of bowhead whales and walruses being wastefully over-hunted in Alaska has now been substantially addressed by setting up a system in which biologists define quota limits and local community associations undertake to operate within them (Huntington 1992).

A recent programme in Papua New Guinea to conserve stocks of grouper fish (*Epinephelus* and *Plectropomus*) has proceeded along similar collaborative lines. It began by first canvassing the views of local fishermen about spawning locations and after these had been confirmed or refuted by professional biologists, went on to define a system of no-fishing zones to be administered by local communities (Hamilton et al. 2001). A comparable success story from South America is provided by the system established for harvesting local mammals for meat in the headwater regions of the Amazon (Constantino et al. 2008). The species in question include white-lipped peccary (*Tayassu pecari*), collared peccary (*Tayassu tajacu*) and brocket deer (*Mazama americama*) and a key part of the process has been to apply a scientific scrutiny to the various hypotheses about game abundance offered by the local *Kaxinawà* people. Important outcomes include: implementing rules against hunting newborns and pregnant females, a prohibition on the use of hunting dogs, except in enclosed garden areas and, very significantly, the creation of reserves in more remote regions. The intention of the latter is to allow excess populations to spill over into more heavily-used areas thereby replenishing them, a little like the spirit bison in the beds of lakes, but with a firmer basis in reality.

The Amazon basin also provides a good working model of the sustainable local utilisation of timber resources. The Mamirauá Sustainable Development Reserve consists mainly of flooded *várzea* forest (Fig. 8.4a). It has previously suffered serious damage from the combined activities of logging contractors and bonded labourers. The new arrangements allow local communities to extract timber for their own use and for commercial sale, but always under the aegis of a sustainable scheme devised by professional foresters from the Mimirauá Institute for Sustainable Development. These arrangements restrict extraction rates to a maximum of 5 trees/hectare over a 25 year period (amounting on average to a removal of less than 1 m^3 of wood/hectare/annum). Additionally, a ban has been placed on felling previously over-exploited species and those serving as important seed sources or supporting key animal populations (Pires 2005; Koziell and Inoue 2006). The trees are generally felled during the dry season when they are more readily accessible, then floated out as rafts as the waters rise again. The main domestic use of timber is for house construction, with large, low-density, logs of *acaçu* (*Hura crepitans*) serving as floating platforms, sawn lengths of *piranheira* (*Piranhea trifoliata*) as wall-planking and trunks of termite-resistant *louro inamuí* (*Octea cymbarum*) as house uprights and cross beams (Fig. 8.4b).

Two other instances of previously unsustainable practices being regularized on the basis of modern knowledge are provided by the harvesting regimes devised for rattans and *ilala* palms. Rattans are climbing forest ferns that provide the raw material for the "cane" furniture popular with Western conservatory owners. Previously, harvesting in Asia took a destructive form with immature shoots being cut prematurely and supporting trees being felled to gain easy access to the climbing stems (Avé 1987; Caldecott 1988). In at least one region, steps have been taken to make operations more sustainable. Following advice from specialists of the International Network for Bamboo and Rattan, harvesting in the *terai* region of Nepal is now carried out on a controlled 5-year cycle, with natural regeneration being augmented with nursery-reared stock (Paudel and Chowdhary 2005).

In southern Africa, *ilala* palms (*Hyphaene coriacea*) are widely used for basket making (Fig. 8.5), but in some areas these operations have run into difficulties. This is because over-cutting has damaged both immature palm shoots and the roots of the bird plum trees (*Berchemia discolor*) providing brown stain for the basket patterns (Cunningham and Milton 1987). Here again a scientific appraisal has shown that sustainability can be achieved by applying the principal of removing only one palm leaf stem in three and by restoring depleted populations through replanting (McKean 2003). Various other instances of appropriate management regimes based on botanical knowledge are documented in Anthony Cunningham's very instructive book *Applied Ethnobotany* (Cunningham 2001).

Fig. 8.4 (**a**) Flooded forest in the Amazonian Mamirauá Sustainable Development Reserve, (**b**) floating house being constructed from local timber

Fig. 8.5 Basket-making enterprises based on the *ilala* palm (*Hyphaene coriacea*). (**a**) Examples of baskets, a Zulu beer-carrying basket in the centre and two *MaShona* produce-carrying baskets on either side, (**b**) mature *ilala* palm tree (**c**) weaving technique in which young palm-leaf fibres are being wound around a core of grass stems

8.5 Augmenting Incomes

8.5.1 Agriculture

Increasing incomes, by whatever means, offers the possibility of addressing life's shortfalls and increasing the sales of agricultural produce has long been seen as the most effective way of doing this. The fact that more food is needed to feed expanding populations adds an extra incentive. Apart from applying the green revolution package in its entirety to achieve these aims, various more specific options are also available. Examples include the introduction of tube wells to increase rice-production in Bangladesh, the liberalisation of seed and fertiliser distribution systems in Kenya and the application of improved techniques to control cassava pests in Sub-Saharan Africa (Spielman and Pandya-Lorch 2009). There is also considerable scope for increasing production by fine-tuning existing traditional techniques (p.33).

Systems involving increased cash-crop production are widely favoured, especially if these can be linked with international market outlets. Such operations are not, however, entirely devoid of pitfalls. World prices of luxury items such as coffee and cocoa can fluctuate widely and farmers who become over-committed to producing them as a sole commodity can run into serious financial difficulties. This is what happened when cocoa prices fell in the 1980s. Growers in countries such as the Ivory Coast found their incomes slashed by more than a half. Moreover, because they had phased out other crops to concentrate on cocoa, the situation caused them considerable hardship, with incomes becoming insufficient to meet basic needs such as clothing, school books and even food. The 2001 fall in coffee prices had similar consequences for growers in Kenya and Mexico.

Also relevant is the phenomenon of "cash-crop malnutrition". This takes the form of diminished nutritional standards amongst the women and children of families heavily involved with cash-cropping (Fleuret and Fleuret 1980; Dewey 1989; DeWalt 1993). Instances have occurred in a variety of agricultural settings, including amongst the families of rice and coffee growers in Kenya, cocoa growers in Nigeria and tea producers in Indonesia (Hughes and Hunter 1970; Haaga et al. 1986; Niemeijer et al. 1988; Husaini et al. 1990). It might be imagined that, whilst converting food growing areas to cash-cropping would inevitably reduce levels of home food production, such deficiencies would be readily made good by wives using the money earned to purchase missing items. The reasons why this frequently fails to happen are sociological in character. Payments for the cash-crops are usually made in lump sums direct to male heads of households and there is commonly a disinclination on their part to make any significant proportion of this money available to their wives (Bryceson 1989; Dewey 1989).

One solution is to reinstate the traditional institution of the women's *home garden* (Niñez 1986). This is usually a plot sited close to the house and dressed with compost made from kitchen wastes. The most effective gardens are ones that are stratified to accommodate plants with complementary growth habits and light requirements, and where species choice is based largely on their capacity to yield essential micro-nutrients (Fig. 8.6). In particular, it is desirable to include plants that contain *carotene*, iron and vitamin C (Pacey 1978; Kuhnlein 2003). As well as providing protection from scurvy, this latter vitamin facilitates the uptake of iron and diminishes the risk of anaemia (Sharma 2003). Recent revelations about the relative inefficiency of the body's mechanism for converting *carotene* to vitamin A (Tang 2010), suggest that there are also advantages in adding at least one animal source to the garden system. Keeping poultry to produce eggs is the most obvious option (Iannotti et al. 2009).

A good example of home gardens doing their job well is to be found amongst small-scale rubber producers in New Guinea (Shack et al. 1990). To avoid the nutritional side-effects of concentrating on a non-edible crop, householders there contrive to obtain much of their food from their gardens, whilst at the same time

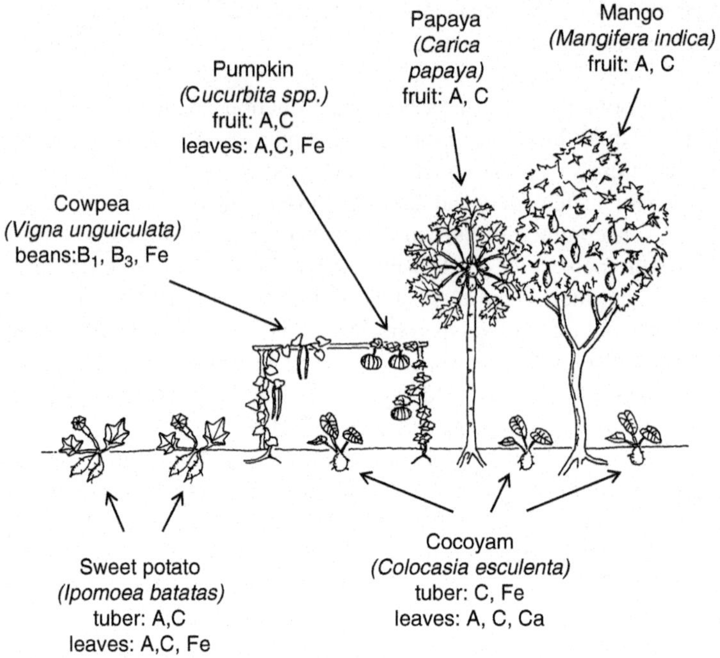

Fig. 8.6 Design for a home garden indicating species in which micro-nutrient contents exceed 25% of daily requirements in a 100 g serving (A = carotene, B_1 = thiamine, B_2 = riboflavin, B_3 = niacin, C = vitamin C, Fe = iron, Ca = calcium)

supplementing these inputs with judicious food purchases. Tinned-fish, for example, are typically brought into service as a suitable protein source. Families following such strategies are noticeably better nourished than those who do not.

In most situations there also remains some scope for reviving the traditional practice of using wild plants as food supplements. Even where the switch to cash-crops has been virtually complete, at least a few edible weeds are likely to survive along field boundaries (Antonsson-Ogle 1984). In Africa Wild amaranth (*Amaranthus hybridus*) is one such species and it can provide useful inputs of carotene, vitamin C, calcium and iron (Oomen and Grubben 1978). Where traditional knowledge about edible wild plants has been lost, it can usually be revived through carefully designed awareness-raising programmes (Opole 1993).

In the matter of coping with trading difficulties, although much faith was once placed in government marketing boards as a means of stabilizing prices, these organisations have generally fallen into disrepute. More attention is now being given to private initiatives, especially ones involving partnerships with sympathetic overseas partners. The *Fairtrade Foundation* is a good example of such a partner. It was set up by a group of charities, and guarantees fair prices for over 1000 items produced by village suppliers. In the Windward Islands of the Eastern Caribbean, for example, the Foundation has brought significant relief to beleagued banana

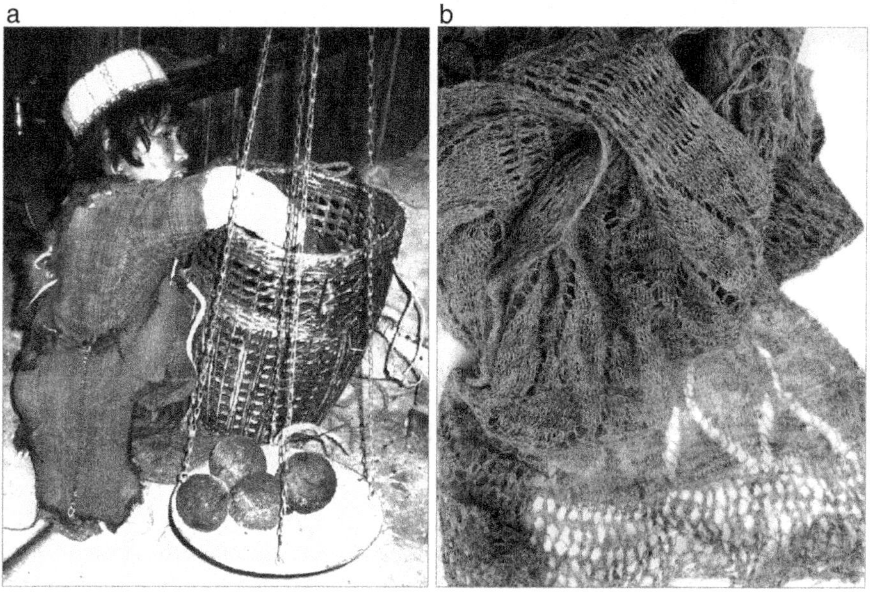

Fig. 8.7 Two examples of unusual locally-traded natural products in Nepal. (**a**) "cannonball" wild tea, (**b**) nettle-fibre scarf

growers, not only by paying them a fair price for their product, but also by raising a fund for community works. Sometimes the collaborative arrangements go very much further. For instance, *Divine Chocolate*, founded in Britain in 1998 numbers amongst its shareholders members of *Kuapa Kokoo,* Ghana's largest cocoa growing co-operative. Likewise *Coffee Pacifica*, a coffee importer in America is one third owned by the Papua New Guinea Coffee Growers Federation representing 12,000 farmers (Anonymous 2007). With all such schemes, the benefits to producers ultimately depend on Western consumers being prepared to pay a small extra premium for ethically-traded products.

8.5.2 Natural Product-Sales

Because village communities have a long history of utilizing natural materials for their own domestic purposes, they are well-positioned to turn these skills to commercial advantage. Trade transactions based on natural products can generate as much as 35% of total household incomes (Kerr 1991). Sometimes this trading takes place on a purely local basis. In Nepal for example, leaves collected from the wild tea plant (*Camellia kissi*) are fermented and compressed into cannonball-shaped masses for sale in local markets (Fig. 8.7a). Other somewhat unusual natural products destined for local sale include fabrics made from stem fibres of the giant nettle

(*Girardinia diversicola*) (Fig. 8.7b) (Dunsmore 1985) and paper manufactured from the bark of the paper daphne shrub (*Daphne bholua*). This paper has the useful property of resisting insect attack and is therefore much in demand for the writing of wills and other contract documents. Comparable enterprises in India include the use of leaves from the *tendu* tree (*Diospyros melanoxylon*) to act as wrappings for *bidi* cigarettes (Rathore 1972) and for the unusually-large leaves of the *sal* tree (*Shorea robusta*) to be brought into service as disposable food "plates" (Basu 1987). Although *bidi* cigarettes achieved something of a cult status amongst the "flower power" generation of California and occasionally, Western tourists buy nettle-fibre jackets and caps, most of this trading is aimed at local consumers.

Other natural products such as rattan furniture and palm baskets have achieved the status of international trade items. Much of this trading has been stimulated by an increasing Western enthusiasm for craft objects made from unusual raw materials or displaying individual craftsmanship (Littrell 1990). There has also been a rekindled enthusiasm for "natural" cosmetics and therapeutic agents, part of the rationale being that these are safer to use than the synthetic offerings of the chemical industry. To an extent, this is a misreading of the situation. For those unfortunate to have especially-sensitive skins, natural fragrances such as cinnamon oil can be just as irritating as their synthetic counterparts (Nater and de Groot 1985; Lovell 1993). It is also the case that otherwise-benign natural products may be rendered less so by having chemical preservatives added to stop them turning rancid during storage. These additives can sometimes themselves act as a source of serious irritation (De Groot and Herxheimer 1989).

Such caveats aside, the trade in cosmetic components can bring significant financial advantages to village communities, a good example being *Body Shop's* purchases of fruit-kernel oils direct from producers in Africa and South America. The oils are used in body lotions, hair conditioners, sun blocks, aromatherapy creams, soaps and lipsticks and derive from such species as the *marula* tree (*Sclerocarya caffra*) in Southern Africa, the shea butter tree (*Butyrospermum parkii*) in West Africa and the Brazil nut tree (*Bertholletia excelsa*) in South America.

Two potential obstacles stand in the way of such trading ventures. One is that careless harvesting can damage the plant populations on which the enterprises depend. The other is that intermediaries in the trading chain are prone to try and extract an unfair proportion of the profits, sometimes resulting in producers receiving as little as 25% of a product's true market value. Ways of countering these problems are on the one hand to devise the kind of sustainable harvesting regimes that have been applied to rattan and *ilala* palm utilisation (p.190). The other is to work with sympathetic trading partners. Non-governmental organisations (NGOs) frequently act in a supportive capacity (Littrell and Dickson 1997). In Zimbabwe, for example, the Women's Bureau has for many years facilitated the trade in *ilala* palm baskets by guaranteeing remuneration at an hourly rate on a par with that of an agricultural worker. In South Africa, basket trading, especially of the much sought after, strikingly-decorated and surprisingly liquid-tight beer-carrying baskets from Zululand, is supported by a community-based "Rural Crafts Association".

In a similar way *Body Shop's* transactions in cosmetic oils form part of their Community Trade Programme. This guarantees fair payments and a continuity of trade links for all participants. An example of a useful intervention by a government agency is provided by an initiative from the Indian Ministry of Tribal Affairs. This has involved setting up a chain of 18 "Tribes" handicraft shops in urban centres, and also organising an annual handicraft fair in a Delhi park. Some 200 craft makers from all over the country are financed to attend the fair and are provided with booths from which to sell their wares. The buyers are a mixture of city residents and visiting tourists and the profits go directly to the craft-makers.

8.5.3 Capitalizing on Ecotourism

Ecotourism is based on the observation of wild species rather than on their extraction and therefore escapes the potential complications of over-harvesting. It may, however, face other difficulties. In its simplest form it involves visitors staying in village houses and enjoying natural-history experiences under the direction of local guides. A fairly typical example is provided by the *Swamp Forest Trail* set up by the village of Bigodi in Western Uganda. After an initial suggestion from an American Peace Corps worker, the villagers constructed a trail through an area of *Phoenix* palm forest and papyrus swamp (Fig. 8.8a). Because of the wet conditions it was necessary to raise part of the trail onto a boardwalk. The main attraction offered is the opportunity to view colobus monkeys at close quarters, particularly the spectacular black and white colobus (*Colobus abyssinicus*), a species rarely encountered in conventional National Parks. Income to the village is generated through charges made for guiding services, simple overnight accommodation and the hire of wellington boots. Over its 20-year history, the project has generated enough money to build and staff a secondary school for the village.

A common ecotourism strategy is to add an extra hut to a village layout to function as a guest house. This is what has been done in a network of nine Mayan Indian villages in Belize (Fig. 8.8b–d). The guest houses are, in practice, village-style huts marginally upgraded to incorporate concrete floors and improved washing and toilet facilities. They provide sleeping facilities for visitors, but meals are taken in various village houses in turn. Activities on offer include observing forest wildlife and making visits to local caves and archaelogical sites. These operations are estimated to raise household incomes by 25%, the money typically being used to meet urgent medical expenses and extra school fees. The original idea for the venture came from an American expatriate, who still lives in Belize and keeps a watching brief on the project. In both these small schemes, communication with potential customers takes place via a web site, an arrangement that has the merit of avoiding the need to pay commission fees to agents and intermediaries.

Bearing in mind the fact that village health awareness is often incomplete; a potential difficulty with home-stays of any kind is that visitors become sick in the process. In particular, poor kitchen hygiene and unclean water supply can cause

Fig. 8.8 Small scale ecotourism projects. (**a**) The Bigodi swamp forest board walk (**b**) guesthouse in a Mayan Indian village (**c**) a meal being prepared for visitors and (**d**) the delivery of drinking and washing water supplies to the guesthouse

gastro-intestinal problems and clearly this is a poor advertisement for future business. The Mayan Indian villages are however exemplary on this score and demonstrate precisely the kind of precautionary measures needed (Edington and Edington 1997). All householders providing meals receive instruction in kitchen hygiene; all drinking water supplied to tourists is boiled (Fig. 8.8c, d) and any household with an illness in the family is temporarily excluded from participation. This is in addition of course to the more routine precaution of equipping sleeping bunks with mosquito nets.

Fig. 8.9 Large scale ecotourism projects. (**a**) Floating ecotourism lodge in the Amazonian Mamirauá Sustainable Development Reserve, (**b**) tourist cottage at the Ngala Game Reserve created by the Conservation Corporation of Africa, (**c**) reconstructed *Shangaan* village at the Londolozi Game Reserve

With higher levels of initial funding, more elaborate facilities are feasible and there are notable examples of sophisticated ecotourism centres being established in Amazonia (Maldonado 2008) and Southern Africa (Curzon 1993) (Fig. 8.9a, b). All these enterprises bring substantial economic benefits to local communities. Wage employment is available both for kitchen staff and other lodge-servicing personnel and also for guides and trackers. Like air-traffic control, English is the *lingua franca* of ecotourism and in Amazonia, where visitors are often avid bird watchers, it has

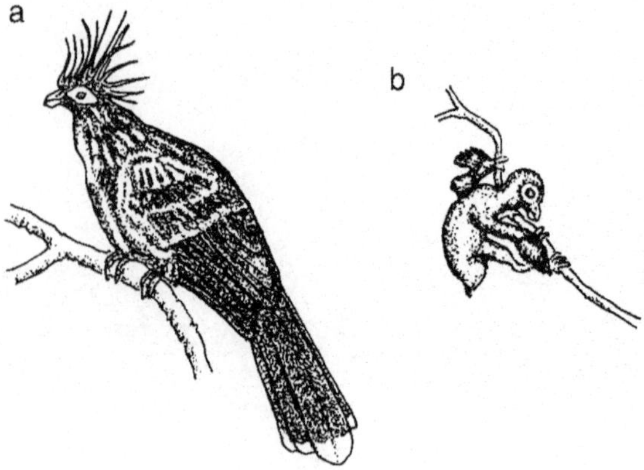

Fig. 8.10 Adult and juvenile hoatzins (**a**, **b**)

proved important for local guides to be able to identify species by their English names. At one facility in the Pantanal region of Brazil a school has been set up to foster such skills. In many schemes, village visits also create opportunities for local people to display and market their handicrafts. Hence, some villages in the flooded *várzea* zone of the Amazon feature, not only floating houses, but also floating gift shops. The Conservation Corporation in South Africa is notable for incorporating into some of its schemes, villages designed to re-accommodate *Shangaan* tribal communities earlier displaced from their traditional territories by the creation of the adjacent Kruger National Park (Fig. 8.9c).

According to the criteria formulated by the *Ecotourism Society*, ecotourism enterprises should, in addition to bringing tangible economic benefits to local communities, also actively contribute to environmental and species conservation (Lindberg and Hawkins 1993). To some extent this latter process happens automatically, once it becomes apparent that wild species are worth more alive and visible, rather than dead or exported. As one observer has pertinently observed, commenting on the desirability of having macaws (large parrots) flying freely in the Amazon forest rather than being lost to the pet trade, "a bird in the bush is worth two in the hand" (Munn 1992). It is also the case that in India, Pakistan and Nepal, ecotourism is making a positive contribution to the conservation of endangered species such as tigers and snow leopards (Dinerstein et al. 1999; Hussain 2000, 2003). In the case of the snow leopards, money from tourism contributes to a fund designed to compensate farmers for stock-losses and to finance the construction of leopard-proof night-time corrals for farm animals. Likewise in Kenya plans are afoot to use tourist revenues to compensate farmers for stock losses to lions, thereby reducing the temptation to mount reprisal attacks.

This is not to imply that ecotourism totally avoids wildlife disturbance problems. Sometimes the very intensity of visitor interest can in itself constitute a threat (Edington 1996, Edington and Edington 1986; Ananthaswamy 2004). A recent example from the Amazon concerns the viewing of the *hoatzin* (*Opisthocomus hoazin*), a bird with a strange prehistoric appearance and juveniles possessing wing-claws that they use to clamber back to the nest after being disturbed (Fig. 8.10). When alarmed, the young birds fascinate visitors by first leaping into the water of the flooded forest then after some minutes swimming, by laboriously climbing back to the nest. Visitors are so fascinated by this unusual behaviour that they stimulate the young birds continually to repeat it. This not only exposes them to attack by aquatic predators such as piranha fish, it also causes stress-induced mortality (Müllner et al. 2004). Clearly in such situations a greater degree of visitor supervision and sense of responsibility are required.

Developing ecotourism projects in desert areas poses particular difficulties, not only because many of viewing targets are night-active and hard to observe but also because local Bedouins, with their mobile life-style and simple tent shelters, make unpromising tourist hosts. At one desert facility in Dubai the natural history attractions on offer combine the viewing of spectacular day-active animals such as the Arabian oryx (*Oryx leucoryx*) and desert gazelle (*Gazella gazella*) with dune excursions to interpret the tracks left by otherwise unseen nocturnal species (Fig. 8.11a–c). Thus far, however, it has proved impossible to involve local Bedouin in these operations in any significant way, in part because of their disinclination to participate and in part because the continued browsing by their large herds of camels would pose a threat to the fragile desert vegetation (Gallacher and Hill 2008). A similar dilemma has arisen with the Dana project in Jordan. Here the solution has been to transfer the pastoralists to an alternative location and to focus ecotourism initiatives on local settled-village communities (Salti 1997; Johnson 1997). These initiatives include encouraging local men to become involved as guides and hotel managers and local women to participate in handicraft production and in the operation of a newly-constructed jam factory that draws its supplies from some rehabilitated fruit orchards (Fig. 8.11d, e).

8.6 Comprehensive Development Plans

Few rural communities can hope to solve all their problems at a single stroke. For most, achieving an acceptable quality of life requires a diversity of new inputs. The most comprehensive framework for providing such inputs is that contained in the United Nations Millennium Project inaugurated in 2000 (Sachs and McArthur 2005). This aimed to achieve eight Millennium Goals by the year 2015 (Table 8.2). Much progress has been made although significant shortfalls are evident. A good illustration of the possibilities is provided by pilot projects in 12 African villages (Sachs 2005). The interventions included: a greater concentration on cash crops; the provision of fertilizers and improved seed varieties; an upgrading of water supplies and household stoves; the provision of mosquito nets; a de-worming programme; a school meals system and the provision of mobile phones. Progress in the first village selected (in Kenya) proved promising. Malaria incidence fell by 50%, food yields more than doubled and the school meals programme (whereby farmers donate a proportion of their harvest directly to the school) produced spectacular improvements in educational attainment (Sachs 2005). Such inputs, which cost a total of

Fig. 8.11 Aspects of desert ecotourism. (**a**) Arabian oryx and desert gazelle at the Al Maha resort in Dubai, (**b**, **c**) sand tracks of a large desert beetle and a night-hunting desert fox, (**d**) ecotourist lodge and craft workshop at the Dana Nature Reserve in Jordan, the open lorry transports visitors into the Reserve (**e**) rehabilitated village fruit orchard at Dana

Table 8.2 Millennium development goals

1.	Reduce extreme poverty and hunger by half, relative to 1990
2.	Achieve universal primary education
3.	Promote gender equality and empowerment of women
4.	Reduce child mortality by two-thirds, relative to 1990
5.	Improve maternal health, including reducing maternal mortality by three-quarters relative to 1990
6.	Prevent the spread of HIV/AIDS virus, malaria and other diseases
7.	Ensure environmental sustainability
8.	Develop a global partnership for development

US$ 350/person over a 5-year period, cannot of course be expected to continue indefinitely and the intention is that the communities involved should ultimately become self-sufficient. In the Kenyan village it is anticipated that increased production and market sale of improved maize varieties will be the main income generator.

Applying this approach to all the needy village communities of the world is clearly well beyond the financial resources of home governments and therefore very substantial Western aid will initially be required. This raises some questions about the general philosophy of aid-giving. A persuasive overall justification is that inhabitants of tropical countries, through no fault of their own, are subject to geographical constraints of a kind that their temperate-zone counterparts escape. Therefore, following this argument, temperate countries should accept an obligation to provide assistance. Those who express doubts often do so on the grounds that corrupt practices prevent the funds reaching their intended targets (Nelson 1994; Ncayiyana 2007). The correct inference must surely be, not that aid programmes should be discontinued, but that they should be designed more carefully and scrutinized more rigorously. A helpful step in this direction has been Transparency International's production of their "Corruption Fighters' Tool Kit". Transparency International is a Berlin-based action group and their booklet describes ways of publicizing corruption issues and also documents specific commitments elicited from governments (Transparency International 2002). Additionally, in a number of countries, including Uganda, Kenya, Pakistan and Indonesia, websites have been created to allow victims of corruption to publicise their experiences.

My own direct experience of a multi-targeted programme relates to projects mounted in Northern India by the Central Himalayan Environment Association (CHEA). This is a non-governmental organisation receiving funding from a variety of official and voluntary sources. Key objectives include the generation of additional income through the marketing of cash crops and natural products, and the implementation of measures to foster the welfare of village women (Fig. 8.12a). The cash crops adopted include improved varieties of potatoes and beans and special arrangements have been made to supplement irrigation water by redirecting supplies from house roofs (Fig. 8.12b). The natural products developed for income generation include *Rhododendron* flowers (Fig. 8.12c) (used to produce a local

Fig. 8.12 Village development initiatives in the Almora district of Uttarakhand in India. (**a**) Targeted village, (**b**) roof-water capture system designed for cash-crop irrigation, (**c**) example of wild plant material being used for income generation (*Rhododendron* flower), (**d**) the village Women's Committee

drink), medicinal herbs and turpentine processed from the resin of local pine trees. The marketing of products of all kinds has been greatly facilitated by a national policy of constructing roads to every village. Special provisions aimed at women include the planting of fodder and firewood crops in more readily accessible places closer to village centres and the upgrading of paths (both measures designed to reduce the burden of carrying these materials); the provision of improved and more private latrine and washing facilities and, in the long term perhaps the most far-reaching step, arranging for a better representation of women on the decision-making assemblies of the villages (Fig. 8.12d). Indeed the point has been reached where some villages are now headed by women. This example illustrates a typical constructive progression for successful village development projects. It builds on a foundation of sound traditional practices and supplements this with external inputs, designed on the one hand to remove residual impediments, and on the other to generate additional economic opportunities.

References

Adongo PB, Phillips JF, Kajihara B, Fayorsey C, Debpuur C, Binka FN (1997) Cultural factors constraining the introduction of family planning among the Kassena-Nankana of Northern Ghana. Soc Sci Med 45:1789–1804

Allen C (1979) Tales from the dark continent. In: Images of British Colonial Africa in the twentieth century. André Deutsch and the British Broadcasting Corporation, London

Altrichter M (2005) The sustainability of subsistence hunting of peccaries in the Argentine Chaco. Biol Conserv 126:351–362

Ananthaswamy A (2004) Massive growth of ecotourism worries biologists. New Sci 4:6–7

Anonymous (2007) Thinking out of the box. How African cocoa-growers are moving upstream into chocolate. *The Economist,* April 7, 2007, 77

Antonsson-Ogle B (1984) Wild plant resources. Ceres 17(5):38–40

Appleton CC (1978) Review of literature on abiotic factors influencing the distribution and life cycles of bilharziasis intermediate host snails. Malacol Rev 11:1–25

Avé W (1987) Small-scale utilization of rattan by a Semai community in West Malaysia. Econ Bot 42:105–119

Basu NG (1987) Forests and tribals. Manisha Granthalaya, Calcutta

Bockstoce J (1980) Battle of the bowheads. Nat Hist 85(5):52–58

Briceño-Garmendia C, Estache A, Shafik N (2004) Infrastructure services in develoing countries: access, quality, costs and policy reform, World Bank policy research working paper 3468. World Bank, Washington, DC

Brosius JP (1991) Foraging in tropical forests: the case of the Penan of Sarawak, East Malaysia (Borneo). Hum Ecol 19(2):123–150

Bryceson DF (1989) Nutrition and the commoditization of food in Sub-Saharan Africa. Soc Sci Med 28:425–440

Bustard HR (1970) A future for crocodiles. Oryx 10(4):249–255

Caldecott J (1988, June) Climbing towards extinction. New Sci 9:62–66

Chin T, Welsby PD (2004) Malaria in the UK: past, present, and future. Postgrad Med J 80:663–666

Clark C (1967) Population growth and land use. Macmillan, London

Colchester M (1997) Salvaging nature: indigenous peoples and protected sites. In: Ghimire KB, Pimbert MP (eds) Social change and conservation, environmental politics and impacts of national parks and protected areas. Earthscan, London pp, pp 97–130

Constantino PAL, Fortini LB, Kaxinawa FRS, Kaxinawa AM, Kaxinawa ES, Kaxinawa AP, Kaxinawa LS, Kaxinawa JM, Kaxinawa JP (2008) Indigenous collaborative research for wildlife management in Amazonia: the case of the Kaxinawá, acre, Brazil. Biol Conserv 141:2718–2729

Cunningham AB (2001) Applied ethnobotany. People, wild plant use & conservation. Earthscan Publications, London

Cunningham AB, Milton SJ (1987) Effects of basket-weaving industry on mokola palm and dye plants in northwestern Botswana. Econ Bot 41:386–402

Curzon C (1993) Ecotourism-conservation ethics-profit. Getting it right in the Eastern Transvaal. Afr Environ Wildl 1(2):36–42

Dasmann RF (1964) African game ranching. Pergamon Press, Oxford

De Groot AC, Herxheimer A (1989) Isothiazolinone preservative: cause of a continuing epidemic of cosmetic dermatitis. Lancet:314–316
DeWalt BR (1993) Nutrition and the commercialization of agriculture: ten years later. Soc Sci Med 36:1407–1416
Dewey KG (1989) Nutrition and the commoditization of food systems in Latin America and the Caribbean. Soc Sci Med 28:415–424
Dinerstein E, Rijal A, Bookbinder M, Kattel B, Rajuria A (1999) Tigers as neighbours: efforts to promote local guardianship of endangered species in rural Nepal. In: Seidensticker J, Christie S, Jackson P (eds) Riding the Tiger. Tiger conservation in human-dominated landscapes. Cambridge University Press, Cambridge, pp 316–333
Drèze J, Sen A (2002) India, development and participation. Oxford University Press, New Delhi
Dunsmore S (1985) The nettle in Nepal, a cottage industry. Land Resources Development Centre, Surbiton
Edington JM (1996) Effects of tourism on natural systems. In: Giller PS, Myers AA (eds) Disturbance and recovery of ecological systems. Royal Irish Society, Dublin, pp 173–181
Edington JM, Edington MA (1986) Ecology, recreation and tourism. Cambridge University Press, Cambridge
Edington JM, Edington MA (1997) Tropical forest ecotourism: two promising projects in Belize. In: Stabler MJ (ed) Tourism & sustainability: from principles to practice. CAB International, Wallingford, pp 163–168
Fewtrell L, Kaufman RB, Kay D, Enanoria W, Haller L, Colford JM (2005) Water, sanitation, and hygiene interventions to reduce diarrhoea in less developed countries: a systematic review and meta-analysis. Lancet Infect Dis 5(1):42–52
Fleuret P, Fleuret A (1980) Nutrition, consumption and agricultural change. Hum Organ 39:250–260
Fong FW (1992) Perspectives for sustainable resource utilization and management of nipa vegetation. Econ Bot 46(1):45–54
Gallacher DJ, Hill JP (2008) Effects of camel grazing on density and species diversity of seedling emergence in the Dubai (UAE) inland desert. Jour Arid Env 72:853–860
Gourou P (1958) The tropical world. Its social and economic conditions and its future status. Longmans & Green, London
Grimwood I (1968) Endangered mammals in Peru. Oryx 9(6):411–422
Haaga J, Mason J, Omoro FZ, Quinn V, Rafferty A, Test K, Wasonga L (1986) Child nutrition in rural Kenya: a geographic and agricultural classification. Ecol Food Nutr 18:297–307
Hamilton RJ, Potuku T, Montambault JR (2001) Community-based conservation results in the recovery of reef spawing aggregation in the Coral Triangle. Biol Conserv 144:1850–1858
Harris DR (1974) Tropical vegetation: an outline and some misconceptions. Geography 59:240–250
Harrison P (1992) The third revolution. Environment, population and a sustainable world. I.B. Tauris/St. Martin's Press, London/New York
Hirschman C, Tan J, Chamratrithirong A, Guest P (1994) The path to below replacement-level fertility in Thailand. Int Fam Plan Perspect 20:82–87
Hughes CC, Hunter JM (1970) Disease and "development" in Africa. Soc Sci Med 3:443–493
Huntington E (1915) Civilisation and climate. Yale University Press, New Haven
Huntington HP (1992) Wildlife management and subsistence hunting in Alaska. Belhaven Press, London
Husaini MA, Suhardjo R, Megawangi R, Nurhadi E, Supardi D, Djojosoebagio S, Karyadi D (1990) Diet, nutritional status, and potential need for home gardens in the tea plantation. In: Landauer K, Brazil M (eds) Tropical home gardens. United Nations University Press, Tokyo, pp 119–125
Hussain S (2000) Protecting the snow leopard and enhancing farmers' livelihood: a pilot insurance scheme in Baltistan. Mt Res Dev 20(3):224–229
Hussain S (2003) The status of the snow leopard in Pakistan and its conflict with local farmers. Oryx 37(1):26–33
Iannotti L, Cunningham K, Ruel M (2009) Diversifying into healthy diets. Homestead food production in Bangladeah. In: Spielman DJ, Pandya-Lorch R (eds) Millions Fed. Proven suc-

cesses in agricultural development. International Food Policy Research Institute, Washington DC, pp 145–151
Innis HA (1956) The fur trade in Canada. University of Toronto Press, Toronto
Janzen DH (1973) Tropical agroecosystems. Science 182:1212–1219
Johnson C (1997) Dana, helping nature, helping people. Al Reen 60:10–15
Kerr K (1991) The economic potential of handicrafts enterprises in rural development: focus on Indonesia. Unasyla 42:31–36
King M (1990) Health is a sustainable state. Lancet 336:664–667
Koziell I, Inoue CYA (2006) Mamirauá sustainable development reserve, Brazil. Lessons learnt in integrating conservation with poverty reduction. International Institute for Environment and Development, London
Krech S (1999) Playing with fire. New Sci 164:56–59
Krech S (2000) The ecological Indian, myth and history. Norton, New York
Kuhnlein HV (2003) Micronutrient nutrition and traditional food systems of indigenous peoples. Food Nutr Agric 32:33–39
Lehman CF (1967) The pet trade and extermination. Oryx 9(2):161–165
Lenon B (1992) The demographic transition model. Geogr Mag 64(5):51
Lindberg K, Hawkins DE (1993) Ecotourism, a guide for planners and managers. The Ecotourism Society, North Bennington
Littrell MA (1990) Symbolic significance of textile crafts for tourists. Ann Tour Res 17:228–245
Littrell MA, Dickson MA (1997) Alternative trading organizations: shifting paradigm in a culture of social responsibility. Hum Organ 56:344–352
Lovell CR (1993) Plants and the skin. Blackwell Scientific Publications, Oxford
Maldonado P (2008) Ecotourism in Peru, rumble in the jungle. De Economist 12:58–59
Manson-Bahr PEC, Apted FIC (1982) Manson's tropical diseases. Baillière Tindall, London
Martens WJM, Jetten TH, Focks DA (1997) Sensitivity of malaria, schistosomiasis and dengue to global warming. Clim Chang 35:145–156
McCord C (1978) Medical technology in developing countries: useful, useless or harmful. Am J Clin Nutr 31:2301–2313
McDonald M, Fleming B (1993) Community-based economic development and resource management in the Hudson Bay area. In: Inglis JT (ed) Traditional ecological knowledge, concepts and cases. International Program on Traditional Knowledge and International Development Research Centre, Ottawa, pp 63–68
McKean SG (2003) Toward sustainable use of palm leaves by a rural community in Kwazulu-Natal, South Africa. Econ Bot 57(1):65–72
McNeil M (1964) Lateritic soils. Sci Am 211(5):96–102
Meggitt MJ (1964) Aboriginal food gatherers of tropical Australia. In: Elliott HFL (ed) The ecology of man in the tropical enviroment. International Union for the Conservation of Nature and Natural Resources, Morges, pp 30–37
Morgan RPC (1979) Soil erosion. Longman, London
Müllner A, Linsenmair KE, Wikelski M (2004) Exposure to ecotourism reduces survival and affects stress response in hoatzin chicks *(Opisthocomus hoazin)*. Biol Conserv 118:549–558
Munn CA (1992) Macaw biology and ecotourism or "when a bird in the bush is worth two in the hand". In: Beissinger SR, Snyder NFR (eds) New world parrots in crisis: solutions from conservation biology. Smithsonian Institution, Washington, DC, pp 47–72
Nakashima DJ (1993) Astute observers on the sea ice edge. Inuit knowledge as a basis for Arctic co-management. In: Inglis JT (ed) Traditional ecological knowledge, concepts and cases. International Program on Traditional Ecological Knowledge and International Development Research Centre, Ottawa, pp 99–110
Nater JP, de Groot AC (1985) Unwanted effects of cosmetics and drugs used in dermatology. Elsevier, Amsterdam
Ncayiyana DJ (2007) Combating poverty: the charade of development aid. Br Med J 335:172–173
Nelson EAS (1994) Development aid: the way forward. Br Med J 308:481

Niemeijer R, Guens M, Kliest T, Ogonda V, Hoorweg J (1988) Nutrition in agricultural development: the case of irrigated rice cultivation in West Kenya. Ecol Food Nutr 22:65–81
Niñez VK (1986) The household garden as a lifeboat. Ceres 19(4):31–36
Oomen HAPC, Grubben GJH (1978) Tropical leaf vegetables in human nutrition. Royal Tropical Institute, Amsterdam
Opole M (1993) Revalidating women's knowledge on indigenous vegetables: implications for policy. In: de Boef W, Amanor K, Welland K, Bebbington A (eds) Cultivating knowledge: genetic diversity, farmer experimentation and crop research. Intermediate Publications, London, pp 157–164
Pacey A (1978) Gardening for better nutrition. Intermediate Technology Publications, London
Parish T (1998) Aerial runways. Enact 6(3):15–19
Paudel SK, Chowdhary CL (2005) Managing rattan as a common property: a case study of community rattan management in Nepal. J Bamboo Rattan 4(1):81–91
Pires A (2005) Community Timber Enterprises in the Mamirauá Sustainable Development Reserve, Brazil. Mamirauá Institute for Sustainable Development, Tefé
Rathore JS (1972) *Diospyros melanoxylon,* a bread-winner tree of India. Econ Bot 26:333–339
Redford KH, Mansour JA (eds) (1996) Traditional peoples and biodiversity conservation in large tropical landscapes. America Verde Publications. The Nature Conservancy, Latin America and Caribbean Division, Arlington
Reeves BOK (1983) Six milleniums of buffalo kills. Sci Am 249(4):92–100
Reeves RR, Mitchell E (1981) The whale behind the tusk. Nat Hist 90(8):50–57
Sachs JD (2005) The end of poverty. How we can make itt happen in our lifetime. Penguin Books, London
Sachs JD, JW MA (2005) The millenium project: a plan for meeting the millennium development goals. Lancet 365:347–373
Salti R (1997) Dana, the human perpective. Al Reen 60:17–19
Sandhu R, Sandler J (1986) The tech and tools handbook. A guide to technologies women are using worldwide. International Women's Tribune Centre/Intermediate Technology Publications, New York/London
Shack KW, Grivetti LE, Dewey KG (1990) Cash cropping, subsistence agriculture, and nutritional status among mothers and children in lowland Papua New Guinea. Soc Sci Med 31:81–88
Shanley P, Galvao J, Luz L (1997) Limits and strengths of local participation: a case-study in eastern Amazonia. PLA Note Int Instit Environ Dev 14:64–67
Sharma KK (2003) Improving bioavailability of iron in Indian diets through food-based approaches for the control of iron deficiency anaemia. Food Nutr Agric 32:51–61
Shepperdson M (1995) Income distribution and health in a North Indian village. Avebury, Aldershot
Sinclair ARE, Fryxell JM, Caughley G (2006) Wildlife ecology, conservation and management. Blackwell Publishing, Oxford
Smith NJH (1976) Spotted cats and the Amazon skin trade. Oryx 13(4):362–371
Spielman DJ, Pandya-Lorch R (eds) (2009) Millions fed. Proven successes in agricultural development. International Policy Research Institute, Washington, DC
Spinage C (1998) Social change and conservation misrepresentation in Africa. Oryx 32:265–276
Stamp LD (1964) The geography of life and death. Collins, London
Strickland D (1981) The eskimo vs. the walrus vs. the government. Nat Hist 90(2):48–97
Sutherland WJ (2001) Sustainable exploitation: a review of principles and methods. Wildl Biol 7:131–140
Tang G (2010) Bioconversion of dietary provitamin A carotenoids to vitamin A in humans. Am J Clin Nutr 91(5):1468–1473
Tiffen M, Mortimore M (1994) Malthus controverted: the role of capital and technology in growth and environment recovery in Kenya. World Dev 22:997–1010
Tiffen M, Mortimore M, Gichuki F (1994) More people, less erosion. John Wiley, Chichester
Transparency International (2002) Corruption fighters toolkit. Civil society experiences and emerging strategies. Transparency International, Berlin

Transparency International (2004) Global corruption report 2004, political corruption. Pluto Press, London

Transparency International (2006) Global corruption report 2006, corruption and health. Pluto Press, London

Wayre P (1969) Wildlife in Taiwan. Oryx 10(1):46–56

Weaver LT, Beckerleg S (1993) Is health a sustainable state? A village study in the Gambia. Lancet 341:1327–1330

Weiyuan C (2008) China's village doctors take great strides. Bull World Health Organ 86(12):914–915

Werner DB, Bower BL (1982) Helping health workers learn. The Hesperian Foundation, Palo Alto

Wheat JB (1967) A Paleo-Indian bison kill. Sci Am 216(1):44–52

Whelan R (1999) Wild in the woods. The Myth of the Noble eco-savage. Institute of Economic Affairs, London

WHO (2009) Neglected tropical diseases, hidden successes, emerging opportunities. World Health Organisation, Geneva

Wilkie DS (2005) Life amidst chaos in a forest filled with its own dangers, DRC violence continues to trample the Efe and Lese. Cult Surv Quart 29(1):1–5

Chapter 9
Lessons for the World at Large

9.1 Contemporary Issues

It is scarcely surprising that the Western World, with all its economic strengths, advanced technologies and sophisticated science base, should be well placed to provide assistance to developing countries. What is less well appreciated is that considerable scope exists for transfers of expertise in the opposite direction. Where might these opportunities lie?

A major dilemma now facing Western nations is the issue of fossil fuels. These have been the mainstay of industrial development for two and a half centuries but are now regarded as posing a threat to climatic stability and human welfare (McCarthy et al. 2001a). This is because their combustion products have the capacity to alter the thermal properties of the atmosphere and in a generally deleterious direction (Houghton 2004). Traditional communities make virtually no demands on fossil fuels and this raises the possibility that some of their simpler techniques for meeting energy needs might have a wider application.

One of the specific challenges offered by climate change is the need to discover (or create) crop varieties able to thrive in the new conditions. Until recently, Western plant breeders have been mainly preoccupied with developing high-yield strains able to respond to enhanced artificial fertilizer inputs. What is required in the new situation is something different and very much akin to the diverse seed repertoires possessed by traditional farmers. This resource allows them to cope with diverse environmental conditions and could make a similar contribution to the strategies required to combat climate change.

A generally welcome outcome of the industrial revolution has been the capacity it created to synthesize new chemical compounds and to mimic and modify naturally existing ones. This facility underpins the production of drugs by the pharmaceutical industry and the processing of raw foodstuffs into consumer products by the food industry. These processes have however prompted certain misgivings. Some of the drugs introduced have been insufficiently tested and have caused

J. Edington, *Indigenous Environmental Knowledge*,
DOI 10.1007/978-3-319-62491-4_9

damaging side effects amongst the patients receiving them (Chetley 1990). In the case of food manufacturers there is a frequently voiced suspicion that commercial gain is given a higher priority than the long-term health interests of consumers (Millstone 1984; Coghlan 1991). In consequence there are those who would advocate a partial reversion to the therapies and dietary procedures of traditional communities. How realistic is this?

Finally an observation about Western attitudes to the natural world; these have in general been strongly influenced by the Judaic-Christian position. This argues that mankind has dominion over the natural world and can deal with it in any way judged convenient (White 1967; Santmire 1985). This is essentially an *anthropocentric* and *utilitarian* stance, which declines to attach value to species failing to contribute to human interests. In this respect it contrasts markedly with the *intrinsic value* notion of many traditional societies, which leads them to cherish even species that make no practical contribution to their enterprises and may even generate problems for them. Some attempt to harmonize these contrasting positions seems long overdue.

9.2 Energy Economies in Agriculture and Food Distribution

Turning first to issues related to the consumption of fossil fuels, Western farmers are lavish consumers of such fuels. Not only are these used to manufacture and run machinery, they are also employed to produce agro-chemicals such as fertilizers, pesticides and herbicides (Pimentel et al. 1973; Pimentel and Pimentel 1996). Traditional farm operations present a very different picture. Most of the motive power comes from the farmer's own physical labour and that of his draught animals. In consequence fossil-energy consumption tends to be very low and often limited to the tiny amounts needed to manufacture simple iron implements such as ploughs and hoes (Leach 1976). Whilst such farmers might legitimately be encouraged to acquire extra technology to lighten their workloads, the reverse is true for their Western counterparts. In their case arguably, the time is ripe for modifying farm operations to reduce fossil-fuel consumption in the interests of planetary health (Cooke 1975; Reganold et al. 1990). Traditional agriculture offers some clues about how this might be achieved.

Savings could be made for example in the energy expenditure associated with the manufacture and application of artificial fertilizers. This could be done by making more use of animal manures and by harnessing the natural process of biological nitrogen fixation. Both strategies usually require a greater degree of farm diversification. A typical rotational sequence on a North American farm pursuing such a course might involve planting maize and oats for 1 year each and then following this with 2 years of the nitrogen-fixing legume, alfalfa (*Medicago sativa*) (Fig. 9.1a). Varying the timing of the rotation on different parts of the farm has the additional merit of producing a crop mosaic that is better equipped to resist pest attack. Furthermore, the replacement of inorganic fertilizers with manure is likely to bring benefits in the form of improved soil structure and reduced susceptibility to erosion

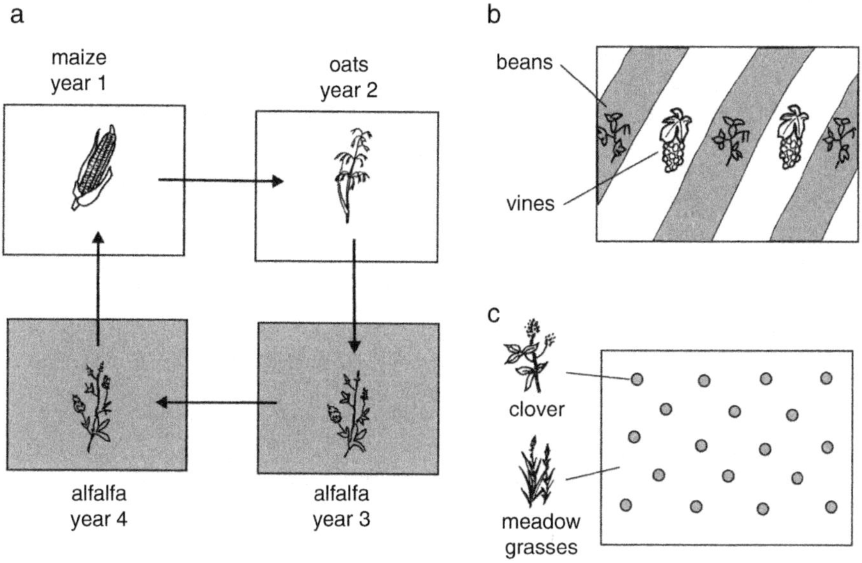

Fig. 9.1 Some strategies for increasing the agricultural use of nitrogen-fixing legumes (*grey shaded*). (**a**) Four-year crop rotation, (**b**) companion planting of vines and beans, (**c**) dispersed clover on pastureland

(Reganold et al. 1987). A parallel approach adopted in some Californian vineyards has been to alternate the lines of vines with rows of legumes (Fig. 9.1b). In Europe, likewise, increasing the proportion of clover in pastures has become a recognized strategy for enhancing nitrogen-fixation (Fig. 9.1c). Farms run along these less-intensive lines need not be less profitable. Yields may be marginally lowered but, because input costs are reduced even more, net financial returns are often greater (Lockeretz et al. 1981; Reganold et al. 1990).

Modern food-distribution systems are also extravagant consumers of fossil energy. A preoccupation with packaging can result in food being transported needlessly long distances. For instance, potatoes grown in Sweden may travel to Italy for packaging before being returned for sale virtually at their place of origin and French walnuts destined for consumption in Britain often make a side excursion to Naples for the same purpose. Such excesses are noticeably absent from most marketing systems in the developing world, and latterly some attempt has been made to recapture these virtues in the West. One such initiative has been the revival of localized "farmers' markets" (Paxton 1994; Festing 1997). Another has been for consumers to make a conscious effort to use locally-grown produce, concentrating for example on items originating from within a hundred-mile radius.

Of course, the most radical way of shortening food-supply lines is to grow some of one's own food either in a garden or urban "allotment" (Crouch and Ward 1997, 1988; Pinkerton and Hopkins 2009). In this task of using small plots effectively, much can be learned from the traditional multi-layered cultivation systems of the tropics. Indeed Robert Hart, the pioneer in this field, drew much of his inspiration from the traditional home gardens of India (de Hart 1996). The underlying principal is to stratify crops according to their respective growth forms and light requirements (p.192). In temperate situations there is the additional need to take account of seasonal variations in sun height, temperature and day-length. This can be achieved with careful planning; the whole operation now frequently going under the name of "permaculture" (Mollison 1978). There are however limits to what can be achieved along these lines. Cities came into existence only when farmers in the countryside managed to produce sufficient food surpluses to free town dwellers for other activities. This general principle still holds good and only a proportion of urban dwellers – estimated at about 1–2% of the total – are likely to have the time, land and inclination to grow a significant proportion of their own food.

9.3 Energy Economies in Architecture and Household Management

9.3.1 Building Design

Modern architects seeking to reduce the energy costs of their constructions can also draw inspiration from their traditional counterparts. The energy expenditure of buildings has two components. The first comes into play when raw materials are being acquired or manufactured, transported to the site and being incorporated into the fabric of a building. The energy expended during this phase is known as *embodied energy* and the amount used is closely related to the type of constructional materials employed. Values for traditional materials such as timber and stone are far smaller than those for modern manufactured ones such as metals and plastics (Friedman and Cammalleri 1995). Average values for these two traditional materials are 7 and 0.4 mega joules/kg respectively and for the modern ones, a very much more substantial 68 and 189 mega joules/kg. Consequently, significant energy savings can be achieved by designing constructions to maximize the use of the natural materials. Using structural steel to support tall buildings is usually a necessity, but cladding their roofs with copper, lead or aluminium is increasingly being seen as a needless extravagance in energy terms (Ramshaw 2003). The second energy-consuming component is more familiar and relates to a building's daily energy expenditure on services such as lighting, heating and cooling.

There are useful traditional precedents on both scores. Village buildings are frequently constructed from locally procured raw materials and opportunities are taken, both to exploit passive energy sources and to recycle materials at the end of the building's useful life. Even mud-bricks are sometimes broken down and reconstituted, and hardwood timbers are frequently passed down from one generation to the next, often achieving the status of family heirlooms. In Indonesia, for example, house posts made from Borneo ironwood (*Eusideroxylon zwageri*) are confidently expected to have a working life of at least 150 years and be readily transferrable from one house to the next (Waterson 1991).

Whilst some modern building projects have sought to adopt these frugal principles, others have chosen to ignore them. For instance the environmentally conscious Findhorn Foundation in Scotland makes a point of obtaining its stone from nearby quarries and its timber from local forests (Talbot 1993) and more recently designers of new village houses in south Somerset have specified that their building-stone should come from the Ham Hill quarry only five miles distant. Other projects have shown no such restraint. There exists, for example an office building in London's Dockland development that is clad with granite wall-facings that originated from a Brazilian quarry, were shipped to Italy for cutting and polishing and thence to Germany for attachment to frames; all before being finally transported to London for installation (Westland and Hinks 1994).

The day-to-day functioning of buildings is also susceptible to energy economies and here again we have some traditional precedents. At Beddington, south of London, there is an interesting residential development for 100 households that seeks to put sustainability concepts into practice. The name BedZED reflects both its location and the intention that it should function as a Zero Energy Development requiring no externally generated energy inputs (BRESCU 2002; Twinn 2003). Although this last objective has not been wholly achieved, the efforts made in this direction frequently echo traditional technologies.

For instance the large roof cowls carry vanes designed to orient their outlet apertures downwind (Fig. 9.2a). As the wind passes over the structures it creates areas of reduced pressure on their lee sides, thereby facilitating the extraction of stale air from the building's interior. Louvres on the opposite sides of the cowls serve the complementary purpose of directing fresh air into the building. These arrangements thus resemble the operation of Middle Eastern wind towers (p.81). Other energy-economic features with traditional precedents include south-facing conservatories to trap winter heat and thick masonry walls designed to provide indoor heat buffering. The latter work in much the same way as the thick mud-brick walls of traditional desert houses.

Echoes of traditional design are also to be found in Nottingham University's new campus buildings. In this case, room interiors are cooled by directing currents of fresh air through them, a process enhanced by positioning lakes in the path of the prevailing winds and by arranging for stale air to be extracted through heat stacks and mobile turrets (Fig. 9.2b). The entire setup, with air being cooled by passage over water and subsequently drawn out on the lee side of roof structures, has clear counterparts in traditional Middle Eastern designs. The relatively few modern embellishments that have been added include photovoltaic cells to drive fans on windless days and rotary heat exchangers designed to lessen temperature differences between the incoming and outgoing air (Fig. 9.2b).

Traditional devices have also started to appear on smaller buildings. At Weobley, a school commissioned by Hereford and Worcester County Council, such features include a sloping roof on its downwind side and a roof overhang above the south-facing wall (Fig. 9.3a). The negative pressure zone created on the lee-side of the

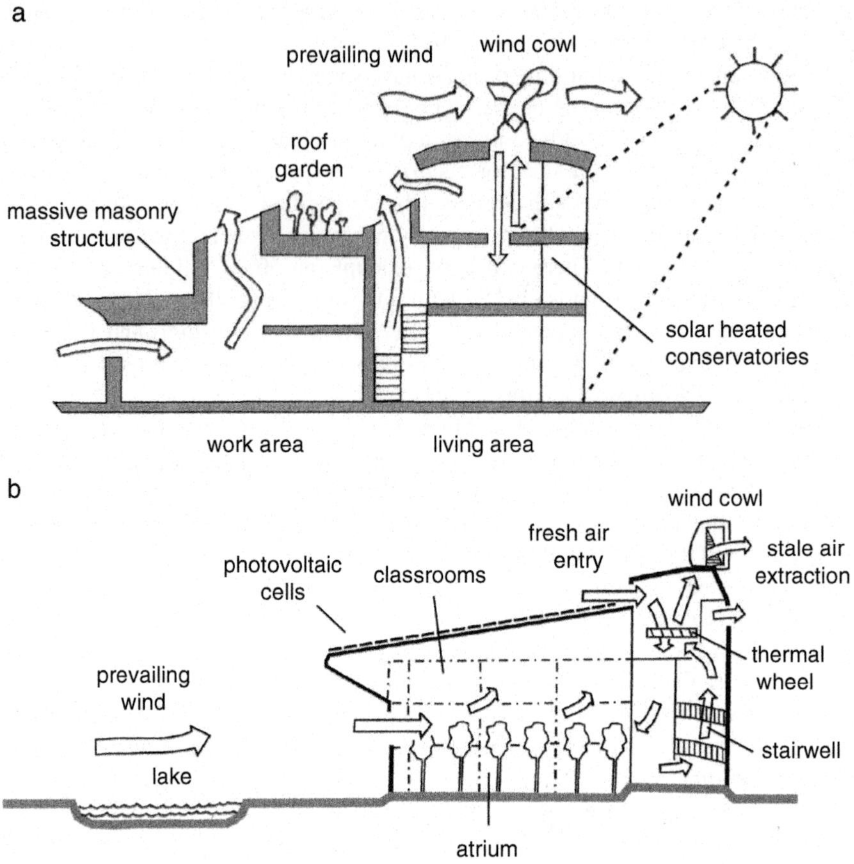

Fig. 9.2 Modern buildings incorporating traditional energy-saving features. (**a**) BedZED residential development (Based on BRECSU (2002) and Twinn (2003)). (**b**) Nottingham University campus building (Based on Pendleton (2000) and McCarthy et al. (2001b))

structure aids the extraction of stale-air from the classrooms, whilst the roof overhang serves to block excessive heat-entry from the high summer sun. Roof overhangs are a long-established traditional device in temperate-zone buildings. They feature conspicuously in the famous *Minka* houses of Japan (Kumara and Saito 1982) and make an early and dramatic appearance in the thirteenth century cliff villages of Arizona. In this latter case it is a rock-overhang rather than a roof that serves as the summer screening device (Fig. 9.3b) (Pillet 1980; Reyman 1982).

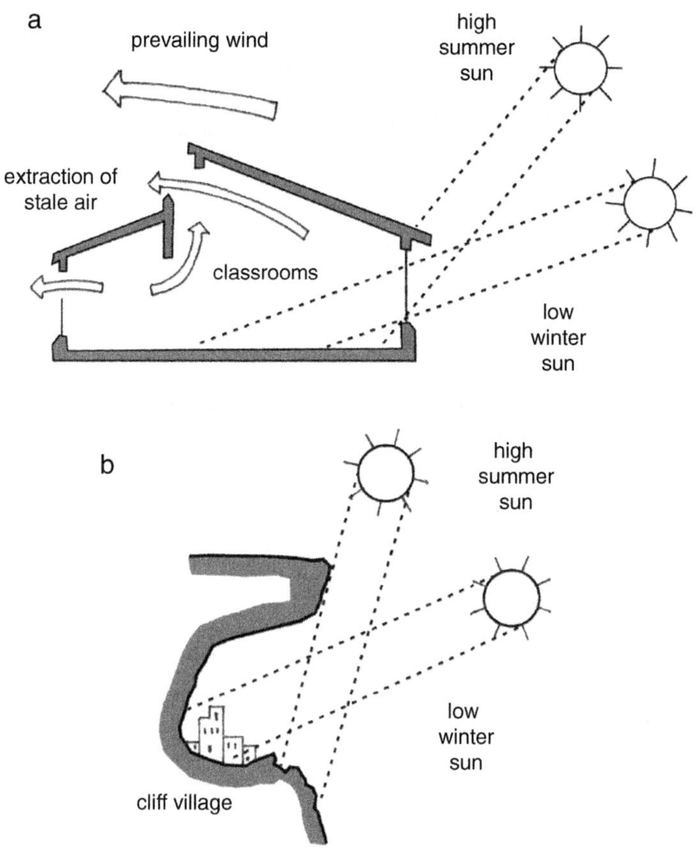

Fig. 9.3 Modern and ancient climatic adaptations from temperate regions. (**a**) Roof design at Weobley School (Based on Lockie 2000), (**b**) rock overhang at Mesa Verde (Based on Pillet 1980)

9.3.2 *Use of Wood-Fuel*

The school at Weobley revisits another traditional strategy by using wood-fuel for supplementary heating, in this case by burning wood chips produced from local coppice woodland. Wood-fuel boilers were also designed into the BedZED scheme although operational and maintenance problems have put them on hold at least for the time being and conventional gas and electrical inputs are currently used to run the facility (Hodge and Haltrecht 2009). In whatever setting, using wood as a fuel has the great merit of releasing no more carbon dioxide on combustion than was originally assimilated during growth. This makes the process effectively carbon neutral. It is for this environmental reason that wood-burning devices are starting to find favour, not only in private houses, but also to help heat public buildings such as hospitals and schools.

There are, however, two reservations to be registered about such systems and they are somewhat reminiscent of the difficulties arising in village situations. One revolves around the issue of atmospheric pollution, not in this case the problem of smoke accumulating indoors (a well-fitted and properly-maintained stove should prevent this), but rather that of emissions polluting the atmosphere at large. For instance, the widespread use of wood stoves on the outskirts of some Canadian cities has been shown to produce significant atmospheric concentrations of such potentially damaging chemicals as *polycyclic aromatic hydrocarbons, dioxins* and *furans* (Environment Canada 2004). Indeed householders with wood-stoves are actively discouraged from using them on days when temperature inversions trap polluted air near to the ground. Similarly, in the San Francisco Bay area of California, it has become the practice under certain meteorological conditions to designate "save-the-air days". At these times domestic wood-burning is forbidden.

Neither are the wood supply problems encountered in village situations entirely absent from Western ones. Even countries with extensive forests are not necessarily well placed to generate the amount of fuel-wood required. The market for stem timber is usually sufficiently buoyant to remove it from consideration; consequently wood-fuel supplies are derived mainly from such secondary sources as discarded branches and stem tips, sawmill wastes and recycled materials. Wood from short-rotation coppices is another option, but one whose use is constrained by shortages of land for the necessary tree planting (Pearce 1995). Consequently in Britain it is predicted that an appreciable gap will open up between fuel-wood consumption and production, with as much as 65% of the supply having to be imported. This raises the prospect of extra supplies being sought from the forests of the developing world, thereby potentially adding to the burdens of village communities already struggling to meet their own firewood needs.

9.3.3 Water-Harvesting

Modern systems for managing water and wastes also involve significant fossil fuel expenditure. Traditional householders offer some interesting pointers in these fields too. For instance they often employ their house roofs as water collecting devices. The simplest arrangement is to install guttering to intercept the rainwater flowing down the roof slope and then to direct it into a suitable storage container (Fig. 9.4a). A minor added sophistication is to arrange the pipe work so that the first flush of water from individual rainstorms – that likely to be contaminated with debris and bird droppings – can be discarded (Pacey and Cullis 1996). In Southern Nigeria, water-collection sometimes involves a traditional "impluvium" courtyard. This is a sunken rectangular basin into which water discharges from surrounding roofs. An underground tank in a corner of the courtyard serves as the water store (Fig. 9.4b).

Similar strategies are by no means out of the question for Western households. In the "autonomous" house designed in Britain by architects Brenda and Robert Vale, roof-water is directed first into underground storage tanks and then passed via a

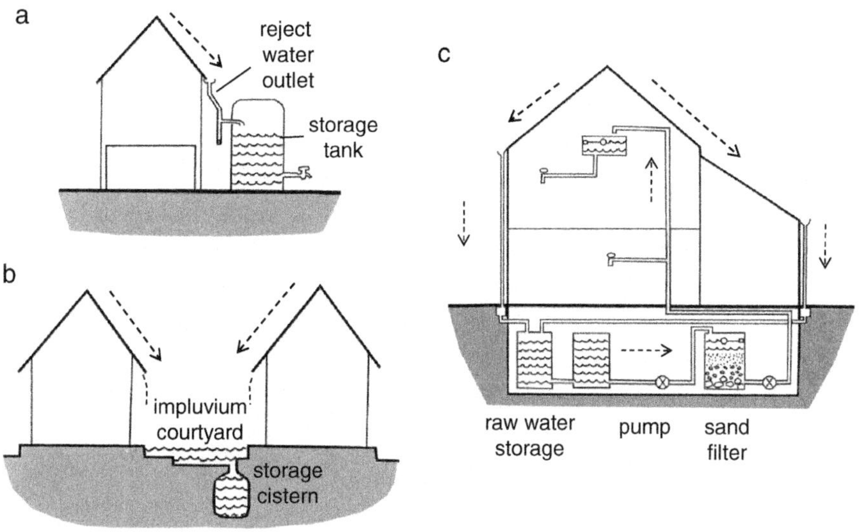

Fig. 9.4 Roof-water harvesting. (**a**) A collection and storage system in Thailand (Based on Pacey and Cullis 1986), (**b**) an impluvium courtyard in Nigeria (Based on Denyer (1978) and Dmochowski (1990)), (**c**) a modern water-collection system from Europe (Based on Vale and Vale 2000)

sand filter to household taps (Fig. 9.4c). In this case, because household toilets are of the composting type not requiring water, the household is in the position of making no demands on the public water supply system (Vale and Vale 2000).

9.3.4 *Waste-Recycling*

The BedZED houses similarly capture roof water and use it to flush toilets and to irrigate gardens. This design aimed initially to integrate water-recycling and waste-disposal in a totally comprehensive way (Fig. 9.5). The basic idea was that drainage from toilets, baths and sinks would not be wastefully discharged to the municipal sewage system but rather into a series of biological treatment tanks. Systems of this type were the original brainchild of John Todd in the United States and have *Living Machine* as a brand name (Todd and Josephson 1996). In order to settle solids and initiate the process of bacterial decomposition, foul drainage is directed first through a device akin to a septic tank and then through a series of aerated tanks incorporating racks of growing plants. The watery end product should be capable, after ultraviolet disinfection, of being combined with rain water and used for various functions not requiring high levels of water purity. An estimated 35% of domestic water use comes into this category and embraces such tasks as flushing toilets, watering roof gardens and irrigating amenity areas (Fig. 9.5). An early Living Machine system has been successfully installed in the Findhorn eco-village in Scotland and although the

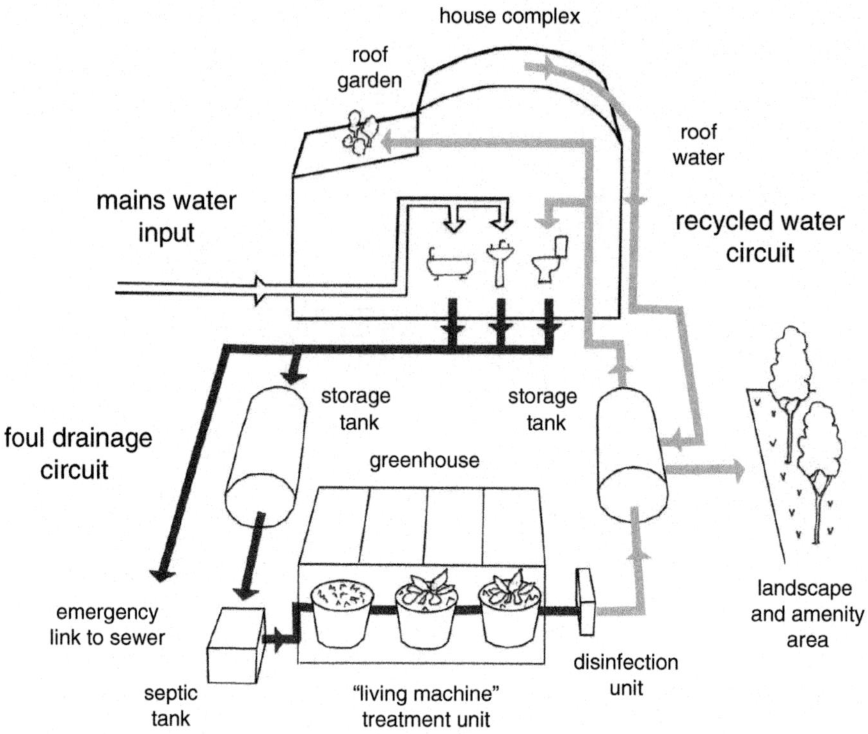

Fig. 9.5 The combined local water treatment and waste disposal system proposed for the BedZED development (Adapted from Shirley-Smith and Butler 2008)

BedZED houses were designed to operate with such units, some technical and managerial issues have still to be ironed out (Shirley-Smith and Butler 2008; Hodge and Haltrecht 2009).

Arguably the time has come for domestic waste systems in general to escape from the stereotyped approach of Victorian engineers. In the wake of nineteenth century urban disease outbreaks, such practitioners were understandably preoccupied with minimizing human contact with faecal wastes. Latterly however we have come to realize that there are numerous possibilities for putting such wastes to practical use. One such scheme in the Brooklyn area of New York involves fertilizing urban landscape areas with wastewater. Thirty ornamental roadside plots have been irrigated with effluents derived from a mixture of storm water and treated sewage. Other possibilities include channelling effluents into fishponds, draining them into willow coppices for biofuel cultivation or directing them into reed beds for the production of thatching materials. In Britain, reed-beds are increasingly being used for the dual purposes of providing the final stage of water treatment and of creating habitats for endangered wetland bird species (Cooper et al. 1996; Hawke and José 1996). These various initiatives do indeed recapture the traditional philosophy of putting wastes to work, but with the crucial difference that the modern schemes use pre-treatment and monitoring procedures to eliminate potential health hazards (WHO 2006).

9.4 Protecting Crop Biodiversity

9.4.1 The Preservation of Seed Stocks

The Western approach to agricultural seed diversity has generally been to focus on a limited number of high-yielding varieties, particularly ones that respond well to fertilizers. Often this is done with little regard for preserving traditional seed repertoires or ensuring that the genetic base of the new varieties is sufficiently wide to cope with new challenges (Cooper et al. 1992; Vellvé 1992). Historically the perils of such oversights have been amply demonstrated by the Irish potato famine of the 1840s – which was a consequence of concentrating on a limited number of disease-susceptible clones – and by the Californian wine-industry collapse of the 1880s that was caused by an over-reliance on a single rootstock vulnerable to attack by the aphid-like insect *Dactylosphaera vitifoliae*. Banana production too, has a history of suffering the penalties of focusing on too limited a range of varieties. The initially favoured commercial variety, *Gros Michel*, proved to be highly susceptible to the fungal pathogen Panama Disease and when it was subsequently replaced by a strain known as *Cavendish* all seemed well until this too has started to be attacked by another variety of the pathogen in the 1950s. Belatedly it has come to be realized that the banana-growing industry needs to draw on a much wider genetic base than has previously been the case and to this end a World Banana Collection comprising some 1400 varieties has been established at Leuven University near Brussels. It is hoped that this will produce the strains required to combat current and future disease problems.

Although the discovery that seeds can be successfully preserved in refrigerated stores seemed to obviate the need to maintain them in the farmer's fields, we now know that not all seeds can be protected in this way. Consequently the only sure way of preserving these varieties is to encourage village farmers to continue growing them (de Boef et al. 1993). These issues have been brought into sharp focus by the threat of climate change. Not only are current geographic zones likely to shift their positions, plant pest and disease problems are likely to appear in new localities. Access to genetically diverse crop materials will be essential to meet both types of challenge (Reusch et al. 2005; Di Falco and Chavas 2008). In countries such as the former Soviet Union, climate change is likely to make some areas warmer and wetter and others warmer and drier. Adjusting to these alterations will mean drawing on stored seed collections such as those accumulated by the notable plant geneticist Nikolai Vavilov at the Institute of Plant Industry in St Petersburg. The stored seeds derive for the most part from collections originally held by traditional farmers.

9.5 Lessons from Traditional Medicine

9.5.1 Success Stories

It is certainly the case that many former village remedies have achieved worldwide acclaim, sometimes by direct transfer, sometimes by synthesizing close copies (p.127). The list continues to lengthen. An interesting new addition is St John's wort (*Hypericum perforatum*) (Fig. 9.6a), a plant formerly used in ancient Greece as a remedy for mental disturbances and recently confirmed as an effective medication for depression (Linde et al. 1996; Szegedi et al. 2005). Another is feverfew (*Tanacetum parthenium*) (Fig. 9.6b), a headache remedy mentioned in Gerard's herbal of 1663 and now recognized as a useful modern treatment for migraine (Murphy et al. 1988). Trials show both these herbal preparations to be just as effective as their modern drug equivalents (Woelk 2000), although care needs to be exercised in the case of self-medication to avoid unwanted interactions with prescription drugs. For instance St John's wort preparations can reduce the effectiveness of oral contraceptives, chemotherapeutic drugs and the widely-used anti-coagulant *warfarin* (Ernst 1999; Yue et al. 2000; Barnes et al. 2002).

Further useful transfers can be anticipated. For example various village remedies for diabetes are currently being actively investigated and a number have already proved capable of lowering blood-sugar levels in experimental animals (Joshi 2011; Khan et al. 2012). The bitter gourd, *Momordica charantia*, looks like a particularly

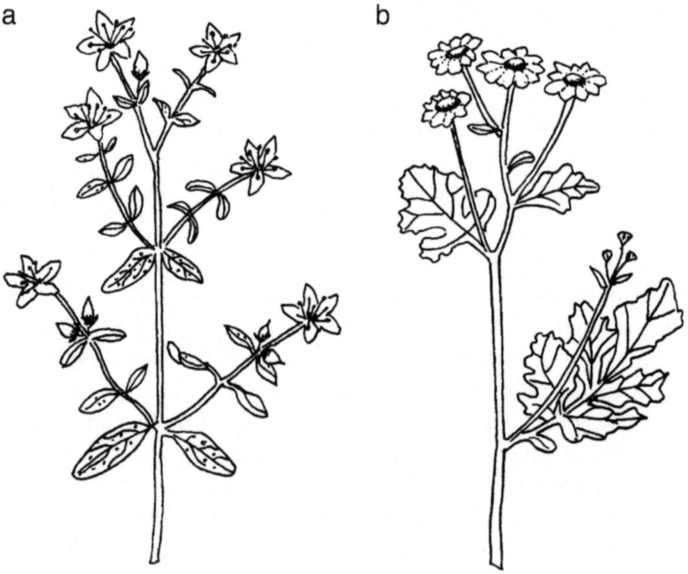

Fig. 9.6 Two traditional herbal remedies in effective modern use. (**a**) St John's wort (*Hypericum perforatum*), (**b**) feverfew (*Tanacetum parthenium*)

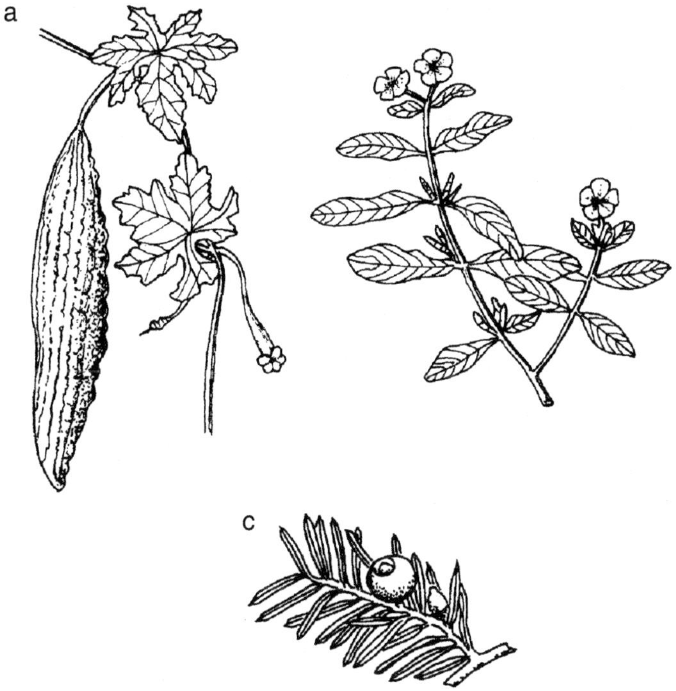

Fig. 9.7 Plants of modern therapeutic interest. (**a**) Bitter gourd (*Mormordica charantia*), (**b**) rosy periwinkle (*Catharanthus roseus*), (**c**) Pacific yew (*Taxus brevifolia*)

promising candidate (Fig. 9.7a). One can also anticipate that there will be further instances of plants used by local herbalists finding a wider and more general application for purposes other than their original ones. This has already happened with the rosy periwinkle (*Catharanthus roseus*) (Fig. 9.7b), a species traditionally used to treat diabetes and only marginally effective in this role (Svoboba et al. 1964), but now recognized as an effective treatment for childhood leukaemia (Lewis and Elvin-Lewis 1977). Another example is *prostratin,* a compound derived from a Samoan tree (*Homalanthus nutans*) and used locally to treat yellow fever. This has now found a wider potential application as an HIV/AIDS remedy (Cox 2001). Yet a further example is *taxol,* a compound obtained from the bark of the Pacific yew tree (*Taxus brevifolia*) (Fig. 9.7c) and used in various minor ways by native North Americans, but now emerging as a useful treatment for ovarian cancer (Markman 1991).

Transfers from traditional practice have, however, some residual pitfalls. It cannot be assumed that "natural" plant remedies are automatically free of side effects. Plants frequently manufacture chemical compounds to repel grazing herbivores (Rosenthal 1986) and it is therefore not surprising that some of these substances should prove equally toxic to humans if they are ingested for medicinal purposes

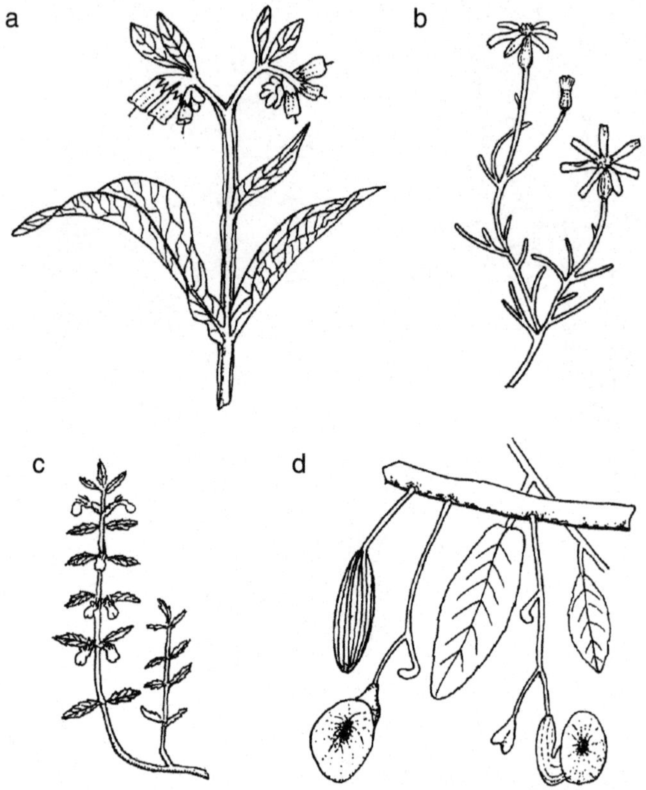

Fig. 9.8 Plants yielding potentially dangerous herbal remedies. (**a**) Comfrey (*Symphytum officiale*), (**b**) threadleaf groundsel (*Senecio longilobus*), (**c**) wall germander (*Teucrium chaemaedrys*), (**d**) birthwort (*Aristolochia fangchi*)

(Huxtable 1992). Neither does a long history of traditional use necessarily guarantee efficacy or safety if the original choice was made on faulty grounds (p.135).

Problem species in this context include comfrey (*Symphytum offinale*), threadleaf groundsel (*Senecio longilobus*) – a form of ragwort – and wall germander (*Teucrium chaemaedrys*) (Fig. 9.8a–c). The first two plants contain *pyrrolizidine alkaloids*; the same substances that are responsible for "bush tea" liver damage in the Caribbean (p.133) (Stillman et al. 1977; Ridker and McDermott 1989). In a typical episode, a 13-year-old boy suffering from intestinal disturbances was, on the advice of a herbalist, dosed with comfrey root extracts and comfrey leaf tea. This did not relieve his condition, but caused the veins in his liver to become dangerously constricted (Weston et al. 1987). In another case, a 47-year-old woman was advised by a homoeopath to treat fatigue and abdominal pain with comfrey preparations and she duly took to consuming as many as ten cups of comfrey tea a day, together with comfrey pills. Sometime later she became ill and was found to have suffered serious liver damage (Bach et al. 1989). With wall germander, the damaging component is probably a compound called *teucrin* and people taking germander capsules or drinking germander tea (usually to lose weight) are similarly prone to develop liver disorders (Larrey et al. 1992; Kouzi et al. 1994).

Sometimes the difficulties seem to stem less from the misapplication of herbs than from their misidentification. In one series of tragic incidents, 48 women in a Belgium health clinic suffered serious kidney damage after being encouraged to take slimming preparations containing Chinese herbs (Vanherweghem et al. 1993; Vanhaelen et al. 1994). The intended active ingredient was a root extract of *Stephania tetrandra* (which has a possible legitimate use as a diuretic), but what the mixture actually contained was root material from *Aristolochia fangchi* (Fig. 9.8d). The roots of the two species are similar in appearance, but those of *Aristolochia* contain compounds capable of causing serious kidney damage. In this particular instance the error had the further disastrous effect of causing a proportion of the patients to develop carcinomas of the urinary system (Cosyns et al. 1999; Kessler 2000; Nortier et al. 2000). In a similar case of mistaken identity, Chinese herbal preparations advertised as containing *Gentian* and *Clematis* extracts were shown subsequently to have become contaminated with root material from *Podophyllum hexandrum*. This relative of the American mayapple has a legitimate, though risky use, as a surface treatment for skin lesions (Duke 1985). However, if taken by mouth it is liable to cause serious nervous-system damage (But et al. 1996).

Such complications aside, there may be a sphere in which the traditional approach serves purposes poorly addressed by conventional medicine. This is where the disorders are largely psychological in origin. In one survey it was concluded that no less than a quarter of attendances at doctors' surgeries and half of all attendances at outpatient departments came into this category (Mayou and Sharpe 1997). The inference is that pressures such as those experienced at work or in fraught family situations (both current and historical) are capable of generating tangible physical symptoms. A recent compilation of case histories has shown how perplexing and disabling these psychosomatic disorders can be (O'Sullivan 2015). In these circumstances the prerequisite for a successful outcome would seem to be an extended exploratory dialogue with a sympathetic practitioner, with any medications given being largely incidental to the process (Payne 1986; Mercer et al. 2001).

This may also help to explain the efficacy of placebos (p.130) where the biochemical potency of any drugs administered seems to be subordinate to the psychological benefits derived from a consultation with a trusted practitioner. The most recent suggestion is that this psychological interplay between healer and patient stimulates the latter's "internal pharmacy" in a useful way (Jonas 2011; Meissner et al. 2011). It may also offer an explanation for the otherwise unlikely successes claimed for *homoeopathy*. The eighteenth century German physician Samuel Hahnemann was the originator of this technique, which he believed to work on the principle that, if a substance employed at full strength reproduces the symptoms of a disease, then a very dilute solution of the same substance will effect a cure. This is a fundamentally implausible proposition. In no other sphere are dilute solutions more active than concentrated ones; neither can water "remember" molecules it once contained but has subsequently lost through dilution (Cowan et al. 2005).

Nonetheless, there are many patients who claim benefits from homoeopathic procedures and one must conclude that the process has little to do with the biochemical content of the preparations and everything to do with the sympathetic manner of the therapists administering them (Ernst 2002; Shang et al. 2005).

As judged by the case histories recounted from Zimbabwe by Michael Gelfand, village healers are fully appreciative of the fact that physical illnesses can have underlying psychological or social causes (Gelfand et al. 1985). Modern Western practitioners on the other hand are often accused of giving insufficient attention to this possibility (Salmon et al. 2004; Rosendal et al. 2005). Possible reasons for this shortfall include time constraints, a poorer knowledge of the family backgrounds of patients than was hitherto the case and a perhaps also a partiality for strictly mechanistic explanations of illness.

There are various ways in which a "whole person" approach might be recaptured. One is to house conventional and alternative therapists under the same roof, with the former acting as portals of entry for the latter (Pietroni 1992), a suggestion that incidentally has raised the objection that this would mean that alternative therapists were being publicly funded within these establishments (McCartney 2011). Another possibility is for hard-pressed general practitioners to avail themselves more readily of the option of referring patients to specialists employing *cognitive behavioural therapy* techniques. In Britain this strategy is already being pursued under the acronym IAPT or *Improving Access to Psychological Therapies*. The main problem apparently is to find sufficient personnel to meet the demand, particularly in view of the large number of sessions required for each patient. A third solution is for the number of existing conventional practitioners to be increased to a point where they have more time to pursue these wider approaches; unfortunately not a very feasible option in an already hard pressed publicly funded health system. Against this background the useful prompts from traditional practice should be seen, not merely in terms of adding herbal remedies to modern medical repertoires, but also as a means of recapturing a more rounded approach to illness.

9.6 Improving Dietary Balance

9.6.1 Food Choices

After much deliberation, the World Health Organization has identified what it considers to be the ideal dietary mix for humans. This includes an appreciable amount of protein, a preponderance of complex carbohydrates over simple sugars, limited quantities of saturated fats and salt, but substantial amounts of fibre (column 4, Fig. 9.9). The WHO further contends that departures from this ideal mix are prone to generate a variety of degenerative disease problems including cardiovascular disorders, diabetes, dental caries, colorectal cancer and diverticulitis (Fig. 9.10). The case is comprehensively made in "*Diet, Nutrition and the Prevention of Chronic*

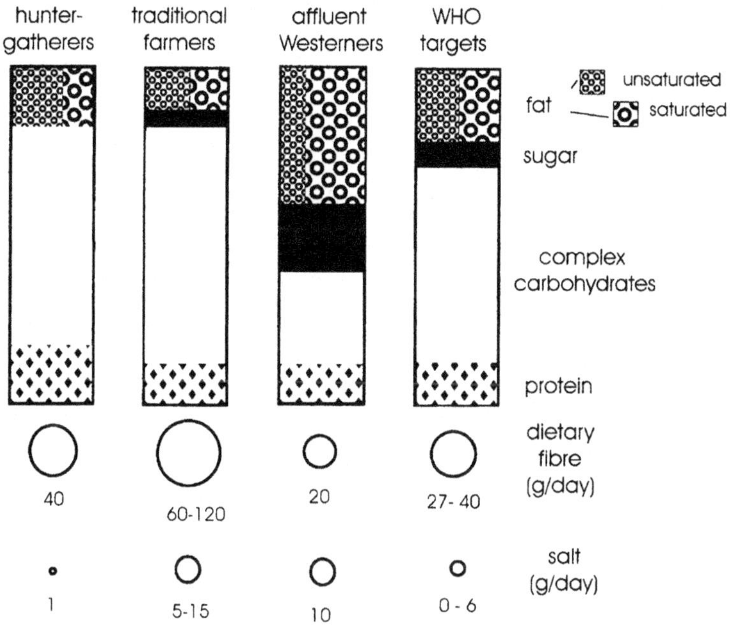

Fig. 9.9 Dietary composition during different phases of human historical development, compared with the "ideal" mix specified by the World Health Organization (Based on Boyden (1988), Eaton et al. (1988) and WHO (2003))

Diseases" (WHO 2003) and subsequently from extra evidence about a link between obesity and cancer (Murthy et al. 2009). The accumulation of calorie-rich materials in cells is thought to interfere with the production of cancer-suppressing agents. The finding that obesity levels in Britain had risen in 2014 to 24.4% amongst men and 25.1% amongst women adds immediacy to this issue.

Bearing these links in mind, it is remarkable that many Westerners follow a diet (column 3 in Fig. 9.9) that departs so radically from the recommended pattern. By contrast, traditional diets generally escape the shortcomings (columns 1 & 2 in Fig. 9.9) and come close to WHO's ideal dietary mix. A likely explanation for this favourable state of affairs is that human nutritional physiology developed early in mankind's history when less healthy foods such as simple sugars and saturated fats were difficult to come by. Whilst traditional communities have often persisted with diets that fit their physiology, Western consumers (aided and abetted by the food industry) have been placed in a position where they can overindulge in foods that, whilst certainly tasty, are potentially injurious to health if consumed in large quantities (Coghlan 1991). The attractiveness of salt as a food flavouring is universally acknowledged. The saturated fat contained in meat likewise contributes significantly to its flavour, and it is for precisely this reason that animal breeders have been working for decades to increase the fat content of domestic animal carcasses (Kyle 1987). In the case of sugar, infants apparently have a natural affinity for this substance (measured by the smiles on their faces), but one that would wane during adulthood were it not being constantly re-activated (Cowart 1981). In the words of one observer, modern humans can physiologically speaking, be rightly regarded as "stone agers attempting to operate in the fast lane" (Eaton et al. 1988).

Various strategies are available to address these problems. One is to require food manufacturers to make their products more wholesome; the other is to persuade consumers to select their foods more carefully. Unfortunately neither objective is readily achievable. Food companies are pressured by their shareholders to maintain profit margins and this they generally do by energetically marketing products favoured by consumers, whether or not these are good for them. Likewise, persuading people to change their diets is not an easy task in the face of entrenched habits and aggressive food advertising. Patent dietary regimes are constantly being proposed, but unfortunately many of these are unrealistic in physiological terms. Interestingly one of the more plausible ones, that advocated by Nathan Pritikin, draws its inspiration from the traditional feeding regime of the Kenyan *Kikuyu* (Pritikin and McGrady 1979). This diet emphasizes the use of cereal grains, potatoes and legumes, and shuns sugar, salt and large quantities of meat. It claims to minimize the risks of heart disease, high blood pressure, diabetes and obesity, in fact all the hazards featured in Fig. 9.10.

A recent dietary initiative has been to encourage consumers to eat 5 (or even 7) portions of fruit and vegetables each day (Oyebode et al. 2014). Again there are interesting parallels with traditional diets, which typically contain large quantities of fruits and vegetables (Table 3.1). The main difficulty with incorporating such a regime into the busy daily lives of Western consumers is the time and effort required to acquire and process the fresh materials. The default position, involving a heavy reliance on processed foods, makes it much more difficult to escape the dietary manipulations of the food industry.

One needs to remind oneself in passing that exercise is an effective way of countering the effects of overconsumption, quite apart from improving cardiovascular fitness (Prentice 1998). Members of village communities generally have such high physical workloads as to need no artificial exercise regimes. The same is not true for most Westerners. One recent set of experiments has shown that the combination of using modern kitchen appliances, driving to work rather than walking and using lifts rather than stairs has the net effect of halving the total energy expended in these daily activities and, in the absence of compensatory measures, can be expected to lead to an annual weight gain of the order of 10 lbs (4.5 kg) (Lanningham-Foster et al. 2003). Not only does exercise boost the power and efficiency of the heart, it releases neurotransmitters and mood-improving *endorphins* into the blood (Farrell et al. 1987). Moreover recent research suggests that it may also stimulate the important bodily process of *autophagy*, a kind of bodily garbage-disposal mechanism evolved to eliminate redundant and harmful components (He et al. 2012).

A cautionary footnote needs to be added to this catalogue of praise for traditional food practices. In the specific matter of food storage, traditional methods usually fall far short of their modern equivalents. Whilst local fermentation techniques are generally sound, there are question marks against some traditional methods of salt-

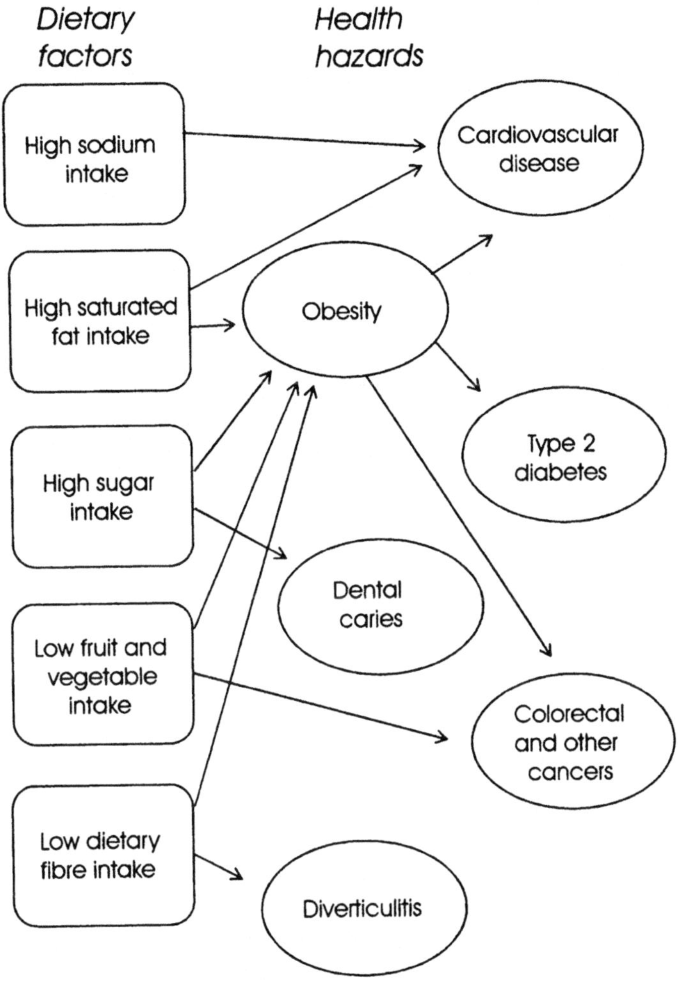

Fig. 9.10 Links between diet and chronic degenerative diseases (Based on WHO (2003) and Murthy et al. (2009))

ing and smoking. For instance there is a high incidence of gastric cancers in communities relying heavily on them (Tsugane and Sasazuki 2007; Murata et al. 2010). Neither are village techniques for storing grain ones to emulate. Granaries are constructed for this purpose and typically receive grain that has previously been dried on mats in the sun. Unfortunately when climatic conditions are constantly humid or become so during the rainy season, the grain in the granary reabsorbs moisture from the air (Giles 1964). This causes the growth of moulds and particular significance attaches to the fungal species *Aspergillus flavus*. This organism releases harmful substances known as *"aflatoxins"* (an acronym derived from its name) and people ingesting these compounds with their food are more susceptible to liver cancer

(Alpert et al. 1971; Shank et al. 1972; Peers and Linsell 1973; Van Rensburg et al. 1985).

The demand of modern consumers for food products with long shelf-lives has provided the impetus for developing efficient storage methods such as canning and freezing. These far out-perform any village procedures, although they can raise extra issues. Because the canning of some foods causes them to lose their colour (strawberries and peas are examples) it is usual to add artificial dyestuffs to retain their consumer appeal. Such "additives" have been criticized on the grounds that their physiological effects are not fully understood (Millstone 1984). An alternative is to make more use of the innocuous plant pigments used in village circles. Prominent amongst these natural agents is the yellow colouring *turmeric* obtained from the rhizome of *Curcuma domestica* and the red colouring *annatto* derived from the seeds of the shrub *Bixa orellana*. The writer well remembers being offered in a Belizean village a pasta-like dish prepared from the leaf shoots of the *jipijapa* tree (*Carlodovica insignis*) that had been attractively coloured pink with a dye obtained from *Bixa* plants growing in the house compound.

9.7 Attitudes to Wild Species

The idea of conservation was born in North America in the late nineteenth century. As land development pushed ever further West on that continent, it came to be realized that reserves and parks were necessary to preserve examples, not only of natural landscapes, but also of wild species and the habitats supporting them. These sentiments have gained ground in modern times and have been especially espoused by town dwellers seeking perhaps to maintain their connections with the natural world (Wilson 1984; Kellert and Wilson 1993). Those who are professionally involved in land management such as farmers and foresters, are less readily persuaded and usually reserve the right to suppress any species remotely suspected of damaging their enterprises. In Britain, for example, there is a long history of house sparrows (*Passer domesticus*) and rooks (*Corvus frugilegus*) being persecuted in spite of a lack of evidence that they cause significant agricultural damage (Murton 1971; Thomas 1983). By the same token, the gray wolf (*Canis lupus*) has been systematically persecuted in most temperate regions because of its reputation for attacking livestock (Anonymous 2012). Many stories of these attacks are certainly apocryphal and DNA saliva testing on killed livestock often shows the real culprits to be wild dogs.

Latterly an ostensibly more liberal stance has been constructed around the notion of "ecosystem services" (Loreau et al. 2002). This postulates that all species in natural communities have worth, because each can be presumed to contribute to the stability of the communities in which they occur and because the ecosystems thus stabilized bring practical benefits to mankind (as for example when forests check floods). With the present state of knowledge, this is a very sweeping assumption, but it is certainly true that instances can be found, in marine fisheries for example, where a full complement of species seems essential for community stability (Worm et al. 2006). Also it can be shown that wild species can contribute positively to farming operations, principally through their pest control or crop pollinating actions and in the case of earthworms through the process of soil conditioning (Stoate 2014).

However the corollary of this approach and one that is far from being universally acceptable is that species, for which no discernible practical function is identifiable, can be allowed to sink into extinction. In this connection the Rio Earth Summit of 1992 reiterated the conventional anthropocentric view, but also fleetingly signalled a departure from it by arguing that wild species can be considered to possess *intrinsic* worth, quite independently of any benefits accruing to mankind (United Nations Conference on Environment and Development 1992; Glowka et al. 1994).

This is a matter where the stance of some traditional groups is again illuminating. For instance the *Bishnoi* of Rajasthan have no difficulty whatsoever with the concept of intrinsic worth and are quite prepared to cherish even species that cause them practical problems (p.30). There are various ways of building on such precedents. Notwithstanding the biblical claim (Genesis 1) that man has dominion over nature and can deal with other species in any way he thinks convenient (White 1967; Santmire 1985), the bible also entertains notions of "stewardship" (Berry 2000). There are similar precedents in the more flexible stance taken by early churchmen such as St. Cuthbert and St. Francis of Assisi (Hughes 1996). A secular approach is equally conceivable and many environmental philosophers support the notion of intrinsic worth on the grounds that wild species (with the possible exception of virulent disease agents) should be allowed to fulfill the propensities of their kind without undue human interference (Taylor 1986; Attfield 1981, 1991; Sterba 1994).

References

Anonymous (2012) Call of the wild. De Economist 22:111–114

Attfield R (1981) The good of trees. J Value Enq 15:35–54

Attfield R (1991) The ethics of evironmental concern. University of Georgia Press, Athens

Bach N, Thung SN, Schaffner F (1989) Comfrey herb tea-induced veno-occlusive disease. Am J Med 87:97–99

Barnes J, Anderson LA, Phillipson JD (2002) Herbal medicines. A guide for healthcare professionals. Pharmaceutical Press, London

Berry RJ (ed) (2000) The care of creation. Focussing concern and action. Inter-Varsity Press, Leicester

Boyden S (1988) Western civilisation in biological perspective. Patterns in biohistory. Oxford University Press, Oxford

BRESCO (2002) BedZED – Beddington Zero Energy Development, Sutton. Building Research Establishment Energy Support Unit, Garston

But PPH, Tomlinson B, Cheung KO, Yong SP, Szeto ML, Lee CK (1996) Adulterants of herb products can cause poisoning. Br Med J 313:117

Chetley A (1990) A healthy business? World health and the pharmaceutical industry. Zed Books, London

Coghlan A (1991) Britain's deadly diet. New Sci 130:26–30

Cooke GW (1975) The energy costs of nitrogen fertilizers used in Britain, the returns received and some savings that are possible. J Sci Food Agric 26:1065–1069

Cooper D, Vellvé R, Hobbelink H (1992) Growing diversity, genetic resources and local food security. Intermediate Technology Publications, London

Cooper PF, Job GD, Green MB, Shutes RBE (1996) Reed beds and constructed wetlands for wastewater treatment. Water Research Centre Publications, Medmenham

Cosyns J-P, Jadoul M, Squifflet J-P, Wese FX, van Ypersele de Strihou C (1999) Urothelial lesions in Chinese herb nephropathy. Am J Kidney Dis 33:1011–1017

Cowan ML, Bruner BD, Huse N, Dwyer JR, Chugh B, Nibbering ETJ, Elsaesser T, Miller RJD (2005) Ultrafast memory loss and energy redistribution in the hydrogen bond network of liquid H_20. Nature 424:199–202

Cowart BJ (1981) Developmmnt of taste perception in humans: sensitivity and preference throughout the life span. Psychol Bull 90:43–73

Cox PA (2001) Ensuring equitable benefits, the Falealupo covenant and the isolation of the antiviral drug prostratin from a Samoan medicinal plant. Pharm Biol 39:33–40

Crouch D, Ward C (1988) A potted geography of the allotment. Geogr Mag 60(8):19–24

Crouch D, Ward C (1997) The allotment: its landscape and culture. Five Leaves Publications, Nottingham

de Boef W, Amanor K, Wellard K, Bebbington A (1993) Cultivating knowledge, genetic diversity, farmer experimentation and crop research. Intermediate Technology Publications, London

de Hart RAJ (1996) Forest gardening. Rediscovering nature and community in a post industrial age. Green Earth Books, Dartington

Denyer S (1978) African traditional architecture. An historical and geographical perspective. Heinemann, London

Di Falco S, Chavas JP (2008) Rainfall shocks, resilience, and the effects of crop diversity on agroecosystem productivity. Land Econ 84:83–96

Dmochowski ZR (1990) An introduction to Nigerian traditional architecture. Volume two, South-West and Central Nigeria. Ethnographica/The National Commission for Museums and Monuments, London/Lagos

Duke JA (1985) CRC handbook of medicinal herbs. CRC Press, Boca Raton

Eaton SB, Konner M, Shostak M (1988) Stone agers in the fast lane: chronic degenerative diseases in evolutionary perpective. Am J Med 84:739–749

Environment Canada (2004) Residential wood heating. Summary of results from 1999–2002. Public Works and Government Services Canada, Ottawa

Ernst E (1999) Second thoughts about the safety of St John's wort. Lancet 354:2014–2015

Ernst E (2002) A systematic review of systematic reviews of homeopathy. Br J Clin Pharmacol 54:577–582

Farrell PA, Gustafson AB, Morgan WP, Pert CB (1987) Enkephalins, catecholamines and psychological mood alterations: effects of prolonged exercise. Med Sci Sports Exerc 19:347–353

Festing H (1997) The potential for direct marketing by small farmers in the UK. Farm Manag 9(8):409–421

Friedman A, Cammalleri V (1995) The environmental impact of building materials in the North American building industry. Build Res Inf 23(3):162–166

Gelfand M, Mavi S, Drummond RB, Ndemera B (1985) The traditional medical practitioner in Zimbabwe. His principles of practice and Pharmacopoeia. Mambo Press, Gweru

Giles PH (1964) The storage of cereals by farmers in Northern Nigeria. Trop Agric Trinidad 41:197–212

Glowka L, Burhenne-Guilmin F, Synge H, McNeely JA, Gündling L (1994) A guide to the convention on biodiversity. IUCN, Gland/Cambridge

Hawke CJ, José PV (1996) Reedbed management for commercial and wildlife interests. Royal Society for the protection of Birds, Sandy

He C, Bassik MC, Moresi V, Sun K, Wei Y, Zou Z, An Z, Loh J, Fisher J, Sun Q, Korsmeyer S, Packer M, May HI, Hill JA, Virgin HW, Gilpin C, Xiao G, Bassel-Duby R, Scherer PE, Levine B (2012) Exercise-induced BCL2-regulated autophagy is required for muscle homeostasis. Nature 481:511–515

Hodge J, Haltrecht J (2009) BedZED seven years on: the impact of the UK's best known eco-village and its residents. Bioregional, Wallingford
Houghton J (2004) Global warming. The complete briefing. Cambridge University Press, Cambridge
Hughes JD (1996) Francis of Assisi and the diversity of creation. Environ Ethic 18:311–320
Huxtable RJ (1992) The myth of beneficient nature: the risks of herbal preparations. Ann Intern Med 117:165–166
Jonas WB (2011) Reframing placebo in research and practice. Philos Trans R Soc Lond B 366:1896–1904
Joshi K (2011) Ethnobotanical study of plants used for the treament of diabetes mellitus in the mountainous regions of Nepal. J Non-timber For Prod 18(1):19–26
Kellert SR, Wilson EO (eds) (1993) The Biophilia hypothesis. Island Press, Washington, DC
Khan V, Najmi AK, Akhtar M, Aqil A, Mujeeb M, Pillai KK (2012) A pharmacological appraisal of medicinal plants with antidiabetic potential. J Pharm BioAll Sci 4(1):27–42
Kouzi SA, McMurty RJ, Nelson SD (1994) Hepatotoxicity of germander *(Teucrium chamaedrys L.)* and one of its constituent neoclerodane diterpenes Teucrin A in the mouse. Chem Res Toxicol 7(6):850–856
Kumara R, Saito Y (1982) Country monograph Japan. In: Bowen A, Vagner R (eds) Passive and low energy alternatives. Pergamon Press, New York pp (15)30–(15) 44
Kyle R (1987) A feast in the wild. Kudu Publishing, Oxford
Lanningham-Foster L, Nysse LJ, Levine JA (2003) Labor saved, calories lost: the energetic impact of domestic labor-saving devices. Obes Res 11:1178–1181
Larrey D, Vial T, Pauwels A, Castot A, Biour M, David M, Michel H (1992) Hepatitis after germander (*Teucrium chamaedrys*) administration, another instance of herbal medicine toxicity. Ann Intern Med 117:129–132
Leach G (1976) Energy and food production. IPC Science and Technology Press, Guildford
Lewis WH, Elvin-Lewis MPF (1977) Medical botany, plants affecting Man's health. John Wiley, New York
Linde K, Ramirez G, Mulrow CD, Pauls A, Weidenhammer W, Melchart D (1996) St John's wort for depression-an overview and meta-analysis of randomised clinical trials. Br Med J 313:253–258
Lockeretz W, Shearer G, Kohl DH (1981) Organic farming in the corn belt. Science 211:540–547
Lockie S (2000) Designs on learning. Energy Environ Manag 2000:28–29
Markman M (1991) Taxol: an important new drug in the management of epithelial ovarian cancer. Yale J Biol Med 64:583–590
Mayou R, Sharpe M (1997) Treating medically unexplained physical symptoms. Br Med J 315:561–562
McCarthy JJ, Canziani O, Leary NA, Dokken DJ, White KS (eds) (2001a) Climate change 2001: impacts, adaptation and vulnerability. Contribution of working group II to the third assessment report of the intergovernmental panel on climate change. Cambridge University Press, Cambridge
McCarthy J, Riddall R, Topp C (2001b) Jubilee Campus, University of Nottingham. Arup J 2001(2):3–10
McCartney M (2011) The scam of integrative medicine. Br Med J 343:160
Meissner K, Kohls N, Colloca L (2011) Introduction to placebo effects in medicine: mechanisms and clinical implications. Philos Trans R Soc Lond B 366:1784–1789
Mercer SW, Watt GCM, Reilly D (2001) Empathy is important for enablement. Br Med J 322:865
Millstone E (1984) Food additives: a technology out of control? New Sci 18:20–24
Mollison B (1978) Permaculture one: a perennial agriculture for human settlements. Transworld Publications, Tyalgum
Murata A, Fujino Y, Pham T-M, Kubo T, Mizoue T, Tokui N, Matsuda S, Yoshimura T (2010) Prospective cohort study evaluating the relationship between salted food and gastrointestinal tract cancer mortality in Japan. Asia Pac J Clin Nutr 19(4):564–571
Murphy JJ, Heptinstall S, Mitchell JRA (1988) Randomised double-blind placebo-controlled trial of feverfew in migraine prevention. Lancet ii:189–192

Murthy NS, Mukherjee S, Roy G, Roy A (2009) Dietary factors and cancer chemoprevention: an overview of obesity-related malignancies. J Postgrad Med 55(1):45–54

Murton RK (1971) Man and birds. Collins, London

Nortier JL, Martinez MC, Schmeiser HH, Arlt VM, Bieler CA, Petein M, Depierreaux MF, De Pauw L, Abramowicz D, Vereerstraeten P, Vanherweghem J-L (2000) Urothelial carcinoma associated with the use of a Chinese herb *(Aristolochia fangchi)*. N Engl J Med 342:686–692

Oyebode O, Gordon-Dseagu V, Walker A, Mindell JS (2014) Fruit and vegetable consumption and all-cause cancer and CVD mortality: analysis of health survey for England. J Epidemiol Community Health. Published online 31st March 2014

Pacey A, Cullis A (1996) Rainwater harvesting, the collection of rainfall and runoff in rural areas. Intermediate Technology Publications, London

Paxton A (1994) The food miles report: the dangers of long distance food transport. SAFE Alliance, London

Payne JJ (1986) Alternative therapy. British Medical Association, Board of Science and Education, London

Pearce F (1995) Seeing the wood for the trees. New Sci 145:12–13

Pendleton K (2000) Green agenda, Hopkins and Partners at Nottingham. Ecotech 1(1):32–36

Pietroni PC (1992) Beyond the boundaries: relationship between general practice and complementary medicine. Br Med J 305:564–566

Pillet MLR (1980) Pueblo house design as a response to the arid-zone climate. In: Golany G (ed) Housing in arid lands, design and planning. The Architectural Press, London, pp 85–90

Pimentel D, Pimentel M (eds) (1996) Food, energy and society. University Press of Colorado, Boulder

Pimentel D, Hurd LE, Bellotti AC, Forster MJ, Oka IN, Sholes OD, Whitman RJ (1973) Food production and the energy crisis. Science 182:443–449

Pinkerton T, Hopkins R (2009) Local food. How to make it happen in your own community. Green Books, Dartington

Prentice A (1998) Sloth, machines that save you energy can also shorten your life. New Sci 157:36–37

Pritikin N, McGrady PM (1979) The pritikin program for diet and exercise. Bantam, New York

Ramshaw J (2003, May 7) Roofing choices – embodied energy and the environment. Ecotech Sustain Architect Today:30–36

Reganold JP, Elliott LF, Unger YL (1987) Long-term effects of organic and conventional farming on soil erosion. Nature 330:370–372

Reganold JP, Papendrick RI, Parr JF (1990) Sustainable agriculture. Sci Am 262(6):72–78

Reusch TBH, Ehlers A, Hammerli A, Worm B (2005) Ecosystem recovery after climatic extremes enhanced by genotypic diversity. Proc Natl Acad Sci 102:2826–2831

Reyman EJ (1982) Southwestern Pueblo architecture: some implications for modern housing design. In: Golany G (ed) Desert planning, international lessons. The Architectural Press, London, pp 105–112

Ridker PM, McDermott WV (1989) Comfrey herb tea and hepatic veno-occlusive disease. Lancet:657–658

Rosendal M, Olesen F, Fink P (2005) Management of medically unexplained symptoms. Br Med J 330:4–5

Rosenthal GA (1986) The chemical defences of higher plants. Sci Am 254(1):76–81

Salmon P, Dowrick CF, Ring A, Humphris GM (2004) Voiced but unheard agendas: qualitative analysis of psychosocial cues that patents with unexplained symptoms present to general practitioners. Br J Gen Pract 54:171–176

Santmire HP (1985) The travail of nature. The ambiguous ecological promise of Christian theology. Fortress Press, Philadelphia

Shang A, Huwiler-Müntener K, Nartey L, Jüni P, Dörig S, Sterne JAC, Pewsner D, Eggar M (2005) Are the clinical effects of homoeopathy placebo effects? Comparative study of placebo-controlled trials of homoeopathy and allopathy. Lancet 336:726–732

Shirley-Smith C, Butler D (2008) Water management at BedZED: some lessons. Eng Sustain 161:113–122

Sterba JP (1994) Reconciling anthropocentric and nonanthropocentric environmental ethics. Environ Value 3:229–244

Stillman AE, Huxtable R, Consroe P, Kohnen P, Smith S (1977) Hepatic veno-occlusive disease due to pyrrolizidine (*Senecio*) poisoning in Arizona. Gastroenterology 73:349–352

Stoate C (2014) Wildlife has its uses. Managing farmland for ecosystem services. Br Wildl 25(3):154–160

Svoboba GH, Gorman M, Root MA (1964) Alkaloids of *Vinca rosea.* A preliminary report on hypoglycaemic activity. Lloydia 27:361–363

Szegedi A, Kohnen R, Dienel A, Kieser M (2005) Acute treatment of moderate to severe depression with hypericum extract WS 5570 (St John's wort); randomised controlled double blind non-inferiority trial versus paroxetine. Br Med J 330:503–506

Talbot JL (1993) Simply Build green. A technical guide to the ecological houses at the Findhorn Foundation. Findhorn Foundation, Findhorn

Taylor PW (1986) Respect for nature. A theory of environmental ethics. Princeton University Press, Princeton

Thomas K (1983) Man and the natural world. Changing attitudes in England 1500–1800. Allen Lane, London

Todd J, Josephson J (1996) The design of living technologies for waste treatment. Ecol Eng 6:109–136

Tsugane S, Sasazuki S (2007) Diet and the risk of gastric cancer, review of epidemiological evidence. Gastric Cancer 10(2):75–83

Twinn C (2003) BedZED. Arup J 2003(1):10–16

Vale B, Vale R (2000) The new autonomous house. Design and planning for sustainability. Thames & Hudson, London

Vanhaelen M, Vanhaelen-Fastre R, But P, Vennerwegnem J-L (1994) Identification of aristolochic acid in Chinese herbs. Lancet 343:174

Vanherweghem J-L, Depierreux M, Tielemans C, Abramowicz D, Dratwa M, Jadoul M, Richard C, Vandervelde D, Verbeelen D, Vanhaelen-Fastre R, Vanhaelen M (1993) Rapidly progressive interstitial renal fibrosis in young women; association with slimming regime including Chinese herbs. Lancet 341:387–391

Vellvé R (1992) Saving the seed. Genetic diversity and European agriculture. Earthscan, London

Waterson R (1991) The living house: an anthropology of architecture in South-East Asia. Oxford University Press, Singapore

Westland P, Hinks J (1994) Minimising environmental impact: lifecycles of building materials. EcoDesign 3(1):14–15

Weston CFM, Cooper BT, Davies JD, Levine DF (1987) Veno-occlusive disease of the liver secondary to ingestion of comfrey. Br Med J 295:183

White L (1967) The historical roots of our ecologic crisis. Science 155:1203–1207

WHO (2003) Diet, nutrition and the prevention of chronic diseases. Technical Report 916, World Health Organization, Geneva

WHO (2006) WHO guidelines for the safe use of wastewater, excreta and grey water. World Health Organization, Geneva

Wilson EO (1984) Biophilia. The human bond with other species. Harvard University Press, Cambridge

Woelk H (2000) Comparison of St John's wort and imipramine for treating depression: randomised control trial. Br Med J 321:536–539

Worm B, Barbier EB, Beaumont N, Duffy JE, Folke C, Halpern BS, Jackson JBC, Lotze HK, Micheli F, Palumbi SR, Sala E, Selkoe KA, Stachowicz JJ, Watson R (2006) Impacts of biodiverity loss on ocean ecosystem services. Science 314:787–790

Yue Q-Y, Bergquist C, Gerdén B (2000) Safety of St John's wort (Hypericum perforatum). Lancet 355:576–577

References

Adetunji JA (1991) Response of parents to five killer diseases among parents in a Yoruba community, Nigeria. Soc Sci Med 32:1379–1387

Adongo PB, Phillips JF, Kajihara B, Fayorsey C, Debpuur C, Binka FN (1997) Cultural factors constraining the introduction of family planning among the Kassena-Nankana of Northern Ghana. Soc Sci Med 45:1789–1804

Agarwal B (1986) Cold hearths and barren slopes. The woodfuel crisis in the third world. Zed Books, London

Agarwal A, Narain S (eds) (1997) Dying wisdom. Rise, fall and potential of India's traditional water harvesting systems. Centre for Science and Environment, New Delhi

Agarwal A, Narain S, Sen S (eds) (2000) State of India's environment 5. *The Citizens Fifth Report.* Centre for Science and Environment, New Delhi

Agyepong IA (1992) Malaria: ethnomedical perceptions and practice in an Adangbe farming community and implications for control. Soc Sci Med 35:31–37

Ahmed S, Grainge M (1986) Potential of the neem tree *(Azadirachta indica)* for pest control and rural development. Econ Bot 40:201–209

Ahmad SS, Rahim S (1986) Response of the drywood termite *Cryptotermes cynocephalus* (Isoptera: Kalotermitidae) to extracts from five Malaysian hardwoods. Malay For 49:470–475

Akerele O (1984) WHO's traditional medicine programme: progress and perspectives. WHO Chron 34(2):76–78

Akerele O (1987) The best of both worlds: bringing traditional medicine up to date. Soc Sci Med 24:177–181

Akerele O, Heywood V, Synge H (1991) Conservation of medicinal plants. Cambridge University Press, Cambridge

Ali S, Ripley SD (1984) Handbook of the birds of India and Pakistan. Oxford University Press, New Delhi

Alkire BH, Tucker AO, Maciarello MJ (1994) Tipo, *Minthostachys mollis* (Lamiaceae) an Ecuadorian mint. Econ Bot 48:60–64

Allen C (1979) Tales from the dark continent. In: Images of British Colonial Africa in the twentieth century. André Deutsch and the British Broadcasting Corporation, London

Alliance for Rural Electrification (2011) Hybrid mini-grid for rural electrification: lessons learned. Alliance for Rural Electrification, Brussels and USAID, Washington, DC

Altieri MA, Merrick LC (1987) In situ conservation of crop genetic resources through maintenance of traditional farming systems. Econ Bot 41:86–96

Altieri MA, Whitcomb WH (1979) The potential use of weeds in the manipulation of beneficial insects. Hortscience 14:12–18

J. Edington, *Indigenous Environmental Knowledge*,
DOI 10.1007/978-3-319-62491-4

Altrichter M (2005) The sustainability of subsistence hunting of peccaries in the Argentine Chaco. Biol Conserv 126:351–362

Ananthaswamy A (2004) Massive growth of ecotourism worries biologists. New Sci 4:6–7

Anonymous (2001) Folk wisdom. Traditional weather forecasting on trial. *The Economist,* November 24, 2001, 102–104

Anonymous (2002) Sad little sorcerers. *The Economist,* December 7, 2002, 68

Anonymous (2007) Thinking out of the box. How African cocoa-growers are moving upstream into chocolate. *The Economist,* April 7, 2007, 77

Anonymous (2008) Fresher cookers. De Economist 6:3–4

Anonymous (2012) Call of the wild. De Economist 22:111–114

Antonsson-Ogle B (1984) Wild plant resources. Ceres 17(5):38–40

Appleton CC (1978) Review of literature on abiotic factors influencing the distribution and life cycles of bilharziasis intermediate host snails. Malacol Rev 11:1–25

Aradeon SB (1987) The architecture of a Hausa master mason. Nigerian Field 52:3–18

Arber A (1953) Herbals, their origin and evolution, A chapter in the history of botany 1470–1670. Cambridge University Press, Cambridge

Armstrong B, Doll R (1975) Enviromental factors and cancer incidence and mortality in different countries with special reference to dietary practices. Int J Cancer 15:617–631

Aronson JK (1985) An account of the Foxglove and its medical uses 1785–1985. Oxford University Press, London

Arvigo R, Balick M (1993) Rainforest remedies. One hundred healing herbs of Belize. Lotus Press, Twin Lakes

Ash C, Crompton D, Keymer A (1984) Nature's unfair food tax. New Sci 8:17–20

Ashton PJ, Walmsley RD (1976) The aquatic fern *Azolla* and its *Anabaena* symbiont. Endeavour 35:39–43

Asibey EOA (1974) Wildlife as a source of protein south of the Sahara. Biol Conserv 6:32–39

Atkinson TC, Briffa KR, Coope GR (1987) Seasonal temperatures in Britain during the past 22,000 years, reconstructed using beetle remains. Nature 375:587–592

Attfield R (1981) The good of trees. J Value Enq 15:35–54

Attfield R (1991) The ethics of evironmental concern. University of Georgia Press, Athens

Avé W (1987) Small-scale utilization of rattan by a Semai community in West Malaysia. Econ Bot 42:105–119

Aya T, Kinoshita T, Imai S, Koizumi S, Mizuno F, Osato T, Satoh C, Oikawa T, Kuzumaki N, Ohigashi H, Koshimizu K (1991) Chromosome translocation and c-MYC activation by Epstein-Barr virus and *Euphorbia tirucalli* in B lymphocytes. Lancet 337:1190

Bach N, Thung SN, Schaffner F (1989) Comfrey herb tea-induced veno-occlusive disease. Am J Med 87:97–99

Bahadori MN (1978) Passive cooling systems in Iranian architecture. Sci Am 238(2):144–152

Bahadori MN, Haghighat F (1985) Passive cooling in hot, arid regions in developing countries by employing domed roofs and reducing the temperature of internal surfaces. Build Environ 20:103–113

Bahti T (1982) Southwestern Indian ceremonials. KC Publications, Las Vegas

Baker ARH (1973) Field systems in medieval England. In: Baker RH, Harley JB (eds) Man made the land, essays in English historical geography. David and Charles, Newton Abbot, pp 59–68

Baldwin S, Geller H, Dutt G, Ravindranath NH (1985) Improved woodburning cookstoves: signs of success. Ambio 14(4–5):280–287

Balée WL (1999) Footprints of the forest. The historical ecology of plant utilization by an Amazonian people. Columbia University Press, New York

Bardach JE, Ryther JH, McLarney WO (1972) Aquaculture: the farming and husbandry of freshwater and marine organisms. John Wiley, New York

Barnard GW, Kristoferson L (1985) Agricultural residues as fuel in the third world. Earthscan, London

Barnes J, Anderson LA, Phillipson JD (2002) Herbal medicines. A guide for healthcare professionals. Pharmaceutical Press, London

Barraclough SL, Ghimire KB (1996) Deforestation in Tanzania. Beyond simplistic generalizations. Ecologist 26:104–109

Bar-Yosef O, Meadow RH (1995) The origins of agriculture in the Near East. In: Gebauer B (ed) Price TD. Last hunters –first farmers. new perspectives on the prehistoric transition to agriculture. School of American Research Press, Santa Fé, pp 39–94

Basu NG (1987) Forests and tribals. Manisha Granthalaya, Calcutta

Beecher HK (1955) The powerful placebo. J Am Med Assoc 159:1602–1606

Beets WC (1982) Multiple cropping and tropical farming systems. Westview Press, Boulder

Beijing Medical College (1984) Dictionary of traditional Chinese medicine. The Commercial Press, Hong Kong

Benedetti F (2008) Placebo effects. Understanding the mechanisms in health and disease. Oxford University Press, Oxford

Bentley JW, Andrews KL (1991) Pests, peasants, and publications: anthropological and entomological views of an integrated pest management program for small-scale Honduran farmers. Hum Organ 50:113–270

Berlin B (1992) Ethnobotanical classification. Principles of categorization of plants and animals in traditional societies. Princeton University Press, Princeton

Berlin B, Breedlove DE, Raven PH (1966) Folk-taxonomies and biological classification. Science 154:273–275

Beroza M, Bottger GT (1954) The insecticidal value of *Tripterygium wilfordii*. J Econ Entomol 47(1):188–189

Berry RJ (ed) (2000) The care of creation. Focussing concern and action. Inter-Varsity Press, Leicester

Bhandari MM (1974) Famine foods in the Rajasthan desert. Econ Bot 28:73–81

Bhandari MM (1978) Flora of the Indian desert. Scientific Publishers, Jodhpur

Bhat RV, Krishnamachari KAVR (1977) Endemic familial arthritis of Malnad – an epidemiological study. Indian J Med Res 66:777–786

Bierlich B (1995) Notions and treatment of Guinea worm in Northern Ghana. Soc Sci Med 41:501–509

Biswas AK (1970) History of hydrology. North Holland Publishing Company, Amsterdam

Black M (1983) Spreading the good news about water and sanitation. UNICEF News 116:13–24

Bledsoe CH, Goubaud MF (1985) The reinterpretation of western pharmaceuticals amongst the Mende of Sierra Leone. Soc Sci Med 3:275–282

Bleek WHI, Lloyd LC (1911) Specimens of bushman folklore. George Allen, London

Blier SP (1987) The anatomy of architecture: ontology and metaphor in Batammaliba architectural expression. Cambridge Univerity Press, Cambridge

Bockstoce J (1980) Battle of the bowheads. Nat Hist 85(5):52–58

Bogardus ST (2006) What do we know about diverticular disease? A brief overview. J Clin Gastroenterol 40(Suppl 3):108–111

Boiteau P, Ratsimamanga AR (1956) Asiaticoside isolé de *Centella asiatica* et ses emplois thérapeutiques. Therapie 11:125–150

Bomba WG (1982) Sanitation aspects of water supply and excreta disposal. In: Schiller EJ, Droste RL (eds) Water supply and sanitation in developing countries. Ann Arbor Science Publishers, Ann Arbor, pp 311–325

Borlaug NE, Dowswell CR (1995) Mobilising science and technology to get agriculture moving in Africa. Dev Policy Rev 13:115–129

Bose PK, Roy AC (1936) The constitution of ayapanin. J Indian Chem Soc 13:586–587

Bottrell DG, Adkisson PL (1977) Cotton insect pest management. Annu Rev Entomol 22:451–481

Bourgeois J-L (1980) Welcoming the wind. Nat Hist 89:70–75

Bowonder B, Prasad SSR, Unni NVM (1987) Deforestation around urban centres in India. Environ Conserv 14:23–28

Boyden S (1988) Western civilisation in biological perspective. Patterns in biohistory. Oxford University Press, Oxford

Brader L (1979) Integrated pest control in the developing world. Annu Rev Entomol 24:225–254
Bradley PN (1991) Woodfuel, women and woodlots. In: The foundations of a Woodfuel Development Strategy for East Africa, vol 1. Macmillan. London
Bradley PN, Chavangi N, Van Gelder A (1985) Development research and energy planning in Kenya. Ambio 14(4–5):228–236
Bras G, Berry DM, Gyorgyi P (1957) Plants as an aetiological factor in veno-occlusive disease of the liver. Lancet:960–962
Bray F (1984) Joseph Needham, science and civilisation in China, vol 6, Biology and biological technology, Part II, Agriculture. Cambridge University Press, Cambridge
BRESCO (2002) BedZED – Beddington Zero Energy Development, Sutton. Building Research Establishment Energy Support Unit, Garston
Briceño-Garmendia C, Estache A, Shafik N (2004) Infrastructure services in develoing countries: access, quality, costs and policy reform, World Bank policy research working paper 3468. World Bank, Washington, DC
Broecker WS, Kennett JP, Flower BP, Teller JP, Trumbore S, Bonani G, Wolfli W (1989) Routeing of meltwater from Lawrentide ice sheet during the Yonger Dryas cold episode. Nature 341:318–321
Bromehead CEN (1942) The early history of water-supply. Geogr J 99(141–151):183–196
Brosius JP (1991) Foraging in tropical forests: the case of the Penan of Sarawak, East Malaysia (Borneo). Hum Ecol 19(2):123–150
Brown RJ (1998) The placebo effect. Sci Am 278(1):90–95
Brown BJ, Marten GG (1986) The ecology of traditional pest management in Southeast Asia. In: Marten GG (ed) Traditional agriculture in Southeast Asia. A human ecology perspective. Westview Press, Boulder, pp 241–272
Bruce N, Perez-Padilla R, Albalak R (2000) Indoor smoke pollution in developing countries: a major environmental and public health challenge. Bull World Health Organ 78(9):1078–1092
Brush SB (1980) Potato taxonomies in Andean agriculture. In: Brokensha D, Warren M, Werner O (eds) Indigenous knowledge systems and development. University Press of America, New York, pp 37–47
Brush SB, Carney HJ, Huamán Z (1981) Dynamics of Andean potato agriculture. Econ Bot 35:70–88
Bryceson DF (1989) Nutrition and the commoditization of food in Sub-Saharan Africa. Soc Sci Med 28:425–440
Bull D (1982) A growing problem. Pesticides and the third world poor. OXFAM, Oxford
Burkitt DP (1969) Epidemiology of cancer of the colon and rectum. Cancer 28:1–13
Burr GO, Burr MM (1929) A new deficiency disease produced by the rigid exclusion of fat from the diet. J Biol Chem 82:345–367
Burr GO, Burr MM (1930) The nature and role of the fatty acids essential in nutrition. J Biol Chem 86:587–621
Bustard HR (1970) A future for crocodiles. Oryx 10(4):249–255
But PPH, Tomlinson B, Cheung KO, Yong SP, Szeto ML, Lee CK (1996) Adulterants of herb products can cause poisoning. Br Med J 313:117
Cain A, Afshar F, Norton J, Daraie MR (1976) Traditional cooling systems in the third world. Ecologist 6(2):60–64
Caldecott J (1988, June) Climbing towards extinction. New Sci 9:62–66
Camino A, Johns T (1988) Laki-laki *(Dennstaedtia glauca,* Polypodiaceae): a green manure used in traditional Andean agriculture. Econ Bot 42:45–53
Campbell BM (1987) The use of wild fruits in Zimbabwe. Econ Bot 41(3):375–385
Campbell BM, Clarke JM, Gumbo DJ (1991) Traditional agroforestry practices in Zimbabwe. Agrofor Syst 14:99–111
Carroll RF (1990) Disease reduction by improved house construction. Overseas building note 194. Building Research Establishment, Garston
Carter WE (1969) New lands and old traditions. Kekchi cultivators in the Guatemalan lowlands. University of Florida Press, Gainesville

Carter AS, Gilmour DA (1989) Increase in tree cover on private land in Central Nepal. Mt Res Dev 9:381–391
Cela-Conde C, Ayala FJ (2007) Human evolution. Trails from the past. Oxford University Press, Oxford
Chacón JC, Gliessman SR (1982) Use of the "non-weed" concept in traditional tropical agroecosystems of south-eastern Mexico. Agro-Ecosyst 8:1–11
Chakravarti AK (1974) Regional preference for food: some aspects of food habit patterns in India. Can Geogr 17:395–410
Chetley A (1990) A healthy business? World health and the pharmaceutical industry. Zed Books, London
Chin T, Welsby PD (2004) Malaria in the UK: past, present, and future. Postgrad Med J 80:663–666
Chiwuzie J, Ukoli F, Okojie O, Isah E, Eriator I (1987) Traditional practitioners are here to stay. World Health Forum 8:240–244
Chopra RN, Nayar SL, Chopra IC (1956) Glossary of Indian medicinal plants. Council of Scientific & Industrial Research, New Delhi
Chopra RN, Nayar SL, Chopra IC (1969) Supplement to glossary of Indian medicinal plants. Council of Scientific and Industrial Research, New Delhi
Clark C (1967) Population growth and land use. Macmillan, London
Cleary GJ, Blackburn RB (1968) Air pollution in native huts in the highlands of New Guinea. Arch Environ Health 17:785–794
Clifford P (1972) Carcinogens in the nose and throat: nasopharyngeal carcinoma in Kenya. Proc R Soc Med 65:682–686
Clifford P, Beecher JL (1964) Nasopharygeal cancer in Kenya, clinical and environmental aspects. Br J Cancer 18:25–43
Codrington HW (1947) A short history of Ceylon. Macmillan, London
Coghlan A (1991) Britain's deadly diet. New Sci 130:26–30
Cohen MND (1977) The food crisis in prehistory: overpopulation and the origins of agriculture. Yale University Press, Newhaven
Colchester M (1997) Salvaging nature: indigenous peoples and protected sites. In: Ghimire KB, Pimbert MP (eds) Social change and conservation, environmental politics and impacts of national parks and protected areas. Earthscan, London pp, pp 97–130
Conklin N (1991) Naga motifs. Sawaddi, Bangkok 37(3):28–31
Constantino PAL, Fortini LB, Kaxinawa FRS, Kaxinawa AM, Kaxinawa ES, Kaxinawa AP, Kaxinawa LS, Kaxinawa JM, Kaxinawa JP (2008) Indigenous collaborative research for wildlife management in Amazonia: the case of the Kaxinawá, Acre, Brazil. Biol Conserv 141:2718–2729
Conway GR (1972) Ecological aspects of pest control in Malaysia. In: Farvar MT, Milton JP (eds) The careless technology: ecology and international development. The Natural History Press, New York, pp 467–488
Cook JW, Duffy E, Schoental R (1950) Primary liver tumours following feeding with alkaloids of *Senecio jacobaea* Lin. Br J Cancer 4:405–410
Cooke GW (1975) The energy costs of nitrogen fertilizers used in Britain, the returns received and some savings that are possible. J Sci Food Agric 26:1065–1069
Cooper MR, Johnson AW (1984) Poisonous plants in Britain and their effects on animals and man. HMSO, London
Cooper D, Vellvé R, Hobbelink H (1992) Growing diversity, genetic resources and local food security. Intermediate Technology Publications, London
Cooper PF, Job GD, Green MB, Shutes RBE (1996) Reed beds and constructed wetlands for wastewater treatment. Water Research Centre Publications, Medmenham
Cosyns J-P, Jadoul M, Squifflet J-P, Wese FX, van Ypersele de Strihou C (1999) Urothelial lesions in Chinese herb nephropathy. Am J Kidney Dis 33:1011–1017
Cowan ML, Bruner BD, Huse N, Dwyer JR, Chugh B, Nibbering ETJ, Elsaesser T, Miller RJD (2005) Ultrafast memory loss and energy redistribution in the hydrogen bond network of liquid H_20. Nature 424:199–202

Cowart BJ (1981) Developmmnt of taste perception in humans: sensitivity and preference throughout the life span. Psychol Bull 90:43–73

Cox PA (1994) The ethnobotanical approach to drug discovery; strengths and limitations. In: Prance G, Chadwick DJ, Marsh J (eds) Ethnobotany and the search for new drugs. Wiley, Chichester, pp 25–36

Cox PA (2001) Ensuring equitable benefits, the Falealupo covenant and the isolation of the anti-viral drug prostratin from a Samoan medicinal plant. Pharm Biol 39:33–40

Crawford MA (1968) Fatty-acid ratios in free-living and domestic animals. Possible implications for atheroma. Lancet i:1329–1333

Croom EM (1983) Documenting and evaluating herbal remedies. Econ Bot 37:13–27

Crouch D, Ward C (1988) A potted geography of the allotment. Geogr Mag 60(8):19–24

Crouch D, Ward C (1997) The allotment: its landscape and culture. Five Leaves Publications, Nottingham

Cuevas VC, Samulde SN, Pajaro PG (1988) *Trichoderma harzianum* Rifai as activator for rapid composting of agricultural wastes. Philipp Agric 71(4):461–469

Cunningham AB (1991) Development of a conservation policy on commercially exploited plants: a case study from South Africa. In: Akerele O, Heywood V, Synge H (eds) Conservation of medicinal plants. Cambridge University Press, Cambridge, pp 337–358

Cunningham AB (2001) Applied ethnobotany. People, wild plant use & conservation. Earthscan Publications, London

Cunningham AB, Milton SJ (1987) Effects of basket-weaving industry on mokola palm and dye plants in northwestern Botswana. Econ Bot 41:386–402

Curzon C (1993) Ecotourism-conservation ethics-profit. Getting it right in the Eastern Transvaal. Afr Environ Wildl 1(2):36–42

Dalby D, Harrison Church RJ, Bezzaz F (1977) Drought in Africa. International African Institute, London

Dang H (1993) Fuel substitution in Sub-Saharan Africa. Environ Manag 17:283–288

Dansgaard W, White JWC, Johnsen SJ (1989) The abrupt termination of the Younger Dryas event. Nature 339:532–534

Dansgaard W, Johnsen SJ, Clausen HB, Dahl-Jensen D, Gundestrup NS, Hammer CU, Hvidberg CS, Steffensen JP, Sveinbjörnsdottir AE, Jouzel J, Bond G (1993) Evidence for general instability of past climate from a 250-kyr ice-core record. Nature 364:218–220

Dasmann RF (1964) African game ranching. Pergamon Press, Oxford

Datta SC, Banerjee AK (1978) Useful weeds of Bengal rice fields. Econ Bot 32:297–310

Dayton L (1991) Saving mud monuments. New Scientist 130, 25 May, 38-42

de Boef W, Amanor K, Wellard K, Bebbington A (1993) Cultivating knowledge, genetic diversity, farmer experimentation and crop research. Intermediate Technology Publications, London

De Groot AC, Herxheimer A (1989) Isothiazolinone preservative: cause of a continuing epidemic of cosmetic dermatitis. Lancet:314–316

de Hart RAJ (1996) Forest gardening. Rediscovering nature and community in a post industrial age. Green Earth Books, Dartington

De Koning HW, Smith KR, Last JM (1985) Biomass fuel combustion and health. Bull World Health Organ 63:11–26

De Schlippé P (1956) Shifting cultivation in Africa. The Zande system of agriculture. Routledge/Kegan Paul, London

De Trey-White S (2006) The witches of Gambaga. Geogr Mag 78(8):48–53

Delange F, Ahluwalia R (eds) (1983) Cassava toxicity and thyroid: research and public health issues. International Development Research Centre, Ottawa

Denevan WM, Padoch C (1988) Introduction: the bora agroforestry project. In: Denevan WM, Padoch C (eds) Swidden-fallow agroforestry in the Peruvian Amazon. Advances in Economic Botany 5. New York Botanical Garden, New York, pp 1–7

Denevan WM, Treacy JM (1987) Young managed fallows at Brillo Nuevo. In: Denevan WM, Padoch C (eds) Swidden-fallow agroforestry in the Peruvian Amazo, Adavances in Economic Botany, vol 5. New York Botanical Garden, New York, pp 8–46

Dentan RK (1979) The Semai, a nonviolent people of Malaya. Holt, Rinehart and Winston, New York

Denyer S (1978) African traditional architecture. An historical and geographical perspective. Heinemann, London

Department of Health (1991) Dietary reference values for food, energy and nutrients for the United Kingdom. HMSO, London

Deressa W, Ali A (2009) Malaria-related perceptions and practices of women with children under the age of five years in Ethiopia. Biomed Centr Public Health 9:259

DeWalt BR (1993) Nutrition and the commercialization of agriculture: ten years later. Soc Sci Med 36:1407–1416

Dewey KG (1989) Nutrition and the commoditization of food systems in Latin America and the Caribbean. Soc Sci Med 28:415–424

Di Falco S, Chavas JP (2008) Rainfall shocks, resilience, and the effects of crop diversity on agro-ecosystem productivity. Land Econ 84:83–96

Dinerstein E, Rijal A, Bookbinder M, Kattel B, Rajuria A (1999) Tigers as neighbours: efforts to promote local guardianship of endangered species in rural Nepal. In: Seidensticker J, Christie S, Jackson P (eds) Riding the Tiger. Tiger conservation in human-dominated landscapes. Cambidge University Press, Cambridge, pp 316–333

Dmochowski ZR (1990) An introduction to Nigerian traditional architecture. Volume two, South-West and Central Nigeria. Ethnographica/The National Commission for Museums and Monuments, London/Lagos

Doutt RL (1964) The historical development of biological control. In: DeBach P (ed) Biological control of insect pests and weeds. Chapman and Hall, London, pp 21–42

Drèze J, Sen A (2002) India, development and participation. Oxford University Press, New Delhi

Drummond JC, Wilbraham A (1958) The Englishman's food. A history of five centuries of English diet. Jonathan Cape, London

D'Souza F (1988) Famine: social security and an analysis of vulnerability. In: Harrison GA (ed) Famine. Oxford University Press, Oxford, pp 1–56

Duby G (1968) Rural economy and country life in the Mediaeval West. University of South Carolina Press, Columbia

Duke JA (1985) CRC handbook of medicinal herbs. CRC Press, Boca Raton

Dunn FL (1968) Epidemiological factors: health and disease in hunter-gatherers. In: Lee RB, DeVore I (eds) Man the hunter. Aldine Publishing Company, Chicago, pp 221–228

Dunsmore S (1985) The nettle in Nepal, a cottage industry. Land Resources Development Centre, Surbiton

Eaton SB, Konner M (1985) Palaeolithic nutrition. A consideration of its nature and current implications. N Engl J Med 312(5):283–289

Eaton SB, Konner M, Shostak M (1988) Stone agers in the fast lane: chronic degenerative diseases in evolutionary perpective. Am J Med 84:739–749

Eckholm EP (1975) The deterioration of mountain environments. Science 189:764–770

Eckholm EP (1976) Losing ground, environmental stress and world food prospects. Worldwatch Institute, W.W. Norton, New York

Edington JM (1996) Effects of tourism on natural systems. In: Giller PS, Myers AA (eds) Disturbance and recovery of ecological systems. Royal Irish Society, Dublin, pp 173–181

Edington JM, Edington MA (1977) Ecology and environmental planning. Chapman and Hall, London

Edington JM, Edington MA (1986) Ecology, recreation and tourism. Cambridge University Press, Cambridge

Edington JM, Edington MA (1997) Tropical forest ecotourism: two promising projects in Belize. In: Stabler MJ (ed) Tourism & sustainability: from principles to practice. CAB International, Wallingford, pp 163–168

Eisenberg DM, Kessler RC, Foster C, Norlock FE, Calkins DR, Delbanco TL (1993) Unconventional medicine in the United States. Prevalence, costs, and patterns of use. N Engl J Med 328:246–252

Eisenberg DM, Davis RB, Ettner SL, Appel S, Wilkey S, Van Rompay M, Kessler RC (1998) Trends in alternative medicine use in the United States, 1990–1997. Results of a follow-up national survey. J Am Med Assoc 280:1569–1575
Eitel EJ (1984) Feng-Shui. The science of sacred landscape in old China, (reprint of 1873 edition). Synergetic Press, London
Ekvall RB (1968) Fields on the Hoof, Nexus of Tibetan Nomadic Pastoralism. Holt, Rinehart and Winston, New York
Elias PM, Brown BE, Ziboh VA (1980) The permeability barrier in essential fatty acid deficiency: evidence for a direct role for linoleic acid in barrier function. J Investig Dermatol 74:230–233
Ellis FP (1953) Tolerable and desirable levels of warmth in warm climates, with special reference to men in the Royal Navy. Ann R Coll Surg Engl 13:369–391
Environment Canada (2004) Residential wood heating. Summary of results from 1999–2002. Public Works and Government Services Canada, Ottawa
Ernst E (1999) Second thoughts about the safety of St John's wort. Lancet 354:2014–2015
Ernst E (2002) A systematic review of systematic reviews of homeopathy. Br J Clin Pharmacol 54:577–582
Escobar GJ, Salazar E, Chuy M (1983) Beliefs regarding the etiology and treatment of infantile diarrhea in Lima, Peru. Soc Sci Med 17:1257–1269
Ettema CH (1994) Indigenous soil classification systems. What is their structure and function and how do they compare to scientific soil classifications? University of Georgia, Athens
Ezzati M, Kammen DM (2001) Indoor air pollution from biomass combustion as a risk factor for acute respiratory infections in Kenya: an exposure response study. Lancet 358:619–624
Falconer J (1990) "Hungry season" food from the forest. Unasylva 160:14–19
Farnsworth NR (1990) The role of medicinal plants in drug development. In: Krogsgaard-Larsen P, Christensen SB, Kofod H (eds) Natural products and drug development. Ballière, Tindall and Cox, London, pp 8–98
Farnsworth NR (1994) Ethnopharmacology and drug development. In: Prance G, Chadwick DJ, Marsh J (eds) Ethnobotany and the search for new drugs. Wiley, Chichester, pp 42–51
Farrell PA, Gustafson AB, Morgan WP, Pert CB (1987) Enkephalins, catecholamines and psychological mood alterations: effects of prolonged exercise. Med Sci Sports Exerc 19:347–353
Fathy H (1986) Natural energy and vernacular architecture. Principles and examples with reference to hot arid climates. University of Chicago Press, Chicago
Feachem RG, Bradley DJ, Garelick H, Mara DD (1983) Sanitation and disease: health aspects of excreta and wastewater management. John Wiley, Chichester
Felker P (1981) Uses of tree legumes in semiarid regions. Econ Bot 35:174–186
Felker P, Bandurski RS (1977) Protein and amino acid composition of tree legume seeds. J Sci Food Agric 28:791–797
Ferretti F, Worm B, Britten GL, Heithaus MR, Lotze HK (2010) Patterns and ecosystem consequences of shark declines in the ocean. Ecol Lett 13:1055–1071
Festing H (1997) The potential for direct marketing by small farmers in the UK. Farm Manag 9(8):409–421
Fewtrell L, Kaufman RB, Kay D, Enanoria W, Haller L, Colford JM (2005) Water, sanitation, and hygiene interventions to reduce diarrhoea in less developed countries: a systematic review and meta-analysis. Lancet Infect Dis 5(1):42–52
Fieldhouse P (1986) Food and nutrition. Customs and culture. Croom Helm, London
Fischer C (1980) Bog bodies of Denmark. In: Cockburn A, Cockburn E (eds) Mummies, disease and ancient cultures. Cambridge University Press, Cambridge, pp 177–193
Fisher P, Ward A (1994) Complementary medicine in Europe. Br Med J 309:107–111
Fleuret A (1979) The role of wild foliage plants in the diet, a case study from Lushoto, Tanzania. Ecol Food Nutr 8:87–93
Fleuret A (1980) Nonfood uses of plants in Usambara. Econ Bot 34:320–333
Fleuret P, Fleuret A (1980) Nutrition, consumption and agricultural change. Hum Organ 39:250–260
Foley G, Barnard G (1984) Farm and community forestry. Earthscan Publications, International Institute for Environment and Development, London

Fong FW (1992) Perspectives for sustainable resource utilization and management of nipa vegetation. Econ Bot 46(1):45–54
Food Standards Agency (2002) McCance and Widdowson's the composition of foods. Sixth Summary Edition. Royal Society of Chemistry, Cambridge
Forde CD (1934) Habitat, economy and society. A geographical introduction to ethnology. Methuen, London
Foster S (1991) Supply and use of essential drugs in sub-Saharan Africa: some issues and possible solutions. Soc Sci Med 32:1201–1218
Fosu GB (1981) Disease classification in rural Ghana: framework and implications for health behaviour. Soc Sci Med 32:1201–1218
Franceschi VR, Horner HT (1980) Calcium oxalate crystals in plants. Bot Rev 46(4):361–427
Friedman A, Cammalleri V (1995) The environmental impact of building materials in the North American building industry. Build Res Inf 23(3):162–166
Fuller GK, Lemma A, Haile T (1979) Schistosomiasis in Omo National Park of Southwest Ethiopia. Am J Trop Med Hyg 28(3):526–530
Funmilayo O (1975) The village weaverbird and the villagers: a protected pest. Nigerian Field 40:183–186
Gadgil M, Vartak VD (1976) The sacred groves of Western Ghats in India. Econ Bot 30:152–160
Gallacher DJ, Hill JP (2008) Effects of camel grazing on density and species diversity of seedling emergence in the Dubai (UAE) inland desert. Jour Arid Env 72:853–860
Garbrecht G (1983) Ancient water works-lessons from history. Impact Sci Soc 1983(1):5–16
Gaur RD, Semwel JK, Negi KS (1981–2) Fodder plants of Garwhal hills and their impact on the quality and quantity of milk. J Himalayan Stud Region Dev, 56, 87–91
Geiger R (1965) The climate near the ground. Harvard University Press, Cambridge, MA
Gelfand M (1964) Medicine and custom in Africa. Livingstone, Edinburgh
Gelfand M, Mavi S, Drummond RB, Ndemera B (1985) The traditional medical practitioner in Zimbabwe. His principles of practice and pharmacopoeia. Mambo Press, Gweru
Gibbs P (1987) Building a Malay house. Oxford University Press, Singapore
Giles PH (1964) The storage of cereals by farmers in Northern Nigeria. Trop Agric Trinidad 41:197–212
Gilmour DA (1988) Not seeing the trees for the forest: a re-appraisal of the deforestation crisis in two hill districts of Nepal. Mt Res Dev 8:343–350
Gilmour DA (1997) Rearranging trees in the landscape in the Middle Hills of Nepal. In: Arnold JEM, Dewees PA (eds) Farms, trees and farmers. Earthscan Publications, London, pp 21–42
Giovannucci E, Willet WC (1994) Dietary factors and risk of colon cancer. Ann Med 26(6):443–452
Glasser M (1988) Accountability of anthropologists, indigenous healers, and their governments: a plea for reasonable medicine. Soc Sci Med 27:1461–1465
Glover IC, Higham CFW (1996) New evidence for early rice cultivation in south, Southeast and East Asia. In: Harris DR (ed) The origins and spread of agriculture and pastoralism in Eurasia. University College London Press, London, pp 413–441
Glowka L, Burhenne-Guilmin F, Synge H, McNeely JA, Gündling L (1994) A guide to the convention on biodiversity. IUCN, Gland/Cambridge
Goldsmith E, Tenakoon M (1982) Traditional agriculture in Sri Lanka. Ecologist 12(5):209–215
Goldsmith E, Allen R, Allaby M, Davoll J, Lawrence S (1972) A blueprint for survival. Penguin Books, Harmondswith
Gomez MI (1982) Sources of vitamin C in the Kenya diet and their stability to cooking and processing. Ecol Food Nutr 12:179–184
Good CM (1978) Man, milieu, and the disease factor, tick-borne relapsing fever in East Africa. In: Hartwig GW, Patterson KD (eds) Disease in African history. Duke University Press, Durham, North Carolina, pp 46–87
Gourou P (1958) The tropical world. Its social and economic conditions and its future status. Longmans & Green, London
Grainge M, Ahmed S (1988) Handbook of plants with pest-control properties. John Wiley, New York

Gray DJ, Ross AG, Li YS, McManus DP (2011) Diagnosis and management of schistosomiasis. Br Med J 342:2651

Green EC, Makhubu L (1984) Traditional healers in Swaziland: toward improved cooperation between the traditional and modern health sectors. Soc Sci Med 18:1071–1079

Grimwood I (1968) Endangered mammals in Peru. Oryx 9(6):411–422

Grivetti LE (1978) Nutritional success in a semi-arid land: examination of Tswana agro-pastoralists of the eastern Kalahari, Botswana. Am J Clin Nutr 31:1204–1220

Grivetti LE (1979) Kalahari agro-pastoral-hunter-gatherers: the Tswana example. Ecol Food Nutr 7:235–256

Guha R (1983) Forestry in British and post-British India. *Economic and Political Weekly,* 29 October, 1882–96

Gunawardene NR, Daniels DAE, Gunatilleke IAUN, Gunatilleke CVS, Karunakaran PV, Nayak KG, Prasad S, Puyravaud P, Ramesh BR, Subramanian KA, Vasanthy G (2007) A brief overview of the Western Ghats-SriLanka biodiversity hotspot. Curr Sci 93(11):1567–1572

Gupta R (1983) Are community biogas plants a feasible proposition? World Health Forum 4:358–361

Haaga J, Mason J, Omoro FZ, Quinn V, Rafferty A, Test K, Wasonga L (1986) Child nutrition in rural Kenya: a geographic and agricultural classification. Ecol Food Nutr 18:297–307

Hall RL (1973) Toxicants occurring naturally in spices and flavors. In: Toxicants occurring naturally in foods. Natural Academy Press, Washington DC, pp 448–463

Hamilton RJ, Potuku T, Montambault JR (2001) Community-based conservation results in the recovery of reef spawing aggregation in the Coral Triangle. Biol Conserv 144:1850–1858

Hanifi Z (1994) Housing design in relation to environmental comfort – a comparison of the traditional Malay house and modern housing. Build Res Inf 22(1):21–33

Hanlon J (1983) Water comes to the "place of the wasps". UNICEF News 116:5–7

Harlan JR (1971) Agricultural origins; centers and non-centers. Science 174:465–473

Harris WV (1971) Termites, their recognition and control. Longman, London

Harris DR (1974) Tropical vegetation: an outline and some misconceptions. Geography 59:240–250

Harris PN, Chen KK (1970) Development of hepatic tumors in rats following ingestion of *Senecio longilobus*. Cancer Res 30:2881–2886

Harrison G (1978) Mosquitoes, malaria and man: a history of the hostilities since 1880. John Murray, London

Harrison P (1992) The third revolution. Environment, population and a sustainable world. I.B. Tauris/St. Martin's Press, London/New York

Hassan N, Huda N, Ahmad K (1985) Seasonal patterns of food intake in rural Bangladesh: its impact on nutritional status. Ecol Food Nutr 17:175–186

Hawke CJ, José PV (1996) Reedbed management for commercial and wildlife interests. Royal Society for the Protection of Birds, Sandy

Hazell PBR (2009) Transforming agriculture: the green revolution in Asia. In: Spielman DJ, Pandya-Lorch R (eds) Millons fed, proven successes in agricultural development. International Food Policy Research Institute, Washington, DC, pp 25–32

He C, Bassik MC, Moresi V, Sun K, Wei Y, Zou Z, An Z, Loh J, Fisher J, Sun Q, Korsmeyer S, Packer M, May HI, Hill JA, Virgin HW, Gilpin C, Xiao G, Bassel-Duby R, Scherer PE, Levine B (2012) Exercise-induced BCL2-regulated autophagy is required for muscle homeostasis. Nature 481:511–515

Helman CG (1990) Culture, health and illness. An introduction for health professionals. Wright, London

Hempel S (2007) The medical detective. John Snow, cholera and the mystery of the broad street pump. Granta Books, London

Herskovits MJ (1926) The cattle complex in East Africa. 28:230-272, 361-380, 494-528, 630-664

Hickin NE (1975) The insect factor in wood decay. An account of wood-boring insects with particular reference to timber indoors. Associated Business Programmes, London

Hill P (1972) Rural Hausa, a village and a setting. Cambridge University Press, Cambridge

Hirschman C, Tan J, Chamratrithirong A, Guest P (1994) The path to below replacement-level fertility in Thailand. Int Fam Plan Perspect 20:82–87
Hodge J, Haltrecht J (2009) BedZED seven years on. The impact of the UK's best known eco-village and its residents. Bioregional, Wallingford
Hoff W, Maseko DN (1986) Nurses and traditional healers join hands. World Health Forum 7:412–416
Holland B, Welch AA, Unwin ID, Buss DH, Paul AA, Southgate AAT (1991) McCance and Widdowson's the comparison of foods, 5th edn. Royal Society of Chemistry and Ministry of Agriculture, Fisheries and Food, Cambridge
Hong ST, Chai JY, Huh S, Rim HJ, Lee SH (2006) A successful experience of soil-transmitted helminth control in the Republic of Korea. Korean J Parasitol 44(3):177–185
Hopkins FG (1912) Feeding experiments illustrating the importance of accessory factors in normal dietaries. J Physiol 44:425–460
Hostetler SW, Bartlein PJ, Clark PU, Small EE, Solomon AM (2000) Simulated influences of Lake Agassiz on the climate of central North America 11,000 years ago. Nature 405:334–337
Houghton J (2004) Global warming. The complete briefing. Cambridge Universty Press, Cambridge
Houmsou RS, Kela SL, Suleiman MM, Ogidi JA (2010) Perceptions and assessment of risk factors in *Schistosoma haematobium* infection in Buruku and Katsina-Ala local government areas of Benue state-Nigeria. Int J Infect Dis 8(1):1–12
Hughes JD (1996) Francis of Assisi and the diversity of creation. Environ Ethic 18:311–320
Hughes CC, Hunter JM (1970) Disease and "development" in Africa. Soc Sci Med 3:443–493
Hugh-Jones S (1982) The Pleiades and Scorpius in Barasana cosmology. In: Aveni AF, Urton G (eds) Ethnoastronomy and archaeoastronomy in the American tropics. New York Academy of Sciences, New York, pp 183–201
Hunter JM (1967) Population pressure in a part of the West African Savanna: a study of Nangodi, Northeast Ghana. Ann Assoc Am Geogr 57:101–114
Huntington E (1915) Civilisation and climate. Yale University Press, New Haven
Huntington HP (1992) Wildlife management and subsistence hunting in Alaska. Belhaven Press, London
Husaini MA, Suhardjo R, Megawangi R, Nurhadi E, Supardi D, Djojosoebagio S, Karyadi D (1990) Diet, nutritional status, and potential need for home gardens in the tea plantation. In: Landauer K, Brazil M (eds) Tropical home gardens. United Nations University Press, Tokyo, pp 119–125
Hussain S (2000) Protecting the snow leopard and enhancing farmers' livelihood: a pilot insurance scheme in Baltistan. Mt Res Dev 20(3):224–229
Hussain S (2003) The status of the snow leopard in Pakistan and its conflict with local farmers. Oryx 37(1):26–33
Huxtable RJ (1992) The myth of beneficient nature: the risks of herbal preparations. Ann Intern Med 117:165–166
Iannotti L, Cunningham K, Ruel M (2009) Diversifying into healthy diets. Homestead food production in Bangladeah. In: Spielman DJ, Pandya-Lorch R (eds) Millions fed. Proven successes in agricultural development. International Food Policy Research Institute, Washington DC, pp 145–151
Idowu AB, Modder WWD (1996) Possible control of the stinking grasshopper, *Zonocerus variegatus* (L) (Orthoptera, Pyrgomorphidae) in Ondo state through human consumption. Nigerian Field 61(1–2):7–14
Ingles AA (1995) Religious beliefs and rituals in Nepal. Their influence on forest conservation. In: Halladay P, Gilmour DA (eds) Conserving biodiversity outside protected areas. The role of traditional agro-ecosystems. IUCN, Gland/Cambridge, pp 205–224
Innis HA (1956) The fur trade in Canada. University of Toronto Press, Toronto
IRRI (1988) Vector-borne disease control in humans through rice agroecosystem management. International Rice Research Institute, Manila
Irvine FR (1969) West African crops. Oxford University Press, London

Isely D (1982) Leguminosae and Homo sapiens. Econ Bot 36:46–70
Ismail M, Rahman AMA (2008) The possibiity of using a hybrid turbine ventilator (HTV) to improve the air velocity (stack ventilation) in Malaysian building interiors. Arch Malaysia 20:60–63
Jacobs AH (1975) Maasai pastoralism in historical perspective. In: Monod T (ed) Pastoralism in Tropical Africa. International African Institute and Oxford University Press, Oxford, pp 406–425
Jacobson M (1989) Botanical pesticides, present and future. In: Arnason JT, Philogene JR, Morand P (eds) Pesticides of plant origin. American Chemical Society, Washington, DC, pp 1–10
Jahoda G (1969) The psychology of superstition. Allen Lane, Penguin, London
Jain S (1984) Standing up for trees, women's role in the Chipko movement. Unasylva 146:12–20
Janzen DH (1973) Tropical agroecosystems. Science 182:1212–1219
Jelliffe DB (1957) Social culture and nutrition. Cultural blocks and protein malnutrition in early childhood in rural West Bengal. J Pediatr 20:128–138
Jelliffe DB, Bennett FJ, Jelliffe EFP, White RHR (1964) Ecology of childhood disease in the Karamajong of Uganda. Arch Environ Health 9:25–36
Jentzch K (1953) Production of alkaloids in Solanaceae. Sci Pharm 21:285–289
Jeyaratnam J, Lun KC, Phoon WO (1987) Survey of acute poisoning among agricultural workers in four Asian countries. Bull World Health Organ 65:521–527
Jiggins J (1989) An examination of the impact of colonialism in establishing negative values and attitudes towards indigenous agricultural knowledge. In: Warren DM, Slikkerveer LJ, Titiola SO (eds) Indigenous knowledge systems: implications for agriculture and international development. Iowa State University, Amco, Iowa, pp 68–78
Johnson C (1997) Dana, helping nature, helping people. Al Reen 60:10–15
Jonas WB (2011) Reframing placebo in research and practice. Philos Trans R Soc Lond B 366:1896–1904
Joshi AR (1987) A study of the environmental relationships of certain village communities in the central development region of Nepal. Doctoral Thesis, University of Wales, Cardiff
Joshi K (2004) Documentation of medicinal plants and their indigenous uses in Likhu Sub-Watershed, Nepal. J Non-Timber Prod 11(2):86–94
Joshi K (2011) Ethnobotanical study of plants used for the treament of diabetes mellitus in the mountainous regions of Nepal. J Non-timber For Prod 18(1):19–26
Joshi AR, Edington JM (1990) The use of medicinal plants by two village communities in the central development region of Nepal. Econ Bot 44:71–83
Joshi AR, Joshi K (2000) Indigenous knowledge and uses of medicinal plants by local communities of the Kali Gandaki Watershed Area, Nepal. Ethnopharmacol 73(1–2):175–183
Joshi AR, Joshi K (2003) Fodder plants of hilly areas of Bagmati and Langtang watershed, Nepal: ethnobotany and future conservation strategy. Ethnobotany 15:107–114
Joshi AR, Joshi K (2007) Ethnomedicinal plants used against skin diseases in some villages of Kali Gandaki, Bagmati and Tadi Likhu Watersheds of Nepal. Ethnobot Leaflets 11:235–246
Joshi AR, Shrestha SL, Joshi K (2003) Environmental management & sustainable development at the crossroad. AnKuS, Kathmandu
Joyce C (1988) The tree that caused a riot. New Sci 117:54–59
Kala CP (1997) Ethnobotanical survey and propagation of rare medicinal herbs for small farmers in the buffer zone of valley of flowers National Park, Garhwal Himalaya. Wildlife Institute of India, Dehradun
Kang BT, Wilson GF, Spikens L (1981) Alley cropping maize *(Zea mays L.)* and leucaena *(Leucaena leucocephala* Lam) in southern Nigeria. Plant Soil 63:165–179
Karyono TH (1996) Thermal comfort in the tropical South East Asia region. Archit Sci Rev 39:135–139
Kayode J (2003) Conservation and Yoruba forest taboos. Nigerian Field 69:53–61
Kean BH, Mott KE, Russell AJ (1978) Tropical medicine and parasitology. Classic Investigations. Cornell University Press, Ithaca
Kellert SR, Wilson EO (eds) (1993) The Biophilia hypothesis. Island Press, Washington, DC

Kennedy KAR (1989) Skeletal markers of occupational stress. In: Işcan MY, Kennedy KAR (eds) Reconstruction of life from the skeleton. Alan R. Liss, New York, pp 129–160

Kerr K (1991) The economic potential of handicrafts enterprises in rural development: focus on Indonesia. Unasyla 42:31–36

Kettle DS (1984) Medical and veterinary entomology. Croom Helm, Beckenham

Khan V, Najmi AK, Akhtar M, Aqil A, Mujeeb M, Pillai KK (2012) A pharmacological appraisal of medicinal plants with antidiabetic potential. J Pharm BioAll Sci 4(1):27–42

Kidd J (1982) The influence of arid-zone climate on house design in the Sudan. In: Golany G (ed) Desert planning: international lessons. The Architectural Press, London, pp 113–123

Kimura K, Saito Y (1982) Country monograph, Japan. In: Bowen A, Vagner R (eds) Passive and low energy alternatives. Pergamon Press, New York, pp (15) 30–(15) 44

King M (1990) Health is a sustainable state. Lancet 336:664–667

Kirkpatrick TW, Simmonds NW (1958) Bamboo borers and the moon. Trop Agric Trinidad 35:299–301

Klayman DL (1985) *Qinghaosu* (Artemisinin): an antimalarial drug from China. Science 228:1049–1055

Kmietowicz Z (2003) Food watchdog warns against high doses of vitamins and minerals. Br Med J 326:1001

Koenigsberger OH, Ingersoll TG, Mayhew A, Szokolay SV (1974) Manual of tropical housing and building. Part one: Climatic design. Longman, London

Kolsky PJ (1993) Water, sanitation and diarrhoea: the limits of understanding. Trans R Soc Trop Med Hyg 87(S3):43–46

Kouzi SA, McMurty RJ, Nelson SD (1994) Hepatotoxicity of germander *(Teucrium chamaedrys L.)* and one of its constituent neoclerodane diterpenes Teucrin A in the mouse. Chem Res Toxicol 7(6):850–856

Koziell I, Inoue CYA (2006) Mamirauá sustainable development reserve, Brazil. Lessons learnt in integrating conservation with poverty reduction. International Institute for Environment and Development, London

Krech S (1999) Playing with fire. New Sci 164:56–59

Krech S (2000) The ecological Indian, myth and history. Norton, New York

Kuhnlein HV (2003) Micronutrient nutrition and traditional food systems of indigenous peoples. Food Nutr Agric 32:33–39

Kumara R, Saito Y (1982) Country monograph Japan. In: Bowen A, Vagner R (eds) Passive and low energy alternatives. Pergamon Press, New York pp (15) 30–(15) 44

Kyle R (1987) A feast in the wild. Kudu Publishing, Oxford

Laderman C (1982) Wild vegetable consumption on the east coast of Peninsular Malaysia. Malay Nat J 35:165–171

Laderman C (1984) Food ideology and eating behaviour: contributions from Malay studies. Soc Sci Med 19:547–559

Lambo TA (1956) Neuropsychiatric observations in the western region of Nigeria. Br Med J ii:1388–1396

Lanningham-Foster L, Nysse LJ, Levine JA (2003) Labor saved, calories lost: the energetic impact of domestic labor-saving devices. Obes Res 11:1178–1181

Larrey D, Vial T, Pauwels A, Castot A, Biour M, David M, Michel H (1992) Hepatitis after germander (*Teucrium chamaedrys*) administration, another instance of herbal medicine toxicity. Ann Intern Med 117:129–132

Larsen CS (1995) Biological changes in human populations with agriculture. Annu Rev Anthropol 24:185–213

Larsen CS (2000) Skeletons in our closet. Revealing our past through Bioarchaeology. Princeton University Press, Princeton

Leach ER (1968) Pul Eliya. A village in Ceylon. A study of land tenure and kinship. Cambridge University Press, Cambridge

Leach G (1976) Energy and food production. IPC Science and Technology Press, Guildford

Leader-Williams N (1992) The world trade in rhino horn: a review. TRAFFIC International, Cambridge
LeBlanc SA, Register KE (2004) Constant battles. Why we fight. St. Martin's Griffin, New York
Lee RB (1972) The !Kung bushmen of Botswana. In: Bicchieri MG (ed) Hunters and gatherers today. Holt, Rinehart and Winston, New York, pp 326–368
Lee RB (1976) !Kung spatial organization, an ecological and historical perspective. In: Lee RB, DeVore I (eds) Kalahari hunter-gatherers. Studies of the !Kung San and their neighbours. Harvard University Press, Cambridge, MA, pp 73–97
Lee RB (1979) The !Kung San: men, women and work in a foraging society. Cambridge University Press, Cambridge
Lee RB, DeVore I (eds) (1968) Man the hunter. Aldine Publishing Company, Chicago
Lee RB, De Vore I (eds) (1976) Kalahari hunter-gatherers. Studies of the !Kung San and their neighbours. Harvard University Press, Cambridge
Lehman CF (1967) The pet trade and extermination. Oryx 9(2):161–165
Lenon B (1992) The demographic transition model. Geogr Mag 64(5):51
Letts JB (1999) Smoke blackened thatch. A unique source of late medieval plant remains from Southern England. English Heritage/University of Reading, London/Reading
Leung AY (1980) Encyclopedia of common natural ingredients used in food, drugs and cosmetics. John Wiley, New York
Levy LF (1968) Porter's neck. Br Med J 1968(2):16–19
Lewcock R, Brans G (1975) The boat as an architectural symbol. In: Oliver P (ed) Shelter, sign and symbol. Barrie and Jenkins, London, pp 107–116
Lewis IM (1965) The Northern pastoral Somali of the horn. In: Gibbs JL (ed) Peoples of Africa. Holt, Rinehart and Winston, New York, pp 321–360
Lewis WH, Elvin-Lewis MPF (1977) Medical botany, plants affecting man's health. John Wiley, New York
Lim JY (1987) The Malay house: rediscovering Malaysia's indigenous shelter system. Institut Masyarakat, Pulau Penang
Lindberg K, Hawkins DE (1993) Ecotourism, a guide for planners and managers. The Ecotourism Society, North Bennington
Linde K, Ramirez G, Mulrow CD, Pauls A, Weidenhammer W, Melchart D (1996) St John's wort for depression-an overview and meta-analysis of randomised clinical trials. Br Med J 313:253–258
Ling B, Den T, Lu Z, Min L, Wang Z, Yuan A (1993) Use of night soil in agriculture and fish farming. World Health Forum 14:67–70
Littrell MA (1990) Symbolic significance of textile crafts for tourists. Ann Tour Res 17:228–245
Littrell MA, Dickson MA (1997) Alternative trading organizations: shifting paradigm in a culture of social responsibility. Hum Organ 56:344–352
Lockeretz W, Shearer G, Kohl DH (1981) Organic farming in the corn belt. Science 211:540–547
Lockie S (2000) Designs on learning. Energy Environ Manag 2000:28–29
Locking M, Cowden J (2009) *Escherichia coli 0187*. Br Med J 339:817–818
Loevinsohn ME (1987) Insecticide use and increased mortality in rural Central Luzon, Philippines. Lancet i:1359–1362
Lokubandra WJM (1991) Policies and organization for medicinal plant conservation in Sri Lanka. In: Akerele O, Heywood V, Synge H (eds) Conservation of medicinal plants. Cambridge University Press, Cambridge, pp 241–252
Lovell CR (1993) Plants and the skin. Blackwell Scientific Publications, Oxford
Lu Z, Qiang G, Dai X (1983) Management of human excreta. World Health Forum 4:279–280
Lumpkin TA, Plucknett DL (1980) Azolla; botany, physiology and use as a green manure. Econ Bot 34:111–153
Lun Z-R, Gasser RB, Lai D-H, Li A-X, Zhu X-Q, Yu X-B, Fang Y-Y (2005) Clonorchiasis: a key foodborne zoonosis in China. Lancet Infect Dis 5:31–41
Mabey R (1972) Food for free. A guide to the edible wild plants of Britain. Collins, Glasgow

Macfarlane A (1981) Death disease and curing in a Himalayan village. In: von Fürer Haimendorf C (ed) Asian highland societies in anthropological perspective. Sterling Publishers, New Delhi, pp 79–130

MacKinnon J, MacKinnon K, Child C, Thorsell J (1986) Managing protected areas in the tropics. International Union for Conservation of Nature and Natural Resources, Gland

Maclean U (1971) Magical medicine, a Nigerian case study. Allen Lane, The Penguin Press, Harmondsworth

Magner LN (1979) A history of the life sciences. Marcel Dekker, New York

Malama A, Sharples S (1997) The performance of traditional and contemporary housing in the cool season of Zambia. Build Environ 32:69–78

Malcolm DW (1953) Sukumaland. An African people and their country. A study of land use in Tanganyika. International African Institute and Oxford University Press, London

Maldonado P (2008) Ecotourism in Peru, rumble in the jungle. De Economist 12:58–59

Malhotra SP (1986) Bishnois – their role in conservation of desert ecosystem. In: Shankarharyan KA, Shankar V (eds) Desert environment, conservation and management. Central Arid Zone Research Institute, Jodhpur, pp 23–24

Manson-Bahr PEC, Apted FIC (1982) Manson's tropical diseases. Baillière Tindall, London

Mara D (1977) Wastewater treatment in hot climates. In: Feachem R, McGarry M, Mara D (eds) Water, wastes and health in hot climates. John Wiley, Chichester, pp 264–283

Mara D, Cairncross S (1989) Guidelines for the safe use of wastewater and excreta in agriculture and aquaculture. World Health Organisation, Geneva

Markman M (1991) Taxol: an important new drug in the management of epithelial ovarian cancer. Yale J Biol Med 64:583–590

Marshall L (1961) Sharing, taking and giving: relief of social tensions among !Kung bushmen. Africa 31:231–249

Marshall L (1964) The !Kung Bushmen of the Kalahari Desert. In: Gibbs JL (ed) Peoples of Africa. Holt, Rinehart and Winston, New York, pp 241–278

Marten GG, Vitakon P (1986) Soil management in traditional agriculture. In: Marten GC (ed) Traditional agriculture in southeast Asia. A human ecology perspective. Westview Press, Boulder, pp 199–225

Martens WJM, Jetten TH, Focks DA (1997) Sensitivity of malaria, schistosomiasis and dengue to global warming. Clim Chang 35:145–156

Martin PS (1984) Prehistoric overkill: the global model. In: Martin PS, Klein RG (eds) Quaternary extinctions, a prehistoric revolution. University of Arizona Press, Tucson, pp 354–403

Martin PS, Klein RG (eds) (1984) Quaternary extinctions. A prehistoric revolution. University of Arizona Press, Tucson

Martin PS, Wright HE (eds) (1967) Pleistocene extinctions: the search for a cause. Yale University Press, New Haven

Mather TH, Trinh TT (1984) Environmental management for vector control in rice fields. Food and Agriculture Organization of the United Nations, Rome

Matteson PC, Altieri MA, Gagne WC (1984) Modification of small farmer practices for better pest management. Annu Rev Entomol 29:383–402

Mattingly PF (1969) The biology of mosquito-borne disease. George Allen and Unwin, London

Maumbe BM, Swinton SM (2003) Hidden health costs of pesticide use in Zimbabwe's smallholder cotton growers. Soc Sci Med 57:1559–1571

Mayou R, Sharpe M (1997) Treating medically unexplained physical symptoms. Br Med J 315:561–562

Mbegu AC (1996) Making the most of compost: a look at *wafipa* mounds in Tanzania. In: Reij C, Scoones I, Toulmin C (eds) Sustaining the soil. Indigenous soil and water conservation in Africa. Earthscan Publications, London, pp 134–138

McBride JR, McBride SM (1987) Tree growth and firewood production near Syabru, Nepal. University of California, Berkeley

McCance RA (1946) Bread. Lancet 6386:77–81

McCance RA, El Neil H, El Din N, Widdowson EM, Southgate DAT, Passmore R, Shirling D, Wilkinson RT (1971) The response of normal men and women to changes in their environmental temperatures and ways of life. Philos Trans R Soc Lond B 259:533–561

McCarthy JJ, Canziani O, Leary NA, Dokken DJ, White KS (eds) (2001a) Climate change 2001: impacts, adaptation and vulnerability. Contribution of working group II to the third assessment report of the intergovernmental panel on climate change. Cambridge University Press, Cambridge

McCarthy J, Riddall R, Topp C (2001b) Jubilee Campus, University of Nottingham. Arup J 2001(2):3–10

McCartney M (2011) The scam of integrative medicine. Br Med J 343:160

McCord C (1978) Medical technology in developing countries: useful, useless or harmful. Am J Clin Nutr 31:2301–2313

McDonald JN (1984) The recorded North American selection regime and late quaternary megafaunal extinctions. In: Martin PS, Klein RG (eds) Quaternary extinctions, a prehistoric revolution. University of Arizona Press, Tucson, pp 404–439

McDonald M, Fleming B (1993) Community-based economic development and resource management in the Hudson Bay area. In: Inglis JT (ed) Traditional ecological knowledge, concepts and cases. International Program on Traditional Knowledge and International Development Research Centre, Ottawa, pp 63–68

McFeters GA (ed) (1990) Drinking water microbiology. Springer, New York

McGarry MG (1976) The taboo resource. The use of human excreta in Chinese agriculture. Ecologist 6(4):150–154

McGarry MG, Stainforth J (1978) Compost, fertilizer and biogas production from human and farm wastes in the People's Republic of China. International Development Research Centre, Ottawa

McKay DA (1980) Foods, illness and folk medicine: insights from Ulu Trengganu, West Malaysia. In: Robson JRK (ed) Food, ecology and culture. Readings in the anthropology of dietary practices. Gordon and Breach Science Publishers, New York, pp 61–66

McKean SG (2003) Toward sustainable use of palm leaves by a rural community in Kwazulu-Natal, South Africa. Econ Bot 57(1):65–72

McMichael JK (1978) The double septic bin in Vietnam. In: Pacey A (ed) Sanitation in developing countries. John Wiley, London, pp 110–114

McNeil M (1964) Lateritic soils. Sci Am 211(5):96–102

Mead M (ed) (1953) Cultural patterns and technical change. United Nations Educational, Scientific and Cultural Organization, Paris

Meggitt MJ (1964) Aboriginal food gatherers of tropical Australia. In: Elliott HFL (ed) The ecology of man in the tropical enviroment. International Union for the Conservation of Nature and Natural Resources, Morges, pp 30–37

Meissner K, Kohls N, Colloca L (2011) Introduction to placebo effects in medicine: mechanisms and clinical implications. Philos Trans R Soc Lond B 366:1784–1789

Melathopoulus AP, Winston ML, Whittington R, Smith R, Lindberg C, Mukai A, Moore M (2000) Comparative laboratory toxicity of neem pesticides to honey bees (Hymenoptera: Apidae), their mite parasites *Varroa jacobsoni* (Acari: Varroidae) and *Acarapis woodi* (Acari: Tarsonemidae), and brood pathogens *Paenibacillus larvae* and *Ascophaera apis*. J Econ Entomol 93(2):199–209

Mendeloff AI (1977) Dietary fiber and human health. N Engl J Med 297(15):811–814

Mercer SW, Watt GCM, Reilly D (2001) Empathy is important for enablement. Br Med J 322:865

Mills JA, Jackson P (1994) Killed for a cure. A review of the worldwide trade in tiger bone. TRAFFIC International, Cambridge

Mills EJ, Foster BC, van Heeswijk RP, Wilson KEP, Kosuge K, Kanfer LB (2005a) Impact of herbal medicine on antiretroviral metabolism. AIDS 19:95–97

Mills EJ, Cooper C, Seely D, Kanfer I (2005b) African herbal medicines in the treatment of HIV: Hypoxis and Sutherlandia. Nutr J 4:19–34

Millstone E (1984) Food additives: a technology out of control? New Sci 18:20–24

Milne G (1936a) Normal erosion as a factor in soil profile development. Nature 138:548–549

Milne G (1936b) A provisional soil map of East Africa. Kenya, Uganda, Tanganyika and Zanzibar with explanatory memoir. Crown Agents, London
Minnaar A, Offringa D, Payze C (1992) To live in fear: witchburning and medicine murder in Venda. Human Sciences Research Council, Pretoria
Miri ES, Hopkins DR, Ruiz-Tiben E, Keana AS, Withers PC, Anagbogu IN, Sadiq LK, Kale OO, Edungbola LD, Braide EI, Ologe JO, Ityonzughul C (2010) Nigeria's triumph: dracunculiasis eradicated. Am J Trop Med Hyg 83(2):215–225
Mishra V, Dai X, Smith KR, Mika L (2004) Maternal exposure to biomass smoke and reduced birth weight in Zimbabwe. Ann Epidemiol 14:740–747
Mithen S (2003) After the ice. A global human history, 20,000–5000 BC. Harvard University Press, Cambridge, MA
Mohapatra SS (1989) Beliefs of rural mothers about diarrhoea in Orissa, India. Dial Diarrhoea 39:7
Molleson T (1994) The eloquent bones of Abu Hureya. Sci Am 270(2):60–65
Mollison B (1978) Permaculture one: a perennial agriculture for human settlements. Transworld Publications, Tyalgum
Moore AMT, Hillman GC (1992) The Pleistocene to Holocene transition and human economy in Southwest Asia: the impact of the Younger Dryas. Am Antiq 5(3):482–494
Moore AMT, Hillman GC, Legge AJ (2000) Village on the Euphrates: from foraging to farming at Abu Hureyra. Oxford University Press, Oxford
Morgan RPC (1979) Soil erosion. Longman, London
Moughtin JC (1982) Settlements and housing in an arid-zone developing country; case study in Hausaland, Northern Nigeria. In: Golany G (ed) Desert planning: international lessons. The Architectural Press, London, pp 145–159
Moulik TK (1985) The biogas program in India and China. Ambio 14(4–5):288–292
Mull JD, Mull DS (1988) Mothers' concepts of childhood diarrhoea in rural Pakistan: what ORT program planners should know. Soc Sci Med 27:53–67
Müllner A, Linsenmair KE, Wikelski M (2004) Exposure to ecotourism reduces survival and affects stress response in hoatzin chicks *(Opisthocomus hoazin)*. Biol Conserv 118:549–558
Munn CA (1992) Macaw biology and ecotourism or "when a bird in the bush is worth two in the hand". In: Beissinger SR, Snyder NFR (eds) New world parrots in crisis: solutions from conservation biology. Smithsonian Institution, Washington, DC, pp 47–72
Murata A, Fujino Y, Pham T-M, Kubo T, Mizoue T, Tokui N, Matsuda S, Yoshimura T (2010) Prospective cohort study evaluating the relationship between salted food and gastrointestinal tract cancer mortality in Japan. Asia Pac J Clin Nutr 19(4):564–571
Murphy JJ, Heptinstall S, Mitchell JRA (1988) Randomised double-blind placebo-controlled trial of feverfew in migraine prevention. Lancet ii:189–192
Murthy NS, Mukherjee S, Roy G, Roy A (2009) Dietary factors and cancer chemoprevention: an overview of obesity-related malignancies. J Postgrad Med 55(1):45–54
Murton RK (1971) Man and birds. Collins, London
Mushita AT (1993) Strengthening the informal seed system in communal areas of Zimbabwe. In: De Boef W, Amanor K, Wellard K, Bebbington A (eds) Cultivating knowledge, genetic diversity, farmer experimentation and crop research. Intermediate Technology Publications, London, pp 85–88
Musvoto C, Mapaure I, Gondo T, Ndeinoma A, Mujawo T (2006) Reality and preferences in community mopane *(Colophospermum mopane)* woodland management in Zimbabwe and Namibia. Int J Hum Soc Sci 1(3):173–177
Nabhan GP (1987) The desert smells like rain. A naturalist in Papago Indian Country. North Point Press, San Francisco
Nagbrahmam D, Sambrani S (1983, January 8) Women's drudgery in firewood collection. *Economic and Political Weekly*
Nakashima DJ (1993) Astute observers on the sea ice edge. Inuit knowledge as a basis for Arctic co-management. In: Inglis JT (ed) Traditional ecological knowledge, concepts and cases.

International Program on Traditional Ecological Knowledge and International Development Research Centre, Ottawa, pp 99–110

Nalin DR, Cash RA, Islam R, Molla M, Phillips RA (1968) Oral maintenance therapy for cholera in adults. Lancet 1968(2):370–373

Nash GD (1995) Timber-framed buildings in Wales. National Museum of Wales, Cardiff

Nater JP, de Groot AC (1985) Unwanted effects of cosmetics and drugs used in dermatology. Elsevier, Amsterdam

Ncayiyana DJ (2007) Combating poverty: the charade of development aid. Br Med J 335:172–173

Ndamba J, Chandiwana SK, Makaza N (1989) Knowledge, attitudes and practices among rural communities in relation to *Phytolacca dodecandra* – a plant molluscicide. Soc Sci Med 28:667–669

Needham J, Lu GD, Huang HT (1986) Science and civilisation in China. Biology and biological technology. Cambridge University Press, Cambridge

Nelson EAS (1994) Development aid: the way forward. Br Med J 308:481

Neuenschwander P, Herren HR (1988) Biological control of the cassava mealybug, *Phenacoccus manihoti*, by the exotic parasitoid *Epidinocarsis lopezi* in Africa. Philos Trans R Soc Lond B 318:319–333

Newbould P (1967) Methods of evaluating the primary production of forests, International biological programme handbook 2. Blackwell Scientific Publications, Oxford

Newman JL (1980) Dimensions of Sandawe diet. In: Robson JRK (ed) Food, ecology and culture. Readings in the anthropology of dietary practice. Gordon and Breach Science Publishers, New York, pp 27–33

Ngowi AB (1997) A hybrid approach to house construction-a case study in Botswana. Build Res Inf 25:142–147

Nichter M (1985) Drink boiled water: a cultural analysis of a health education message. Soc Sci Med 21:667–669

Nicol JF, Humphreys MA (2001) Adaptive thermal comfort and sustainable standards for buildings. In: McCartney K (ed) Moving thermal comfort standards into the 21st century. Oxford Brookes University, Oxford, pp 45–59

Niemeijer R, Guens M, Kliest T, Ogonda V, Hoorweg J (1988) Nutrition in agricultural development: the case of irrigated rice cultivation in West Kenya. Ecol Food Nutr 22:65–81

Niñez VK (1986) The household garden as a lifeboat. Ceres 19(4):31–36

Nortier JL, Martinez MC, Schmeiser HH, Arlt VM, Bieler CA, Petein M, Depierreaux MF, De Pauw L, Abramowicz D, Vereerstraeten P, Vanherweghem J-L (2000) Urothelial carcinoma associated with the use of a Chinese herb *(Aristolochia fangchi)*. N Engl J Med 342:686–692

Norton J (1986) Building with earth: a handbook. Intermediate Technology Development Group, Rugby

Nweke F (2009) Resisting viruses and bugs: cassava in Sub-Saharan Africa. In: Spielman DJ, Pandya-Lorch R (eds) Millions fed, proven successes in agricultural development. International Food Policy Research Institute, Washington, DC, pp 41–46

Nye PH, Greenland DJ (1960) The soil under shifting cultivation. Commonwealth Agricultural Bureaux, Farnham Royal

O'Dea WT (1958) The social history of lighting. Routledge and Kegan Paul, London

Ogbeide O (1974) Nutritional hazards of food taboos and preferences in Mid-West Nigeria. Am J Clin Nutr 27:213–216

Oguntoyimbo JS, Richards P (1978) Drought and the Nigerian farmer. J Arid Environ 1:165–194

Okafor FC (1987) Population pressure and land resource depletion in southeastern Nigeria. Appl Geogr 7:243–256

Olgyay V (1992) Design with climate. Bioclimatic approach to architectural regionalism. Van Nostrand Reinhold, New York

Oliveira JFS, de Çarvalho JP, de Sousa RFXB, Simao MM (1976) The nutritional value of four species of insects consumed in Angola. Ecol Food Nutr 5:91–97

Oliver P (1987) Dwellings. The house across the world. University of Texas Press, Austin

Oliver-Bever B (1986) Medicinal plants in tropical West Africa. Cambridge University Press, Cambridge
Oomen HAPC, Grubben GJH (1978) Tropical leaf vegetables in human nutrition. Royal Tropical Institute, Amsterdam
Opole M (1993) Revalidating women's knowledge on indigenous vegetables: implications for policy. In: de Boef W, Amanor K, Welland K, Bebbington A (eds) Cultivating knowledge: genetic diversity, farmer experimentation and crop research. Intermediate Publications, London, pp 157–164
Orraca-Tetteh R (1963) The giant African snail as a source of food. In: Ovington JD (ed) The better use of the world's fauna for food. Institute of Biology, London, pp 53–61
Osato T, Mizuno F, Imai S, Mizuno F, Aya T, Koisumi S, Kinoshita T, Tokuda H, Ito Y, Hirai N, Hirota M, Ohigashi H, Koshimizu K, Kofi-Tsekpo WM, Were JBO, Muganbi M (1987) African Burkitt's lymphoma and an Epstein-Barr virus-enhancing plant *Euphorbia tirucalli*. Lancet 1987(i):1257–1258
Osborne T, Mendel LB (1911) Feeding experiments with isolated food substances. Carnegie Institute of Washington, Publication No 156, Washington
Ostrander GK, Cheng KC, Wolf MJ, Wolfe MJ (2004) Shark cartilage, cancer and the growing threat of pseudoscience. Cancer Res 64:8485–8491
Oyebode O, Gordon-Dseagu V, Walker A, Mindell JS (2014) Fruit and vegetable consumption and all-cause cancer and CVD mortality: analysis of health survey for England. J Epidemiol Community Health. Published online 31st March 2014
Oyebola DDO, Elegbe RA (1975) Cow's urine poisoning in Nigeria. Trop Geogr Med 27:194–202
Pacey A (1978) Gardening for better nutrition. Intermediate Technology Publications, London
Pacey A, Cullis A (1996) Rainwater harvesting, the collection of rainfall and runoff in rural areas. Intermediate Technology Publications, London
Painter NS, Burkitt DP (1971) Diverticular disease of the colon: a deficiency disease of western civilisation. Br Med J ii:450–454
Palgrave KC (1983) Trees of Southern Africa. Struik Publishers, Cape Town
Palm O, Sandell K (1989) Sustainable agriculture and nitrogen supply: farmers' and scientists' perspective. Ambio 18(8):442–448
Pande PN (1996) Drudgery of the hill women. Indus Publishing Company, New Delhi
Pandey MR (1984a) Prevalence of chronic bronchitis in rural commmunities of the hill region of Nepal. Thorax 39:331–336
Pandey MR (1984b) Domestic smoke pollution and chronic bronchitis in a rural community of the hill region of Nepal. Thorax 39:337–339
Parish T (1998) Aerial runways. Enact 6(3):15–19
Parodi JL (1988) The use of palms and other native plants in non-conventional, low cost rural housing in the Peruvian Amazon. Adv Bot 6:119–129
Patel VL, Eisemon TO, Arocha JF (1988) Causal reasoning and the treatment of diarrhoeal disease by mothers in Kenya. Soc Sci Med 27:1277–1286
Patten M (1985) We'll eat again. A collection of recipes from the war years selected by Marguerite Patten. Imperial War Museum and Hamlyn Publishing, London
Paudel SK, Chowdhary CL (2005) Managing rattan as a common property: a case study of community rattan management in Nepal. J Bamboo Rattan 4(1):81–91
Paxton A (1994) The food miles report: the dangers of long distance food transport. SAFE Alliance, London
Payawal PC, Paderon EM (1986) Sporocarp formation and spore germination of *Azolla* strains in the Philippines. Philipp Agric 69:633–643
Payne JJ (1986) Alternative therapy. British Medical Association, Board of Science and Education, London
Pearce F (1995) Seeing the wood for the trees. New Sci 145:12–13
Pendleton K (2000) Green agenda, Hopkins and Partners at Nottingham. Ecotech 1(1):32–36
Persson J (1986) Trees, plants and a rural community in the southern Sudan. Unasylva 154:32–43

Pettifor JM (2004) Nutritional rickets: deficiency of vitamin D, calcium or both? Am J Clin Nutr 80(6):1725–1729

Pietroni PC (1992) Beyond the boundaries: relationship between general practice and complementary medicine. Br Med J 305:564–566

Pillet MLR (1980) Pueblo house design as a response to the arid-zone climate. In: Golany G (ed) Housing in arid lands, design and planning. The Architectural Press, London, pp 85–90

Pimentel D, Pimentel M (eds) (1996) Food, energy and society. University Press of Colorado, Boulder

Pimentel D, Hurd LE, Bellotti AC, Forster MJ, Oka IN, Sholes OD, Whitman RJ (1973) Food production and the energy crisis. Science 182:443–449

Pinkerton T, Hopkins R (2009) Local food. How to make it happen in your own community. Green Books, Dartington

Pires A (2005) Community Timber Enterprises in the Mamirauá Sustainable Development Reserve, Brazil. Mamirauá Institute for Sustainable Development, Tefé

Platt BS (1964) Biological ennoblement: improvement of the nutritive value of foods and dietary regimes by biological agencies. Food Technol 18:662–670

Pluknett DL, Beemer HL (eds) (1981) Vegetable farming systems in China. Westview Press, Boulder

Polunin NVC (1985) Traditional marine practices in Indonesia and their bearing on conservation. In: McNeely JA, Pitt D (eds) Culture and conservation: the human dimension in environmental planning. Croom Helm, London, pp 155–179

Poschen P (1986) An evaluation of the *Acacia albida*-based agroforestry practices in the Haranghe Highlands of Ethiopia. Agrofor Syst 4(2):129–143

Prance G, Chadwick D, Marsh J (eds) (1994) Ethnobotany and the search for new drugs. Wiley, Chichester

Prentice A (1998) Sloth, machines that save you energy can also shorten your life. New Sci 157:36–37

Pretty JN (1995) Regenerating agriculture. Policies and practice for sustainability and self-reliance. Earthscan Publications, London

Prevett PF (1962) The reduction of Bruchid infestation of Cowpeas by postharvest storage methods. Reports of the West African Stored Products Research Unit, Nigeria

Price LL (1997) Wild plant food in agricultural environments: a study of occurrence, management, and gathering rights in Northeast Thailand. Hum Organ 56:209–221

Prill-Brett J (1986) The Bontok: traditional wet-rice and swidden cultivators of the Philippines. In: Marten GG (ed) Traditional agriculture in southeast asia. A human ecology perspective. Westview Press, Boulder, pp 54–84

Prior J, Cutler D (1992) Trees to fuel Africa's fires. New Sci 135:35–39

Prior J, Tuohy J (1987) Fuel for Africa's fires. New Sci 115:48–51

Pritikin N, McGrady PM (1979) The pritikin program for diet and exercise. Bantam, New York

Purseglove JW (1968) Tropical crops, dicotyledons. Longman, London

Purseglove JW (1972) Tropical crops, monocotyledons. Longman, London

Quin PJ (1959) Food and feeding habits of the Pedi. Witwatersrand University Press, Johannesburg

Rackham O (1976) Trees and woodland in the British landscape. J.M. Dent, London

Rácz-Kotilla E, Rácz G, Solomon A (1974) The action of *Taraxacum officinale* extracts on the body weight and diuresis of laboratory animals. Planta Med 26:212–217

Rahman AMA (1995) House design in relation to environmental comfort. Build Res Inf 23(1):49–54

Rajalakshmi R (1981) Applied nutrition. Oxford and IBH Publishing, New Delhi

Ramshaw J (2003, May 7) Roofing choices – embodied energy and the environment. Ecotech Sustain Architect Today:30–36

Rathore JS (1972) *Diospyros melanoxylon,* a bread-winner tree of India. Econ Bot 26:333–339

Ravindranath NH, Murali KS, Malhotra KC (eds) (2000) Joint Forest management and community forestry in India. An ecological and institutional assessment. Oxford & IBH Publishing Company, New Delhi

Reddy KS (2002) Cardiovascular diseases in the developing countries:dimensions, determinants, dynamics and directions for public action. Public Health Nutr 5:231–237

Redford KH, Mansour JA (eds) (1996) Traditional peoples and biodiversity conservation in large tropical landscapes. America Verde Publications. The Nature Conservancy, Latin America and Caribbean Division, Arlington

Reeves BOK (1983) Six milleniums of buffalo kills. Sci Am 249(4):92–100

Reeves RR, Mitchell E (1981) The whale behind the tusk. Nat Hist 90(8):50–57

Reganold JP, Elliott LF, Unger YL (1987) Long-term effects of organic and conventional farming on soil erosion. Nature 330:370–372

Reganold JP, Papendrick RI, Parr JF (1990) Sustainable agriculture. Sci Am 262(6):72–78

Reid HF, Smith KR, Sherchand B (1986) Indoor smoke exposures from traditional and improved cookstoves comparisons among rural Nepali people. Mt Res Dev 6(4):293–304

Reissig WH, Heinrichs EA, Litsinger JA, Moody K, Fiedler L, Mew TW, Barrion AT (1986) Illustrated guide to integrated pest management in rice in tropical asia. International Rice Research Institute, Los Baños

Reusch TBH, Ehlers A, Hammerli A, Worm B (2005) Ecosystem recovery after climatic extremes enhanced by genotypic diversity. Proc Natl Acad Sci 102:2826–2831

Reyman EJ (1982) Southwestern Pueblo architecture: some implications for modern housing design. In: Golany G (ed) Desert planning, international lessons. The Architectural Press, London, pp 105–112

Reynolds JEF (ed) (1996) Martindale. The extra pharmacopoeia, 31st edn. London, Royal Pharmaceutical Society

Rhoades R, Bebbington A (1995) Farmers who experiment: an untapped resource for agricultural research and development. In: Warren DM, Slikkerveer LJ, Brokensha D (eds) The cultural dimension of development. Indigenous knowledge systems. Intermediate Technology Publications, London, pp 296–307

Ribeiro BG, Kenhíri T (1989) Rainy seasons and constellations: the Desâna economic calendar. In: Posey DA, Balée W (eds) Resource management in Amazonia: indigenous and folk strategies, Advances in economic botany 7. New York Botanical Gardens, New York, pp 97–114

Richards P (1985) Indigenous agricultural revolution. Ecology and food production in West Africa. Hutchinson, London

Ridker PM, McDermott WV (1989) Comfrey herb tea and hepatic veno-occlusive disease. Lancet:657–658

Roaf S (1983) The windcatchers of the Middle East. In: Germen A (ed) Islamic architecture and urbanism. King Faisal University, Damman, pp 257–268

Robbins WW, Harrington JP, Freire-Marreco B (1916) The ethnobotany of the Tewa Indians. Smithsonian Institution Bureau of American Ethnology, Government Printing Office, Washington, DC

Robert V, Carnevale P (1991) Influence of deltamethrin treatment of bed nets on malaria transmission in the Kou valley, Burkina Faso. Bull World Health Organ 69(6):735–740

Rodahl K (1949) Toxicity of polar bear liver. Nature 164(4169):530–531

Rodahl K, Moore T (1943) The vitamin A content and toxicity of bear and seal liver. Biochem J 37:166–168

Rodger A (1974) The Sudanese heat trap. Ecologist 4(3):102–106

Roonwall ML, Chatterjee PN, Thapa RS (1966) Effect of leaching out water soluble substances in the bamboo, *Bambusa nutans* Wall., as protection against insect borers. Indian For Rec Entomol 10:249–257

Rosendal M, Olesen F, Fink P (2005) Management of medically unexplained symptoms. Br Med J 330:4–5

Rosenthal GA (1986) The chemical defences of higher plants. Sci Am 254(1):76–81

Roundy RW (1985) Clean water provision in rural areas of less developed countries. Soc Sci Med 20:293–300

Ruddle K, Furtado J, Zhong GF, Deng HZ (1983) The mulberry dike-carp pond resource system of the Zhujiang (Pearl River) Delta, People's Republic of China. 1. Environmental context and system overview. Appl Geogr 3:45–62

Ruebush TK, Weller SC, Klein RE (1992) Knowledge and beliefs about malaria on the Pacific coastal plain of Guatemala. Am J Trop Med Hyg 46(4):451–459
Ruthenberg H (1976) Farming systems in the tropics. Clarendon Press, Oxford
Sachs JD (2005a) The end of poverty. How we can make itt happen in our lifetime. Penguin Books, London
Sachs JD, McArthur JW (2005b) The millenium project: a plan for meeting the millennium development goals. Lancet 365:347–373
Salazar R (1992) Community plant genetic resources management: experiences in Southeast Asia. In: Cooper D, Vellvé R, Hobbelink H (eds) Growing diversity. genetic resources and local food security. Intermediate Technology Publications, London, pp 17–29
Salti R (1997) Dana, the human perpective. Al Reen 60:17–19
Salmon P, Dowrick CF, Ring A, Humphris GM (2004) Voiced but unheard agendas: qualitative analysis of psychosocial cues that patents with unexplained symptoms present to general practitioners. Br J Gen Pract 54:171–176
Sandhu R, Sandler J (1986) The tech and tools handbook. A guide to technologies women are using worldwide. International Women's Tribune Centre/Intermediate Technology Publications, New York/London
Sankhala KS, Jackson P (1985) People, trees and antelope in the Indian Desert. In: McNeely JA, Pitt D (eds) Culture and conservation: the human dimension in environmental planning. Croom Helm, London, pp 205–210
Santmire HP (1985) The travail of nature. The ambiguous ecological promise of Christian theology. Fortress Press, Philadelphia
Sarkinfada F, Oyebanji A-A, Sadiq IA, Iyasu Z (2009) Urinary schistosomiasis in the Danjarima community in Kano, Nigeria. J Infect Dev Ctries 3(6):452–457
Sawaya WN, Daghir NJ, Khan P (1983) Chemical characterization and edibility of the oil extracted from *Citrullus colocynthis* seeds. J Food Sci 48(1):104–106
Saxena SK (1986) Desert plants used as human food during scarcity and famines. In: Shankarnarayan KA, Shankar V (eds) Desert environment – conservation and management. Central Arid Zone Research Institute, Jodhpur, pp 43–47
Scher AT (1978) Injuries to the cervical spine sustained while carrying loads on the head. Paraplegia 16:94–101
Schlegel SA (1979) Tiruray subsistence: from shifting cultivation to plow agriculture. Ateneo de Manila University Press, Quezon City
Schoental R (1968) Toxicology and carcinogenic action of pyrrolizidine alkaloids. Cancer Res 28:2237–2246
Schoental R (1972) Prevention or cure? Trop Geogr Med 24:194–198
Schreck CE, Self LS (1985) Bed nets that kill mosquitoes. World Health Forum 6:342–344
Sdei A (2006) Thermal comfort in the traditional japanese house. Martin Centre for Architectural and Urban Studies. Cambridge University, Cambridge
Secoy DM, Smith AE (1983) Use of plants in control of agricultural and domestic pests. Econ Bot 37:28–57
Shack KW, Grivetti LE, Dewey KG (1990) Cash cropping, subsistence agriculture, and nutritional status among mothers and children in lowland Papua New Guinea. Soc Sci Med 31:81–88
Shang A, Huwiler-Müntener K, Nartey L, Jüni P, Dörig S, Sterne JAC, Pewsner D, Eggar M (2005) Are the clinical effects of homoeopathy placebo effects? Comparative study of placebo-controlled trials of homoeopathy and allopathy. Lancet 336:726–732
Shanley P, Galvao J, Luz L (1997) Limits and strengths of local participation: a case-study in eastern Amazonia. PLA Note Int Instit Environ Dev 14:64–67
Sharland RW (1995) Using indigenous knowledge in a subsistence society of Sudan. In: Warren DM, Slikkerveer LJ, Brokensha D (eds) The cultural dimension of development. Indigenous knowledge systems. Intermediate Technology Publications, London, pp 385–395
Sharma KK (2003) Improving bioavailability of iron in Indian diets through food-based approaches for the control of iron deficiency anaemia. Food Nutr Agric 32:51–61

Sharples S, Malama A (1997) A thermal comfort field survey in the cool season of Zambia. Build Environ 32(3):237–243
Shellard EJ (1966) Medicinal plants and the pharmacist. Pharm J 237:495
Shepard M, Carner GR, Turnipseed SG (1977) Colonization and resurgence of insect pests of soybean in response to insecticides and field isolation. Environ Entomol 6:501–506
Shepherd G (1992) Managing Africa's tropical dry forests: a review of indigenous methods. Overseas Development Institute, London
Shepperdson M (1995) Income distribution and health in a North Indian village. Avebury, Aldershot
Sherratt A (1997) Climatic cycles and behavioural revolutions: the emergence of modern humans and the beginning of farming. Antiquity 71:271–287
Shirley-Smith C, Butler D (2008) Water management at BedZED: some lessons. Eng Sustain 161:113–122
Shiva V, Bandyopadhyay J (1986) The evolution, structure and impact of the Chipko movement. Mt Res Dev 6:133–142
Shiva V, Sharatchandra HC, Bandyopadhyay J (1981) Social, economic and ecological impact of social forestry in Kolar. Indian Institute of Management, Bangalore
Shiva V, Sharatchandra HC, Bandyopadhyay J (1982) Social forestry – no solution within the market. Ecologist 12(4):158–168
Silberbauer GB (1972) The G/wi Bushmen. In: Bicchieri MG (ed) Hunters and gatherers today. Holt, Rinehart and Winston, New York, pp 271–326
Silva S (2011) Frutas da Amazônia Brasileira. Metalivros, São Paulo
Simmonds NW (1962) The evolution of bananas. Longman, London
Simoons FJ (1974) Rejection of fish as human food in Africa: a problem in history and ecology. Ecol Food Nutr 3:89–105
Sinclair ARE, Fryxell JM, Caughley G (2006) Wildlife ecology, conservation and management. Blackwell Publishing, Oxford
Skjønsberg E (1989) Change in an African Village. Kefa speaks. Kumarian Press, West Hartford
Smartt J, Simmonds NW (1995) Evolution of crop plants. Longman Scientific and Technical, Harlow
Smith KGV (ed) (1973) Insects and other arthropods of medical importance. British Museum (Natural History), London
Smith NJH (1976) Spotted cats and the Amazon skin trade. Oryx 13(4):362–371
Sofowara A (1982) Medicinal plants and traditional medicine in Africa. John Wiley, Chichester
Spielman DJ, Pandya-Lorch R (eds) (2009) Millions fed. Proven successes in agricultural development. International Policy Research Institute, Washington, DC
Spinage C (1998) Social change and conservation misrepresentation in Africa. Oryx 32:265–276
Sreeby LM (1982) Sugar availability, sugar consumption and dental caries. Community Dent Oral Epidemiol 10(6):1–7
Stamp LD (1964) The geography of life and death. Collins, London
Stamp LD (1969) Man and the land. Collins, London
Stanley NF, Alpers MP (eds) (1975) Lakes and human health. Academic, London
Stapleton MC (1989) Diarrhoeal diseases: perceptions and practices in Nepal. Soc Sci Med 28:593–604
Stephenson LS, Holland C (1987) The impact of helminth infections on human nutrition. Schistosomes and Soil-transmitted Helminths. Taylor & Francis, London
Sterba JP (1994) Reconciling anthropocentric and nonanthropocentric environmental ethics. Environ Value 3:229–244
Stillman AE, Huxtable R, Consroe P, Kohnen P, Smith S (1977) Hepatic veno-occlusive disease due to pyrrolizidine (*Senecio*) poisoning in Arizona. Gastroenterology 73:349–352
Stoate C (2014) Wildlife has its uses. Managing farmland for ecosystem services. Br Wildl 25(3):154–160
Stoll G (1988) Natural crop protection, based on local farm resources in the tropics and subtropics. AGRECOL & Margraf Publishers, Weikersheim

Strickland D (1981) The eskimo vs. the walrus vs. the government. Nat Hist 90(2):48–97
Stringer C (2002) The morphological and behavioural origins of modern humans. Proc Br Acad 106:23–30
Sunquist F (1984) New approach to conservation in the third world. Smithsonian 14(12):164–169
Sutherland WJ (2001) Sustainable exploitation: a review of principles and methods. Wildl Biol 7:131–140
Svoboba GH, Gorman M, Root MA (1964) Alkaloids of *Vinca rosea*. A preliminary report on hypoglycaemic activity. Lloydia 27:361–363
Symington CF (1974) Forester's manual of dipterocarps. Penerbit Universiti Malaya, Kuala Lumpur
Szegedi A, Kohnen R, Dienel A, Kieser M (2005) Acute treatment of moderate to severe depression with hypericum extract WS 5570 (St John's wort); randomised controlled double blind non-inferiority trial versus paroxetine. Br Med J 330:503–506
Talbot JL (1993) Simply Build green. A technical guide to the ecological houses at the Findhorn Foundation. Findhorn Foundation, Findhorn
Tan SP, Wenlock RW, Buss DH (1985) Immigrant foods, second supplement to McCance and Widddowson's the composition of foods. HMSO, London
Tang G (2010) Bioconversion of dietary provitamin A carotenoids to vitamin A in humans. Am J Clin Nutr 91(5):1468–1473
Taylor CE (1951) A medical survey of the Kali Gandak and Pokhara Valleys of Central Nepal. Geogr Rev 41:421–437
Taylor PW (1986) Respect for nature. A theory of environmental ethics. Princeton University Press, Princeton
Taylor JM, Jones WI, Hagan EC, Gross MA, Davis DA, Cook EL (1967) Toxicity of oil of calamus (jammu variety). Toxicol Appl Pharmacol 10:405
Taylor P, Chandiwana SK, Govere JM, Chombo F (1987) Knowledge attitudes and practices in relation to schistosomiasis in a rural community. Soc Sci Med 24:607–611
Temphel KJ, Beukeboom HJJ (2007) Community forests: supporting Bhutan's national and MDG goals while protecting forests. In: Oberndorf R, Mahanty S, Burslem K, Suzuki R (eds) A cut for the poor. Food and Agriculture Organization Bangkok, Regional Community Forestry Training Centre Bangkok, Netherlands Development Organisation Hanoi, Hanoi, pp 114–125
Thomas EM (1959) The harmless people. Secker & Warburg, London
Thomas K (1983) Man and the natural world. Changing attitudes in England 1500–1800. Allen Lane, London
Tiffen M, Mortimore M (1994) Malthus controverted: the role of capital and technology in growth and environment recovery in Kenya. World Dev 22:997–1010
Tiffen M, Mortimore M, Gichuki F (1994) More people, less erosion. John Wiley, Chichester
Tindale NB (1933) Report on field trip. Oceania Sydney 4:101–105
Tindale NB (1972) The Pitjandjara. In: Bicchieri MG (ed) Hunters and gatherers today. Holt, Rinehard and Winston, New York, pp 217–268
Tishkoff SA, Varkonyi R, Cahinhinan N, Abbes S, Argyropoulos G, Destro-Bisol G, Drousiotou A, Dangerfield B, Lefranc G, Loiselet J, Piro A, Stoneking M, Tagarelli A, Tagarelli G, Touma EH, William SW, Clark AG (2001) Halotype diversity and linkage disequilibrium at human G6PD: recent origin of alleles that confer malarial resistance. Science 293:455–462
Todd J, Josephson J (1996) The design of living technologies for waste treatment. Ecol Eng 6:109–136
Transparency International (2002) Corruption fighters toolkit. Civil society experiences and emerging strategies. Transparency International, Berlin
Transparency International (2004) Global corruption report 2004, political corruption. Pluto Press, London
Transparency International (2006) Global corruption report 2006, Corruption and health. Pluto Press, London
Tredgold MH (1986) Food plants of Zimbabwe. Mambo Press, Gweru

Trowell H, Burkitt D, Heaton K (1985) Dietary fibre, fibre-depleted foods and disease. Academic, London

Tsey K (1997) Traditional medicine in contemporary Ghana: a public policy analysis. Soc Sci Med 45:1065–1074

Tsugane S, Sasazuki S (2007) Diet and the risk of gastric cancer, review of epidemiological evidence. Gastric Cancer 10(2):75–83

Turnbull C (1961) The forest people. Jonathon Cape, London

Twinn C (2003) BedZED. Arup J 2003(1):10–16

Ulluwishewa R (1991) Modernization versus sustainability: disintegrating village agroecocomplexes in the dry zone of Sri Lanka. Environ Conserv 18(2):103–109

Upvall MJ (1992) Nursing perceptions of collaboration with indigenous healers in Swaziland. Int J Nurs Stud 29(1):27–36

Vale B, Vale R (2000) The new autonomous house. Design and planning for sustainability. Thames & Hudson, London

Van Buren A (1979) A chinese biogas manual. Intermediate Technology Publications, London

van der Geest S (1987) Self-care and the informal sale of drugs in South Cameroon. Soc Sci Med 25:293–305

van der Post L (1958) The lost world of the Kalahari. Hogarth Press, London

van der Post L (1961) The heart of the hunter. Hogarth Press, London

Vanhaelen M, Vanhaelen-Fastre R, But P, Vennerwegnem J-L (1994) Identification of aristolochic acid in Chinese herbs. Lancet 343:174

Vanherweghem J-L, Depierreux M, Tielemans C, Abramowicz D, Dratwa M, Jadoul M, Richard C, Vandervelde D, Verbeelen D, Vanhaelen-Fastre R, Vanhaelen M (1993) Rapidly progressive interstitial renal fibrosis in young women; association with slimming regime including Chinese herbs. Lancet 341:387–391

Velimirovic H, Velimirovic B (1980) Do traditional plant medicines have a future in third world countries? Curare 3:173–191

Vellinga M, Oliver P, Bridge A (2007) Atlas of vernacular architecture of the world. Routledge, Abingdon

Vellvé R (1992) Saving the seed. Genetic diversity and european agriculture. Earthscan, London

Vijayalalaksmi K, Sridhar S, Damodharan E (1998) Rice, non-chemical pest control. Centre for Indian Knowledge Systems, Chennai

Vogel VJ (1970) American Indian medicine. University of Oklahoma Press, Norman

Volkman SK, Barry AE, Lyons EJ, Nielsen KM, Thomas SM, Choi M, Thakore SS, Day KP, Wirth DF, Hartl DL (2001) Recent origin of *Plasmodium falciparum* from a single progenitor. Science 293:482–484

Wada Y (1988) Strategies for control of Japanese encephalitis in rice production systems in developing countries. In: Smith WH (ed) Vector-borne disease control in humans through rice agroecosystem management. International Rice Research Institute, Manila, pp 153–160

Wade A (ed) (1982) Martindale, the extra pharmacopoeia, 28th edn. London, The Pharmaceutical Press

Walton GA (1962) The *Ornithodorus moubata* superspecies problem in relation to human relapsing fever epidemiology. In: Arthur DR (ed) Aspects of disease transmission by ticks. Zoological Society of London Symposium 6, London, pp 83–156

Warren J (1999) Conservation of earth structures. Buttterworth-Heinemann, Oxford

Warren DM, Bova GS, Tregoning MA, Kliewer M (1982) Ghanian national policy towards indigenous healers. The case of primary health training for indigenous healers (PRHETIH) program. Soc Sci Med 16:1873–1881

Waterkeyn J, Cairncross S (2005) Creating demand for sanitation and hygiene through community health clubs: a cost-effective intervention in two districts in Zimbabwe. Soc Sci Med 61:1958–1970

Waterson R (1990) The living house: an anthropology of architecture in South-East Asia. Oxford University Press, Singapore

Watt JM, Breyer-Bandwijk MG (1962) The medicinal and poisonous plants of Southern and Eastern Africa. Livingstone, Edinburgh
Wayre P (1969) Wildlife in Taiwan. Oryx 10(1):46–56
Weaver LT, Beckerleg S (1993) Is health a sustainable state? A village study in the Gambia. Lancet 341:1327–1330
Wehmeyer AS (1966) The nutrient composition of some edible wild fruits found in the Transvaal. S Afr Med J 40:1102–1104
Weiss MG (1988) Cultural models of diarrheal illness; conceptual framework and review. Soc Sci Med 27:5–16
Weiyuan C (2008) China's village doctors take great strides. Bull World Health Organ 86(12):914–915
Wellin E (1955) Water boiling in a Peruvian town. In: Paul BD (ed) Health, culture and community. Russel Sage Foundation, New York, pp 71–103
Wells J (1995) Population, settlements and the environment: the provision of organic materials for shelter. Habitat Int 19:73–90
Wells J, Haddar F, Uhde M-L, Sy A (1998) Shelter provision in the context of deforestation: some evidence from Mali. Habitat Int 22:463–475
Werner R (1986) Bomah-Poyang. Traditional medicine and ceremonial art of the aborigines of Malaysia. University of Malaya, Kuala Lumpur
Werner DB, Bower BL (1982) Helping health workers learn. The Hesperian Foundation, Palo Alto
Westland P, Hinks J (1994) Minimising environmental impact: lifecycles of building materials. EcoDesign 3(1):14–15
Weston CFM, Cooper BT, Davies JD, Levine DF (1987) Veno-occlusive disease of the liver secondary to ingestion of comfrey. Br Med J 295:183
Wheat JB (1967) A Paleo-indian bison kill. Sci Am 216(1):44–52
Whelan R (1999) Wild in the woods. The myth of the noble eco-savage. Institute of Economic Affairs, London
White L (1967) The historical roots of our ecologic crisis. Science 155:1203–1207
White AU (1977) Patterns of domestic water use in low-income communities. In: Feachem R, McGarry M, Mara D (eds) Water, wastes and health in hot climates. Wiley, Chichester, pp 96–112
WHO (1985a) Guidelines for drinking-water quality, vol 3. Drinking-water quality control in small-community supplies. World Health Organization, Geneva
WHO (1985b) The use of essential drugs. Technical Report 722, World Health Organization, Geneva
WHO (1990) Diet, nutrition, and the prevention of chronic diseases. Technical Report 797, World Health Organization, Geneva
WHO (1993) Guidelines for Drinking-water quality. World Health Organization, Geneva
WHO (1995). Control of foodborne trematode infections. Technical Report 849, World Health Organization, Geneva
WHO (2002) Guidelines for drinking-water quality. Addendum, microbiological agents in drinking water. World Health Organization, Geneva
WHO (2003) Diet, nutrition and the prevention of chronic diseases. Technical Report 916, World Health Organization, Geneva
WHO (2006) WHO guidelines for the safe use of wastewater, excreta and grey water. World Health Organization, Geneva
WHO (2009) Neglected tropical diseases, hidden successes, emerging opportunities. World Health Organisation, Geneva
Wijewardene R, Waidyanatha P (1984) Conservation farming for small farmers in the humid Tropics. Department of Agriculture, Sri Lanka and the Commonwealth Consultative Group for the Asia Pacific Region, Colombo
Wilkie DS (2005) Life amidst chaos in a forest filled with its own dangers, DRC violence continues to trample the Efe and Lese. Cult Surv Quart 29(1):1–5

Willet WC, MacMahon B (1984) Diet and cancer – an overview. N Engl J Med 310(10–11):633–638
Williams CD (1935) Kwashiorkor: a nutritional disease of children associated with a maize diet. Lancet 226:1151–1152
Williams PH (1979) Vegetable crop protection in the People's Republic of China. Annu Rev Phytopathol 17:311–324
Williamson G, Payne WJA (1978) An introduction to animal husbandry in the tropics. Longman, London
Wilson EO (1984) Biophilia. The human bond with other species. Harvard University Press, Cambridge
Winch PJ, Makemba AM, Kamazima SR, Lurie M, Lwihula GK, Prmji Z, Minjas JN, Shiff CJ (1996) Local terminolgy for febrile illnesses in Bagamoyo District, Tanzania and its impact on the design of a community-based malaria control programme. Soc Sci Med 42:1057–1067
Woelk H (2000) Comparison of St John's wort and imipramine for treating depression: randomised control trial. Br Med J 321:536–539
Wolf-Gould CS, Taylor N, Horwitz SM, Barry M (1991) Misinformation about medications in rural Ghana. Soc Sci Med 33:83–89
Wood BJ (1971) Development of integrated control programmes for pests of tropical perennial crops in Malaysia. In: Huffaker CB (ed) Biological control. Plenum Press, New York, pp 422–457
Worede M, Mekbib H (1993) Linking genetic resource conservation to farmers in Ethiopia. In: Boef W d, Amanor K, Wellard K, Bebbington A (eds) Cultivating knowledge: genetic diversity, farmer experimentation and crop research. Intermediate Technology Publications, London, pp 78–84
World Bank (1995) China: nightsoil management improvement study in Hubei. Regional water and sanitation group for East Asia and the Pacific environmental sanitation study. World Bank, Washington, DC
Worm B, Barbier EB, Beaumont N, Duffy JE, Folke C, Halpern BS, Jackson JBC, Lotze HK, Micheli F, Palumbi SR, Sala E, Selkoe KA, Stachowicz JJ, Watson R (2006) Impacts of biodiverity loss on ocean ecosystem services. Science 314:787–790
Wratten S (1992, August 22) Farmers weed out the cereal killers. *New Scientist*, 31–35
Wright G (1985) The stone villages of Britain. David & Charles, Newton Abbot
Wulff HE (1968) The qanats of Iran. Sci Am 218(4):94–105
Yang RZ, Tang CS (1988) Plants used for pest control in China: a literature review. Econ Bot 42:373–400
Yap HH, Tan HT, Abdul Malik Y, Rohaizat B, Loh PY, Chong NL (1990) Field efficacy of mosquito coil formulations containing *d*-allethrin and *d*-transallethrin against indoor mosquitoes especially *Culex quinquefasciatus* say. Southeast Asian J Trop Med Public Health 21:558–563
Yatazawa M, Tomomatsu N, Hosoda N, Nunome K (1980) Nitrogen fixation in *Azolla-Anabaena* symbiosis as affected by mineral nutrient status. Soil Sci Plant Nutr 26:415–426
Yue Q-Y, Bergquist C, Gerdén B (2000) Safety of St John's wort *(Hypericum perforatum)*. Lancet 355:576–577
Zandstra BH, Motooka PS (1978) Beneficial effects of weeds in pest management – a review. PANS 24(3):333–338
Zennie TM, Ogzewalla CD (1977) Ascorbic acid and vitamin a content of edible wild plants of Ohio and Kentucky. Econ Bot 31:76–79
Zeuner FE (1963) A history of domesticted animals. Hutchinson, London
Zuidema PJ (1981) The Katayama syndrome; an outbreak in Dutch tourists to the Omo National Park, Ethiopia. Tropic Geogr Magaz 33:30–35
Zweig RD (1985) Freshwater aquaculture in China: ecosystem system management for survival. Ambio 14(2):66–74

Index

J. Edington, *Indigenous Environmental Knowledge*,
DOI 10.1007/978-3-319-62491-4

Printed by Printforce, the Netherlands